Phytoplankton Productivity

Dedication

This book is dedicated to the ship and boat crews who have helped us in our work, for without their skills, dedication and consummate patience there would be very little to write about the productivity of lakes and oceans.

Phytoplankton Productivity

Carbon Assimilation in Marine and Freshwater Ecosystems

Edited by

Peter J. le B. Williams,* David N. Thomas* and
Colin S. Reynolds**

*School of Ocean Sciences, University of Wales, Bangor, UK
**Centre for Ecology and Hydrology, Windermere, UK

Blackwell
Science

© 2002 by Blackwell Science Ltd,
a Blackwell Publishing Company
Editorial Offices:
Osney Mead, Oxford OX2 0EL, UK
　Tel: +44 (0)1865 206206
Blackwell Science, Inc., 350 Main Street,
Malden, MA 02148-5018, USA
　Tel: +1 781 388 8250
Iowa State Press, a Blackwell Publishing
Company, 2121 State Avenue, Ames, Iowa
50014-8300, USA
　Tel: +1 515 292 0140
Blackwell Science Asia Pty, 54 University Street,
Carlton, Victoria 3053, Australia
　Tel: +61 (0)3 9347 0300
Blackwell Wissenschafts Verlag,
Kurfürstendamm 57, 10707 Berlin, Germany
　Tel: +49 (0)30 32 79 060

First published 2002 by Blackwell Science Ltd

Library of Congress
Cataloging-in-Publication Data
is available

ISBN 0-632-05711-4

A catalogue record for this title is available from
the British Library

Set in 10/13 Times
by DP Photosetting, Aylesbury, Bucks
Printed and bound in Great Britain by
MPG Books Ltd, Bodmin, Cornwall

For further information on
Blackwell Science, visit our website:
www.blackwell-science.com

Contents

Contributors

Richard T. Barber
Marine Laboratory
Nicholas School of the Environment and
    Earth Sciences
Duke University
Beaufort, NC
USA

Michael J. Behrenfeld
National Aeronautic and Space
    Administration
Goddard Space Flight Center
Greenbelt, MD
USA

Robert R. Bidigare
Department of Oceanography
School of Ocean and Earth Science and
    Technology
University of Hawaii
Honolulu, HI
USA

Wayne E. Esaias
National Aeronautic and Space
    Administration
Goddard Space Flight Center
Greenbelt, MD
USA

Paul G. Falkowski
Environmental Biophysics and Molecular
    Biology Program
Institute of Marine and Coastal Science, and
    Dept of Geology
Rutgers University
New Brunswick, NJ
USA

G.E. Fogg
Emeritus Professor
School of Ocean Sciences
University of Wales, Bangor
UK

Richard J. Geider
Department of Biological Sciences
University of Essex
Colchester
UK

D. Glen George
Centre for Ecology and Hydrology
Ambleside
Windermere
UK

Anna K. Hilting
Marine Laboratory
Nicholas School of the Environment and
    Earth Sciences
Duke University
Beaufort, NC
USA

David M. Karl
Department of Oceanography
School of Ocean and Earth Science and
    Technology
University of Hawaii
Honolulu, HI
USA

Ricardo M. Letelier
College of Oceanic and Atmospheric
    Sciences
Oregon State University
Corvallis, OR
USA

Marlon R. Lewis
Department of Oceanography
Dalhousie University
Halifax, NS
Canada

Hugh L. MacIntyre
Horn Point Laboratory
University of Maryland
University of Maryland Center for
    Environmental Science
Cambridge, MD
USA

John Marra
Lamont-Doherty Earth Observatory of
    Columbia University
Palisades, NY
USA

Marina Montresor
Stazione Zoologica 'A. Dohrn'
Napoli
Italy

Colin S. Reynolds
Centre for Ecology and Hydrology
Ambleside
Cumbria
UK

Ulf Riebesell
Alfred Wegener Institute for Polar and
    Marine Research
Bremerhaven
Germany

Wilhelm Ripl
Technische Universität Berlin
Department of Limnology
Berlin
Germany

Victor Smetacek
Alfred-Wegener Institute for Polar and
    Marine Research
Bremerhaven
Germany

Morten Søndergaard
Freshwater Biological Laboratory
University of Copenhagen
Denmark

Kevin Turpie
SAIC-General Sciences Corporation
Laural, MD
USA

Peter Verity
Skidaway Institute of Oceanography
Savannah, GA
USA

Dieter A. Wolf-Gladrow
Alfred Wegener Institute for Polar and
    Marine Research
Bremerhaven
Germany

Klaus-Dieter Wolter
Technische Universität Berlin
Department of Limnology
Berlin
Germany

Foreword

G.E. Fogg

'I should rather incline to believe, that that wonderful power of nature, of changing one substance into another, and of promoting perpetually that transmutation of substances, which we may observe every where, is carried on in this green vegetable matter in a more ample and conspicuous way.' (Jan Ingenhousz, 1779)

This generalisation by Ingenhousz elegantly expresses the essence of this book. He based it on a single experiment which showed that the green matter in spring water, probably the unicellular alga *Chorella*, produced 'very fine dephlogisticated air' under the influence of light. However, the picture is now much more extensive and complicated. We find ourselves dealing with a distinct biological community, worldwide in distribution, sustaining the fertility of oceans and lakes alike and involving an intricate web of chemical and biological processes extending from the hydrosphere into the atmosphere and lithosphere, and both responding to and modifying changes in the physical environment.

The basic structure of this picture has been provided as much by chemists as by biologists. Jean Senebier in 1782 showed that fixed air (carbon dioxide) is necessary for the production of dephlogisticated air (oxygen), leading Ingenhousz to propose that the carbon in organic matter comes from this source. The quantitative investigations of Thèodore de Saussure published eight years later proved that water is the other ingredient necessary to give stoichiometric balance. At this time biologists seemed not to worry too much about the ultimate sources of animal energy. The sailor scientist William Scoresby found that animals such as whales, seals and fish feed on small plankton such as medusae and crustaceans and these, in turn, on minute animalcules, forming a dependent chain of existence, but he did not speculate on the means by which life at the bottom of the chain was sustained.

It was the physicist Julius Robert von Mayer in the middle of the nineteenth century who, in propounding the doctrine of the conservation of energy, emphasised the key position of photosynthesis by plants in the maintenance of life. The botanist Joseph Hooker (1847), in an almost throw-away paragraph in his *Flora Antarctica* around the same time, asserted that diatoms are plants, not animals as believed then by the zoological establishment, and that in the oceans, in which they are ubiquitous, they play a similar role in supporting life as do green plants on land. Following this, the introduction of the fine-meshed silk plankton net gave a great stimulus to studies of microscopic organisms in both the sea and freshwater. However, this allowed only semi-quantitative assessment of standing crops and biased interest towards taxonomy.

Towards the end of the nineteenth century, investigations of marine plankton were mainly carried out by German scientists, led by Victor Hensen and these, fortunately one might think, were not distracted by the prevailing obsession with Darwinism, but based their studies on the concepts of plant nutrition put forward by agricultural chemists of the Justus von Liebig school. The comprehensive picture of phytoplankton dynamics which they provided (see Chapter 13) was refined in the early decades of the twentieth century into something like its present form by Allen, Atkins and Harvey at Plymouth.

Direct measurement of the driving process, photosynthesis, in this dynamic system was first achieved using the Winkler technique for estimating dissolved oxygen concentrations, introduced by Gaarder and Gran in 1927. In its original form this method was rather insensitive. After Calvin and his colleagues in the late 1940s had used radiocarbon (^{14}C) with spectacular success to trace the path of carbon in photosynthesis, Einer Steemann Nielsen on the Danish Deep-Sea Expedition of 1950–1952 (also known as the *Galathea* Expedition) applied it to the measurement of the rate of photosynthesis. Soon it was used in freshwaters by the Swedish limnologist Rodhe. This radiocarbon technique is relatively simple, extremely sensitive and precise, and for a few years the necessary radiochemicals could be taken worldwide with nothing more than a little scratching of chins and shrugging of shoulders on the part of custom officials.

Steemann Nielsen's introduction of the ^{14}C technique gave an unprecedented stimulus to phytoplankton studies. This book, written by leading marine and freshwater scientists, celebrates the tremendous development in the subject since, and largely because of, Steemann Nielsen's innovation. The book takes for the basis of its structure space and time-scales. It begins with a biography of Steemann Nielsen and the history of the subject and then ascends the size scales from the physiological to the global. It then considers ecosystem change and variability on a series of time-scales from the diurnal to geological. The book chapters conclude with a look into the future.

The first chapter, a biography by Søndergaard, traces the development of Steemann Nielsen's interest in primary productivity, in which trouble with his eyesight and the impact of the Second World War seem to have played crucial roles in deflecting his activities from taxonomy of oceanic phytoplankton to photosynthesis by aquatic plants. Chapter 2, by Barber and Hilting, amplifies the sketch of the pioneer work before 1950 given above and also places the recent major trends in the study of primary productivity into broad historical context.

It has been, and must continue to be if the subject is to develop to its full potential, essential to have a general understanding of the physiology and biochemistry of carbon assimilation by algae. After the initial delirious piling up of radiocarbon data, questions inevitably arose as to the meaning of these results. Assimilation of radiocarbon-labelled carbon can take place in the dark as well as in the light – so is gross or net photosynthesis being measured? This question is still with us and is grappled by Marra in Chapter 4. To a greater or lesser extent some of the dark fixation will be carried out by heterotrophs contained in samples of natural waters. Furthermore, laboratory experiments with pure cultures suggest that relationships between photosynthesis and respiration may vary according to species, the physiological conditions of cells, and environmental factors.

Another possibility is that some of the fixed carbon may leak out of healthy photosynthesising cells and be missed if, as is usually the case, the radioactivity of particulate matter only is measured. Tests with radiocarbon tracers have shown that, particularly under conditions of high light or nutrient deficiency, excretion of products of photosynthesis *in situ* in the sea or in freshwaters can be considerable. In Chapter 3, Geider and MacIntyre relate the photosynthetic process to the complex of metabolic activities in the cell and give a guide to answering these questions. The controlling role of inorganic nutrients, recognised by the German school a century ago, is considered on the micro-scale by Riebesell and Wolf-Gladrow (Chapter 5). Study of this has been greatly assisted by application of Steemann Nielsen's tracer idea to elements other than carbon together with better understanding of the chemical kinetics involved supplied by marine chemists. In view of possible changes in surface water chemistry brought about by global changes, such studies may be of critical importance.

The first radiocarbon determinations of primary productivity were made in the course of cruises not specifically planned for the purpose and as a result were widely scattered in space and usually separated by long time intervals. Intelligent guesswork might fit such data into a plausible picture of the production processes in a given sea area but it is not surprising that controversy often arose when results were extrapolated to larger scales and different sea areas. Limnologists were at an advantage here and around this time were able to produce more exact studies of relationships between primary production and light and other physical and chemical conditions in the water column. Valid comparisons could be made between different lakes. In Chapter 6, Lewis describes how observations of ocean reflectance spectra from space and the tardy recognition that mesoscale fluctuations play an important part in determining long-term averages have contributed to more accurate assessment of primary productivity in the sea. Behrenfeld, Esaias and Turpie in Chapter 7 consider how this, together with greater understanding of phytoplankton physiology, has enabled modelling of the oceanic carbon budget and estimation of its role in global climate change.

Apart from salinity, which is of no great importance in this context, the physicochemical environment of phytoplankton is essentially the same in freshwaters and the sea and the organisms themselves, although taxonomically different, are physiologically and biochemically similar. There has not been as much collaboration between oceanographers and limnologists in the study of primary production as might be desirable but it is to be welcomed that three of the chapters in this volume deal with freshwater production. Reynolds (Chapter 8) considers year-to-year variability in phytoplankton production, some lakes showing little, but others considerable, differences. The parts played by variability in environmental forcing and patterns of phytoplankton response are examined by means of long-term data sets. These, which have been obtained by standard methods over a period of more than 50 years in the English Lake District, are also used by George (Chapter 10). Among the quasi-cyclic weather events influencing the flux of nutrients and phytoplankton dynamics, he finds the feature known as the North Atlantic Oscillation to have most effect. Ripl and Wolter (Chapter 11) discuss the complexity of ecosystem development and the emergent tendency towards stability through the conservation of matter. Fundamental to the development of all ecosystems is the extent to which they regulate the through-flow of energy and water. According to this concept, these essentially molecular processes restrict the ecological dynamics to proceeding in a non-random way, with an increasing role for cyclical processes and the truncation of losses. This, indeed, is what makes natural systems sustainable. For the most part, human interference with the dynamics of water balance and vegetation works in the opposite direction, leading to accelerated degradation and loss of enduring, functional landscapes. The authors make clear links between the understanding of biochemical synthesis and the design of the measures that are now necessary to restore landscapes and aquatic ecosystems to sustainability.

Chapter 9, by Karl, Bidigare and Letelier, gives a fascinating picture of the North Pacific Subtropical Gyre. It has proved to be an unexpectedly dynamic system – not an ocean desert – with high rates of gross primary production and complex, decoupled, seasonal patterns of carbon production, consumption and export. A particular point to which attention should be drawn – because users of the radiocarbon technique have often used filters with larger pores than those originally used by Steemann Nielsen, or indeed neglected to specify pore size – is that of the role of picoplankton. For the first three decades of the period we are considering, most students of plankton devoted themselves to species that were most obvious under their microscopes, ignoring the mathematical fact that the total biomass, and therefore the ecological importance, of many tiny cells can outweigh that of large cells in fewer numbers. Improvements in microscopy eventually enabled the demonstration around 1979 of the ubiquity and abundance of picoplankton. The term 'microlitersphere' was proposed for the community of organisms – including heterotrophic bacteria, fungi and protozoa, as well as phototrophic algae, flagellates, cyanobacteria and prochlorophytes – having a maximum size of 2 μm. It is perhaps more satisfactory to include micro-organisms of size up to 20 μm since they have similar properties to picoplankton, and to use the inclusive term 'ultraplankton'. Recent work having shown that viruses are ubiquitous, extremely abundant and highly active in both sea waters and freshwaters, these must be included under this heading as well. The ultraplankton appears to be a highly dynamic, self-contained equilibrium community, adapted to an environment in which molecular diffusion is dominant, and to a large extent independent and distinct from the non-steady-state community (adapted to turbulent conditions) of the microplankton. Karl and his colleagues conclude that gross primary production in the sea area studied is dominated by the picoplankton whereas export production and net carbon sequestration are controlled by the microplankton. The interactions between the two systems are probably climate-sensitive, which may explain the interannual and interdecadal fluctuations observed.

In Chapter 12, Falkowski looks at the history of productivity in the aquatic system on the grand scale, using molecular biological and geological evidence to outline the evolution of the major phytoplankton and their roles in mediating biogeochemical cycles from the Archean to the present day. He concludes with a consideration of how productivity and carbon burial, both inorganic and organic, relate to the taxonomic richness in the ocean. The final chapter by Smetacek, Montresor and Verity returns to the history of quantitative biological oceanography, emphasising that measurements themselves are not the final objective. Integrated knowledge of the whole system is needed to understand what goes on and to carry out the central task of biological oceanography at the present time, namely to integrate organismal biology and life cycles with biogeochemical cycles, and so turn the spectre of global change into something more comprehensible and, possibly, manageable.

For many of us, interest in phytoplankton began with a first look down a microscope at a strange world of minutely exquisite forms or a feeling of direct involvement with the environment when sampling in all weathers at sea or on a lake. Now, our subject, as this volume amply shows, has made tremendous, unexpected advances. It is rash to make predictions as to how it will develop in the future. My own inclinations, were I given the opportunity to start again, would be to collaborate with an enthusiastic rheologist to investigate the relationships between phytoplankton form and water movements: micro-organisms seem to have subtle ways of exploiting turbulence and can certainly modify it to some extent. We need, too, to have much more than the patchy knowledge we have at present about the effects – biological, physical and even, it seems, meteorological – which the huge variety of extracellular products released by phytoplankton exert. Unexpected lines of investigation will certainly emerge and, always, we can be sure that, in the words of the astronomer Fred Hoyle:

'If when by patient inquiry we learn the answer to any problem, we always find, both as a whole and in detail, that the answer thus revealed is finer in concept and design than anything we could ever have arrived at by a random guess.' (Hoyle, 1950, p. 118)

References

Hooker, J.D. (1847) *Flora Antarctica*. Reeve Bros, p. 505.
Hoyle, F. (1950) *The Nature of the Universe*. Blackwell, Oxford.
Ingenhousz, J. (1779) *Experiments upon Vegetables*. Elmsly & Payne, London.

Acknowledgements

We are grateful to various organisations for their generous support towards the compilation of this book as well as towards the associated conference 'An Appreciation of 50 years of the Study of Primary Production in Oceans and Lakes' at the University of Wales, Bangor, March 2002. These are: British Phycological Society; American Society for Limnology and Oceanography; Niels Bohr Foundation; Freshwater Biological Association, UK; Marine Biological Association, UK; Scottish Association for Marine Science; University of Wales, Bangor; and Amersham Pharmacia Biotech.

We thank Tracey Bentley, David Roberts and David Bowers for their help at various stages of the production of the text.

David Thomas is also grateful to the Hanse Institute for Advanced Study for its support.

Chapter 1
A Biography of Einer Steemann Nielsen: the Man and his Science

Morten Søndergaard

Steemann published about 200 papers and three books during his very active academic life. A short curriculum vitae is presented in Box 1.1. The application of $^{14}CO_2$ to quantify aquatic primary production and thus the invention of the ^{14}C method in 1949–1950 (see appendix references 61, 66)[1] is without doubt his most outstanding scientific achievement. The method has after 50 years proven its value and must be considered one of the most important methodological developments within marine ecology and limnology in the past century. However, other aspects of his research are of importance to aquatic ecologists, although these may not be well-known or appreciated by international audiences.

Steemann (Fig. 1.1) began his academic career as a botanist and specialised in phytoplankton taxonomy. At the age of 21 he became a member of the Dana round-the-world expedition 1928–1930 and his doctoral thesis at the age of 27 was on the genus *Ceratium* in the South Pacific (see appendix reference 6). His publications from the 1930s show that at this time he had developed an interest in quantitative ecology and specifically plankton production and the relationship between phytoplankton and zooplankton (see appendix references 2, 5, 8, 11, 12). His second scientific paper published in 1932 was on plankton production. Problems with his eyesight forced him to give up the many hours at the microscope and he changed, in his own words, 'to an experimental biologist', where marine primary production, plant physiology and specifically photosynthesis, light adaptation and inorganic carbon assimilation became his 'pet areas'.

His penchant for simple experiments, a thorough understanding of the importance of the physico-chemical environment and an almost pedantic attitude to experimental details were developed during the early 1930s, when he was learning the experimental approach to biological questions in the very creative scientific environment at the Zoophysiological Laboratory of August Krogh (Nobel laureate). This period became crucial to his career within experimental ecology and when he changed from the descriptive to an experimental approach.

His first experimental studies were focused on inorganic carbon uptake by

---

[1] The full publications of Steemann Nielsen are listed in the Appendix at the end of this book.

Box 1.1 A short curriculum vitae of E. Steemann Nielsen.

Steemann Nielsen was born on 13 June 1907 and baptised Halfdan Einer Steemann Nielsen, but never used his first given name. His high school education was within classical languages, art and history. Actually his parents had wanted him to study theology and become a priest. With an education without much mathematics, chemistry and physics, Steemann had to work hard when, after a rather short period within descriptive and taxonomical botany, he was forced to change to more hardcore natural science. He started to study botany at the University of Copenhagen in 1925, graduated with a Master in Botany 1931 and defended his DPhil thesis in 1934. He died on 17 April 1989.

Professor Pétur M. Jónasson presented Steemann Nielsen's obituary in 1991 (The Royal Danish Academy of Sciences and Letters Annual Report for 1990–1991) and in *Archiv für Hydrobiologie* **131**, 253–254, 1994. Steemann (Fig. 1.1) was honoured for his academic achievements by many international societies and academics.

Academic positions:

1932–1944: Research assistant at the Plankton Laboratory, the Commission for Danish Fisheries and Sea Research.
1935–1944: Research assistant at the Royal Danish School of Pharmacy.
1944–1969: Professor in Botany at the Royal Danish School of Pharmacy.
1962–1969: External lecturer in Oceanography, University of Copenhagen.
1969–1977: Professor in Freshwater Biology, University of Copenhagen.
1977:   Retired and left science.

During the years 1937–1939 he carried out studies in Monaco, Plymouth and Hull. He became member of the Royal Danish Academy of Sciences and Letters from 1958.

Major cruises:

- Botanist on the Dana round-the-world expedition 1928–1930.
- Dana research cruises in 1932 and 1934.
- Thor's expedition to the North Atlantic 1935.
- *Galathea* round-the-world expedition 1950–1952.

---

submerged macrophytes. The results showed that some macrophytes can take advantage of the huge pool of bicarbonate present in most natural waters and continue to photosynthesise at high pH. Steemann proposed a conceptual chemical model to explain how the bicarbonate was utilised (see appendix references 33, 35, 43, 45). However, his model did not survive the later finding of a pH-lowering proton pump and other documented mechanisms for bicarbonate utilisation. The studies on inorganic carbon uptake were carried out at the same time as those of F. Ruttner in Austria, but the results of Steemann have in my opinion not been recognised in the international literature according to their merits. Most of these studies were carried out during or shortly after World War II. In the 1960s, quantitative study of the annual cycle of plankton carried out in the outer Kiel Fjord by Hans Lohmann in 1905–1906 who also followed in Hensen's footsteps (Lohmann, 1908). In addition to nets, he introduced a new method – counting live plankton concentrated by centrifugation – and demonstrated that biomass was high during the warm summer months. He attributed this to the favourable effects of light and temperature on plankton growth, implying that nutrients were replete. Brandt disagreed strongly because he felt that his nutrient-control view of plankton productivity was under attack.

Brandt should have known better (Smetacek, 1985a). He was aware that nutrients entered the sea through rivers and that Lohmann's site in the Fjord was susceptible to eutrophication (the contemporary word for the concept was 'Verjauchung' or over-manuring) hence should have maintained higher biomass than the open sea where Brandt's studies were carried out. In the light of regional differences in plankton cycles, Lohmann's high biomass could have been assimilated into a bigger picture without strife, instead, the footsteps diverged. Brandt's attack was all the more unfortunate because Lohmann's monograph was as much a milestone in plankton ecology as Brandt's own papers on the metabolism of the sea (Brandt, 1899, 1902). Lohmann, the father of rigorous plankton ecology, compared the annual cycles of biomass of the main components and, amongst many other details, pointed out the abundance hence important role of protozoa in grazing phytoplankton and even documented sinking out of the spring bloom. However, presumably because of his nutrient-replete view, he did not reflect on their respective impacts on nutrient recycling and depletion. Both findings and their implications for system functioning were rediscovered 70 years later (Sieburth *et al.*, 1978; Smetacek, 1985b). Lohmann did not attempt to estimate productivity but was absorbed in unravelling organism interactions and life cycles as the basis of quantitative plankton species succession. His approach would have culminated in compilation of a systematic autecology of plankton species of Kiel Bight but he was not able to continue this line of study (Mills, 1989).

Mills (1989) has vividly portrayed the scientific interaction that took place within the Kiel school and how an outsider, Alexander Nathansohn who had been based in Leipzig but also worked in Oslo with Gran, brought gravity and hence structure into the developing picture of metabolism of the sea. By postulating that nutrients were lost by sinking particles and that new nutrients were injected back into the surface by vertical water movement, he connected all the essential links in the big picture. Hensen's grand view now had its mechanisms and its quantities and he should have basked in the vindication of his pioneering approach. Instead he and his colleagues were reluctant to modify parts of their views in order to grasp the big picture, whose coherence should have been obvious. The founding triumvirate – Hensen, Brandt and Lohmann – did not celebrate the advances of their school but preferred to wallow in nit-picking dissent.

13.3   Steps to the future

The legacies of the separated schools championed by Haeckel and established by Hensen have taken very different courses. The study of comparative morphology, exemplified by taxonomy and phylogeny, lost its philosophical attraction after the concept of biogenetic basic law became obsolete and Darwinian evolution (in its modern terms) came to be accepted as fact. Long after biogenetic basic law was forgotten, morphogenesis remained a mystery until its mechanisms were laid bare in the past decade by the new science of molecular biology. But Haeckel's illustrations continue to mesmerise lay persons and scientists alike. In the case of production biology, microscopical examination of the phytoplankton continued to be the method of choice until it was relegated to the background by the combination of the ^{14}C method with chemical analyses, including pigments. These new methods refined the big picture but also brought confusion, which was finally resolved with acceptance of the distinction between new and regenerated production in the 1980s. At about the same time, in the aftermath of the sink-link debate on the fate of the microbial loop, the justification for productivity studies shifted from living resources to biogeochemical cycles within the context of earth system research. The concepts, however, have remained the same, because the shift in focus was brought about by external forces: the collapse of the fisheries by industrial fishing and the spectre of global warming respectively, and not by solution of problems within the field. It is now time to take stock and reflect on what progress has been made since Hensen's days and what homework needs to be completed.

Hensen's vision of the role of phytoplankton is now accepted and balancing the budget by comparing average values of production and its fate in a given water column should no longer be an end to itself. Numbers by themselves mean little, and even if coupled with one another in one-dimensional box models, the whole model is not more than the sum of its parts. This is exemplified by the North Sea model of Steele (1974): since it was published, estimates of North Sea production have tripled, it has been shown that *Calanus finmarchicus* is imported and not home-grown, and that more nutrients enter from the Atlantic than are introduced via the eutrophied rivers. The new insights underline the overriding role of physics in driving pelagic ecosystems. The challenge now is to explain how the North Atlantic Oscillation exerts its significant influence on the North Sea ecosystems as also on the cod and herring stocks. On the other hand, the extent to which ocean physics drives the patterns of species succession of phytoplankton in the North Sea remains unclear. The fact that this phenomenon also occurs at comparable time-scales in lakes speaks for a greater role of driving forces operating within species populations, but these have yet to be elucidated. Clearly there is a need for more information on the biology of individual species. Integrating this knowledge on the basis of annual cycles is required if we are to tease out the crucial links and pathways that determine the functional dynamics of the North Sea ecosystem.

Biogeochemistry is interested in species that accumulate, in those that sink, and those that remove specific elements from the cycle. This field is not interested in growth rate *per se* but in functional groups such as DMS producers, opal exporters, calcifiers, N_2-fixers, etc. Which sets of environmental and internal factors determine their respective temporal and regional dominance patterns needs to be elucidated. We need to explore alternative approaches to better characterise pelagic ecosystems. Thus it is becoming increasingly apparent that the competition paradigm does not explain survival of the many disparate phyla, ongoing endosymbiosis, the relationship between form and function, or species succession patterns. For example, the Sverdrup model applied to the spring water column is as much dependent on growth as it is on loss rates (Smetacek and Passow, 1990). The latter have received too little systematic attention but could well play a key role in explaining species composition (Geider *et al.*, 2001). The current explanation for the status of high-nutrient–low-chlorophyll waters reflects grazing losses as much as iron limitation (Landry *et al.*, 1997) but what determines which species 'take over' when iron limitation is removed?

Sources of mortality are manifold and include not only zooplankton and protozoa, but also viruses (Fuhrman, 1999), pathogenic bacteria, parasites and parasitoids (Geider *et al.*, 2001; Smetacek, 2001). One can only understand defence strategies if one knows the attackers and their weapons. This information cannot be gained from unialgal cultures which yield only information on growth physiology. Physiologies devoted to defence are poorly known but some interesting mechanisms are coming to light. Thus doubt is now being cast on the post-Hensenian view of diatoms as pastures of the sea. Miralto *et al.* (1999) have reported the presence of aldehydes in diatoms that reduce the viability of copepod eggs. It has since been shown that these aldehydes are produced from fatty acids by enzymes activated in the algal cells on sustaining damage (Pohnert, 2000). An analogous mechanism whereby the innocuous molecule DMSP is converted to acrylic acid which deters protozoan grazers has been reported for *Emiliania huxleyi* (Wolfe *et al.*, 1997). These examples suggest that the presence of herbivore deterrents – widespread in terrestrial plants – also occur among phytoplankton (Shaw *et al.*, 1997; Wolfe, 2000). An argument against their presence might be that the cells are killed before having an effect, but are not the cells of an asexual reproducing phytoplankton species population like the leaves on a tree? Much work needs to be carried out in this field. Production diverted to defence slows growth rate, but by reducing mortality it can increase survival, particularly of the non-growth stage of the life cycle.

13.4   Concluding comments

The science of marine productivity has come a long way since the grey beginnings a century ago, but some issues have remained as grey as ever. There have been significant advances, for example flow cytometers (developed for blood research and now successfully applied to 'this blood of the sea') and molecular genetics have considerably advanced our grasp of the plankton and revealed new groups in the ocean. In the process, enumeration under the microscope has been relegated to the back bench. The introduction of the ^{14}C method by Steemann Nielsen revolutionised measurements of aquatic production and, in combination with remote sensing, we now have a very good impression of how much is being produced where, and why (Behrenfeld and Falkowski, 1997). Now we have methods that, in terms of accuracy as well as depth and breadth of coverage, surpass even those available to terrestrial ecologists. However, although our numbers are much better, the management of fish stocks on the basis of productivity studies envisioned by Hensen seems as remote as ever. For the time being, industrial fishing and the ensuing wholesale depletion of commercial fish stocks have rendered the effort irrelevant, so it has not been noticed that productivity studies have so far failed to keep the promise of Hensen's original agricultural paradigm. But the situation is likely to change when today's call for sustainability is eventually put into practice. Hensen's goal to 'subjugate the sea' should be reworded to 'utilise the sea' but the challenge for the next generation remains.

So, how many generations will we have to wait before Hensen's demand for a scientific basis for utilisation of the sea is fulfilled? Although we have a good understanding of the physico-chemical driving forces of marine ecosystems, we are still ignorant of the factors that shape species succession, and the relationship between form and function in the plankton. We must prepare ourselves to open new conceptual avenues, as our community will some day again be called upon to estimate potential yields and carry out environmental manipulations. This requires predictive understanding which goes beyond nutrient control of phytoplankton production. We need to understand why species occur where and when they do (Verity and Smetacek, 1996). More research will have to be directed towards elucidating annual cycles and their dependence upon life history strategies of individual species, of the type carried out on taxa such as *Pfiesteria* (Burkholder and Glasgow, 1997) and *Phaeocystis* (Verity *et al.*, 1988). To this end we will need to understand the evolutionary pressures that shaped the phytoplankton, because, as Dobzhansky (1973) pointed out, 'Nothing makes sense in biology except in the light of evolution'. Needless to say, only sense can be predicted.

Fifty years ago Steemann Nielsen had a vision which left us a legacy. What is our current vision? Integrating organism biology and life cycles with the great biogeochemical cycles of our planet is the central task facing biological oceanographers today. Our science is now being challenged by the spectre of global change: we need to radically improve our ability to understand the plankton if we are to join the other earth system scientists in successfully assessing, modelling and predicting interactions between the components that drive the climate of our planet. It is our opinion that comparative studies of plankton ecology combined with *in situ* experiments will prove to be the most fruitful approaches to identifying crucial processes and mechanisms driving pelagic ecosystems.

References

Behrenfeld, M.J. & Falkowski, P.G. (1997) Photosynthetic rates derived from satellite-based chlorophyll concentration. *Limnology and Oceanography* **42**, 1–20.

Brandt, K. (1899) Über den Stoffwechsel im Meere. *Wissenschaftliche Meeresuntersuchungen, Abteilung Kiel, Neue Folge* **4**, 215–30.

Brandt, K. (1902) Über den Stoffwechsel im Meere. 2. Abhandlung. *Wissenschaftliche Meeresuntersuchungen, Abteilung Kiel, Neue Folge* **6**, 23–79.

Breidbach, O. (1997) Entphysiologisierte Morphologie – Vergleichende Entwicklungsbiologie in der Nachfolge Haeckels. *Theory in Biosciences* **116**, 328–48.

Breidbach, O. (1998) Brief instructions to viewing Haeckel's pictures. In: *Art Forms in Nature* (with contributions by O. Breidbach and I. Eibl-Eibensfeldt) (ed. E. Haeckel). Prestel Verlag, München, New York.

Burkholder, J.M. & Glasgow, J.B., Jr. (1997) The ichthyotoxic dinoflagellate *Pfiesteria piscicida*: behavior, impacts, and environmental controls. *Limnology and Oceanography* **42**, 1052–75.

Dobzhansky, T. (1973) Nothing in biology makes sense except in the light of evolution. *American Biology Teacher* **35**, 125–9.

Frenzel, J. (1897) Die Diatomeen und ihr Schicksal. *Naturwissenschaftliche Wochenschrift* **XII**, Nr. 14.

Fuhrman, J.A. (1999) Marine viruses and their biogeochemical and ecological effects. *Nature* **399**, 541–8.

Geider, R.J., DeLucia, E.H., Falkowski, P.G. *et al.* (2001) Forum. Primary productivity of planet earth: biological determinants and physical constraints in terrestrial and aquatic habitats. *Global Change Biology* **7** (in press).

Haeckel, E. (1862) *Die Radiolarien (Rhizopoda Radiaria). Eine Monographie*, (2 Vols) G. Reimer, Berlin.

Haeckel, E. (1866) *Generelle Morphologie der Organismen*, 2 Vols. G. Reimer, Berlin.

Haeckel, E. (1873) *Natürliche Schöpfungsgeschichte. 4. Auflage*. G. Reimer, Berlin.

Haeckel, E. (1904) *Kunstformen der Natur*. Biliographisches Institut, Leipzig and Vienna.

Haeckel, E. (1998) *Art Forms in Nature* (With contributions by O. Breidbach and I. Eibl-Eibensfeldt). Prestel Verlag, München, New York.

Hardy, A.C. (1956) *The Open Sea, its Natural History: the World of Plankton*. Collins, London.

Hensen, V. (1887) Über die Bestimmung des Planktons oder des im Meere treibenden Materials an Pflanzen und Thieren, pp. 1–107. *Kommission zur wissenschaftlichen Untersuchungen der deutschen Meere in Kiel, 1992–1886. V. Bericht, Jahrgang 12–16.*

Hensen, V. (1911) Das Leben im Ozean nach Zahlung seiner Bewohner. Übersicht und Resultate der quantitativen Untersuchungen, pp. 1–406. *Ergebnisse der Plankton-Expedition der Humboldt-Stiftung. Band V.O.*

Hensen, V. (1926) Die Biologie des Meeres. In: *Die Wunder des Meeres* (ed. G. Gellert), pp. 129–39. Peter J. Oestergaard Verlag, Berlin Schöneberg.

Huisman, J. & Weissing, F.J. (1999) Biodiversity of plankton by species oscillations and chaos. *Nature* **402**, 407–10.

Karsten, G. (1899) Die Diatomeen der Kieler Bucht. *Wissenschaftliche Meeresuntersuchungen, Abteilung Kiel, Neue Folge* **4**, 17–205.

Landry, M.R., Barber, R.T. Bidigare, R.R. *et al.* (1997) Iron and grazing constraints on primary production in the central equatorial Pacific: an EQPAC synthesis. *Limnology and Oceanography* **42**, 405–18.

Lohmann, H. (1908) Untersuchungen zur Feststellung des vollständigen Gehaltes des Meeres an Plankton. *Wissenschaftliche Meeresuntersuchungen, Abteilung Kiel, Neue Folge* **10**, 129–370.

Lötsch, B. (1998) Gibt es Kunstformen der Natur? Radiolarien, Haeckel's biologische Ästhetik und ihre Überschreitung. In: *Welträtsel und Lebenswunder. Ernst Haeckel – Werk, Wirkung und Folgen*, pp.339–72. Stapfia 56, zugleich Kataloge des OÖ. Landesmuseums Neue Folge 131.

Mills, E.L. (1989) *Biological Oceanography: an Early History, 1970–1960.* Cornell University Press, Ithaca and London.

Miralto, A., Barone, G., Romano, G. *et al.* (1999) The insidious effect of diatoms on copepod reproduction. *Nature* **402**, 173–6.

Pohnert, G. (2000) Wound-activated chemical defense in unicellular planktonic algae. *Angewandte Chemie-International Edition* **39**, 4352–4.

Rauschenplat, H. (1900) Über die Nahrung von Thieren aus der Kieler Bucht. *Wissenschaftliche Meeresuntersuchungen, Abteilung Kiel, Neue Folge* **5**, 85–151.

Richardson, M.K., Hanken, J., Gooneratne, M.L. *et al.* (1997) There is no highly conserved embryonic stage in the vertebrates: implications for current theories of evolution and development. *Anatomy and Embryology* **196**, 91–106.

Shaw, B.A., Anderson, R.J. & Harrison, P.J. (1997) Feeding deterrent and toxicity effects of apo-fucoxanthinoids and phycotoxins on a marine copepod (*Tigriopus californicus*). *Marine Biology* **128**, 273–80.

Sieburth, J.McN., Smetacek, V. & Lenz, J. (1978) Pelagic ecosystem structure: heterotrophic components of the plankton and their relationship to plankton size fractions. *Limnology and Oceanography* **23**, 1256–63.

Smetacek, V. (1985a) The annual cycle of Kiel Bight plankton: A long term analysis. *Estuaries* **8**, 145–57.

Smetacek, V.S. (1985b) Role of sinking in diatom life-history cycles: ecological, evolutionary and geological significance. *Marine Biology* **84**, 239–51.

Smetacek, V. (1999) Revolution in the ocean. *Nature* **401**, 647.

Smetacek, V. (2001) A watery arms race. *Nature* **411**, 745.

Smetacek, V, von Bodungen, B., Knoppers, B. *et al.* (1984) Seasonal stages characterizing the annual cycle of an inshore pelagic system. *Rapports et Procès-Verbaux des Réunions. Conseil International pour l'Exploration de la Mer* **183**, 126–35.

Smetacek, V. & Passow, U. (1990) Spring bloom initiation and Sverdrup's critical-depth model. *Limnology and Oceanography* **35**, 228–234.

Steele, J.H. (1974) *The Structure of Marine Ecosystems.* Harvard University Press, Harvard.

Steemann Nielsen, E. (1934) *Die Verbreitung, Biologie und Variation der Ceratien im südlichen Stillen Ozean*, pp. 1–67. Dana Report n. 4, Carlsberg Foundation, Copenhagen.

Steemann Nielsen, E. (1939a) *Die Ceratien des Indischen Ozeans und der Ostasiatischen Gewässer mit einer allgemeinen Zusammenfassung über die Verbreitung der Ceratien in den Weltmeeren*, pp. 1–33. Dana Report n. 17, Carlsberg Foundation, Copenhagen.

Steemann Nielsen, E. (1939b) Über die vertikale Verbreitung der Phytoplankton im Meere. *Internationale Revue der gesamten Hydrobiologie und Hydrographie* **38**, 421–40.

Steemann Nielsen, E. (1952) The use of radioactive carbon (C^{14}) for measuring organic production in the sea. *Journal du Conseil International pour l'Exploration de la Mer* **18**, 117–40.

Verity, P.G. & Smetacek, V. (1996) Organism life cycles, predation, and the structure of marine pelagic ecosystems. *Marine Ecology Progress Series* **130**, 277–93.

Verity, P.G., Villareal, T.A. & Smayda, T.J. (1988) Ecological investigations of blooms of colonial *Phaeocystis pouchetii*. II. The role of life cycle phenomena in bloom termination. *Journal of Plankton Research* **10**, 749–66.

Wolfe, G.V. (2000) The chemical defence ecology of marine unicellular plankton: constraints, mechanisms, and impacts. *Biological Bulletin* **198**, 225–44.

Wolfe, G.V., Steinke, M. & Kirst, G.O. (1997) Grazing-activated chemical defence in a unicellular marine alga. *Nature* **387**, 894–7.

Appendix
Steemann Nielsen's Publications
Compiled by Professor Morten Søndergaard

(Note Steemann Nielsen's name has been abbreviated to SN.)

Titles in Danish and Swedish have been translated. Articles communicated to the general public or of a more popular character are preceded by an asterisk (selected by MS). It has not been possible to evaluate all details of some of the papers published in the 'grey' literature. Number 26 and 27 are identical, but published in both Danish and English. Number 28, 32 and 56 are just very short notes to tell the readers of Botanisches Zentralblatt and the members of The Nordic Society of Plant Physiology that Steemann had published a paper. By today's standards a number of the papers from the 1950s are rather alike and identical tables and long phrases can be found in more than one paper. Most of the words and results published in four of his *Nature* papers can be found in other papers; however, this is not that uncommon for the follow-up of seminal breakthrough papers, and a forgivable redundancy.

(1)   SN, E. (1931) Einige Planktonalgen aus den warmen Meeren. *Dansk Botanisk Arkiv* **6** (9), 1–13.

(2)   SN, E. (1932) Einleitende Untersuchungen über die Stoffproduktion des Planktons. *Meddeleser fra Kommissionen for Danmarks Fiskeri-og Havundersøgelser. Serie: Plankton* **2** (4), 1–14.

*(3)   SN, E. (1932) Mangroven og dens biologi (in Danish; The mangrove and its biology). *Natur og Videnskab* **1**, 148–51.

(4)   SN, E. (1933) Phytoplanktonet i Ringkøbing Fjord 1915–31. With an English summary (in Danish; The phytoplankton in Ringkøbing Fjord). In: *Ringkøbing Fjords naturhistorie i brakvandsperioden 1915–1931* (eds A.C. Johansen & H. Blegvad), pp. 36–48. Reitzels forlag, Copenhagen.

(5)   SN, E. (1933) Über quantitative Untersuchung von marinen Plankton mit Utermöhls umgekehrten Mikroskop. *Journal du Conseil International pour l'Exploration de la Mer* **8**, 201–10.

(6)   SN, E. (1934) Untersuchungen über die Verbreitung, Biologie und Variation der *Ceratien* im südlichen Stillen Ozean. *Dana-Report*, **4**, 1–68.

(7)   SN, E. & Brand, T. (1934) Quantitative Zentrifugenmethoden zur Planktonbestimmung. *Rapports et Procès-Verbaux des Réunions* **12**, App. III, 99–100.

(8)   SN, E. (1935) Eine Methode zur exakten quantitativen Bestimmung von Zooplankton. Mit allgemeinen Bemerkungen über quantitative Planktonarbeiten. *Journal du Conseil International pour l'Exploration de la Mer* **10**, 302–14.

(9)   SN, E. (1935) The production of phytoplankton at the Faroe Isles, Iceland, East

Greenland and the waters around. *Meddelelser fra Kommisionen for Danmarks Fiskeriog Havundersøgelser, Serie: Plankton,* **3** (1), 1–94.

(10) SN, E. (1937) Undersøgelser om stofproduktionen i havet ved Island (in Danish; Investigations of the production in the sea at Island). In, *Beretning om M.S. Thor's Havundersøgelsestogt 1935. Publikationer om Østgrønland* Nr. **5**: 63–72.

(11) SN, E. (1937) The annual amount of organic matter produced by the phytoplankton in the Sound off Helsingör. *Meddelelser fra Kommisionen for Danmarks Fiskeri-og Havundersøgelser, Serie: Plankton* **3** (3), 1–37.

(12) SN, E. (1937) On the relation between the quantities of phytoplankton and zooplankton in the sea. *Journal du Conseil International pour l'Exploration de la Mer* **12**, 147–54.

(13) SN, E. (1937) Om planktonproduktionen under molérdannelsen (in Danish; About the plankton production during the formation of moler-clay). In: *Et vulkanomraades Livshistorie* (ed. I.S.A. Andersen). Geologisk Förening Stockholm Förhandlinger **59**: 336–40.

(14) SN, E. (1938) Über die Anwendung von Netzfängen bei quantitativen Phytoplanktonuntersuchungen. *Journal du Conseil International pour l'Exploration de la Mer* **13**, 197–205.

(15) Gabrielsen, E.K. & SN, E. (1938) Kohlensäureassimilation und Lichtqualität bei den marinen Planktondiatomeen. (Vorläufige Mitteilung). *Rapports et Procès-Verbaux des Réunions* **108**, App. II, **4**, 19–21.

*(16) SN, E. (1938) Lysforholdene i havet og den marine planteverden (in Danish; The light climate of the ocean and plant life in the marine environment). *Naturens Verden* **22**, 24–35.

*(17) SN, E. (1938) Om betydningen of den såkaldte 'xeromorphe' bladstruktur hos planter på næringsfattig bund (in Danish; About the importance of the 'xeromorphe' leaf structure of plants growing on nutrient deprived soils). *Naturhistorisk Tidskrift* **2**, 151–3.

(18) SN, E. (1938) De danske farvandes hydrografi i Litorinatiden. Mit einer Zusammenfassung (in Danish; The hydrography of the Danish coastal waters during the Litorina age). *Meddelelser fra Dansk Geologiske Forening* **9**, 337–50.

(19) SN, E. (1939) Die Ceratien des Indischen Ozeans und der Ostasiatischen Gewässer. Mit einer allgemeinen Zusammenfassung über die Verbreitung der Ceratien in den Weltmeeren. *Dana Report* **17**, 1–33.

(20) SN, E. (1939) Letter to the editor. (Über quantitative Netzunterschugungen.) *Journal du Conseil International pour l'Exploration de la Mer* **14**, 106–107.

(21) SN, E. (1939) Über die vertikale Verbreitung der Phytoplanktonten im Meere. *Internationale Revue der gesamten Hydrobiologie* **38**, 421–40.

*(22) SN. E. (1940) Produktionen af fytoplankton i de danske farvande inden for Skagen (in Danish: The production of phytoplankton in Danish waters within the Skagen region). *Naturhistorisk Tidskrift* **4**, 102–104.

*(23) SN, E. (1940) Vækststoffer i havet og deres betydning for planteplanktonet (in Danish; Growth elements in the sea and their importance for the phytoplankton). *Naturens Verden* **24**, 465–70.

(24) SN, E. (1940) Über die Bedeutung der sogenanten xeromorphen Struktur im Blattbau der Pflanzen auf nährstoffarmen Boden. *Dansk Botanisk Arkiv* **10** (2), 1–28.

(25)   SN, E. (1940) Die Produktionsbedingungen des Phytoplanktons im Übergangsgebiet zwischen der Nord – und Ostsee. *Meddelelser fra Kommisionen for Danmarks Fiskeri-og Havundersøgelser, Serie: Plankton* **3** (4), 1–55.

(26)   SN, E. & Otterstrøm, C.V. (1940) Two cases of extensive mortality in fishes caused by the flagellate *Prymnesium parvum* Carter. *Report of the Danish Biological Stations* 1939, **44**, 1–24.

(27)   SN, E. & Otterstrøm, C.V. (1940) To tilfælde af omfattende dødelighed hos fisk forårsaget af flagellaten *Prymnesium parvum* Carter. *Beretning fra de Danske Biologiske Stationer* 1939, **44**, 1–23.

(28)   SN, E. (1939/40) Author ref. *Botanisches Zentralblatt N.F.* **33**, 379.

(29)   SN, E. (1940) Book review. *Naturhistorisk Tidskrift* **4**.

*(30) SN, E. (1941) Øresundskanalen – Et Østersøproblem (in Danish; The Oeresund canal – a problem for the Baltic?). *Naturhistorisk Tidskrift* **5**, 87–9.

(31)   SN, E. (1941) Über das Verhältnis zwischen Verwandtschaft und Verbreitung von Organismen in Beziehung zu ökologischen Studien auf Grundlage der Verbreitung. *Royal Danish Society of Sciences and Letters. Biologiske Meddelelser* **16** (4), 1–25.

(32)   SN, E. (1941/42) Author ref. *Botanisches Zentralblatt N.F.* **35**, 88.

(33)   SN, E. (1942) Der Mechanismus der Photosynthese. Versuche mit *Fucus serratus* und anderen submersen. *Dansk Botanisk Arkiv* **11** (2), 1–95.

(34)   SN, E. (1943) Über das Frühlingsplankton bei Island und den Faröer-Inseln. *Meddelelser fra Kommisionen for Danmarks Fiskeri- og Havundersøgelser, Serie: Plankton* **3** (6), 1–14.

(35)   SN, E. (1944) Dependence of freshwater plants on quantity of carbon dioxide and hydrogen ion concentration. Illustrated through experimental investigations. *Dansk Botanisk Arkiv* **11**, 1–25.

(36)   SN, E. (1944) Havets planteverden i økologisk og produktionsbiologisk belysning (in Danish; The plants of the oceans with respect to ecology and production). *Skrifter udgivet af Kommisionen for Danmarks Fiskeri- og Havundersøgelser* **13**, 1–108.

*(37) SN, E. (1944) Hvordan ernæres vandplanterne? (in Danish; How do waterplants acquire nutrition?). *Akvariebladet* **10**, 128–31.

*(38) SN, E. (1945) Om forandringer i Roskildefjords saltholdighed (in Danish; About the salinity changes in Roskilde Fjord). *Naturens Verden* **29**, 108–14.

(39)   SN, E. (1945) Book review. *Naturhistorisk Tidskrift* **9**.

*(40) SN, E. (1946) Carbon sources in the photosynthesis of aquatic plants. *Nature* **158**, 594–9.

*(41) SN, E. (1946) Vandplanternes C-kilde under fotosyntese (in Danish; The carbon sources of waterplants during photosynthesis). *Naturhistorisk Tidskrift* **10**, 23–24.

(42)   SN, E. (1947) Professor dr. phil. Ove Paulsen. Et par mindeord. (Obituary). *Naturens Verden* **31**, 33–7.

(43)   SN, E. (1947) Diffusion of dissolved substances through thalli and leaves of aquatic plants. *Nature* **160**, 376–7.

(44)   SN, E. (1947) Hovedtrækkene af planterigets system (in Danish; The main taxonomic system of the plant kingdom). *Farmaceuten* **10**, 39–46.

(45)   SN, E. (1947) Photosynthesis of aquatic plants with special reference to the carbon sources. *Dansk Botanisk Arkiv* **12** (8), 1–71.

(46) SN, E. (1948) Plant Ecology. *The Humanities and the Sciences in Denmark during the Second World War*, pp. 306–308. Munksgaards Forlag, Copenhagen.

(47) SN, E. (1948) Biography of J. Boye Petersen. In: *Salomons Leksidale Tidskrift*, (Danish encyclopedia) Vol. **8**, p. 1056. Salomon, Copenhagen.

*(48) SN, E. (1948) Diffusion og fotosyntese hos vandplanter (in Danish; Diffusion and photosynthesis of aquatic plants). *Naturhistorisk Tidskrift* **12**, 55.

(49) SN, E. (1949) A reversible inactivation of chlorophyll *in vivo*. *Physiologia Plantarum* **2**, 247–65.

*(50) SN, E. (1949) Stofproduktionen i de grønlandske farvande (in Danish; The organic production in the waters off Greenland). *Grønlandsposten* **8**, 56–7.

(51) SN, E. (1949) Ove Paulsen, 1874–1947 (Obituary). *Journal du Conseil International pour l'Exploration de la Mer* **11** (1), 14–15.

(52) SN, E. & Kristiansen, J. (1949) Carbonic anhydrase in submersed autotrophic plants. *Physiologia Plantarum* **2**, 325–31.

(53) SN, E. & Mikkelsen, V. (1949) *Nøgle til bestemmelse af de vigtigste grupper af blomsterplanterne* (in Danish; A key to the most important groups of angiosperms). Copenhagen, pp. 1-19. 2nd edn 1953, pp. 1–15. (Lecture notes.)

(54) SN, E. (1949) Book review. *Oikos* **1**, 153.

(55) SN, E. (1949) Book review. *Botanisk Tidskrift* **48**, 131.

(56) SN, E. (1950) Author ref. Meddelelser fra Nordisk Förening for Fysiologisk Botanik **2**.

*(57) SN, E. (1950) Samspillet mellem planter og dyr i akvarier (in Danish; The interaction between plants and animals in aquarias). *Dansk Akvarieblad* **4**, 25–8.

(58) SN, E. & Mikkelsen, V. (1951) *De lavere planters systematik* (in Danish, The taxonomy of the 'lower' plants), p. 30. (Lecture notes.)

(59) SN, E. (1951) The marine vegetation of the Isefjord – a study on ecology and production. *Meddelelser fra Kommisionen for Danmarks Fiskeri- og Havundersøgelser, Serie: Plankton* **5** (4), 1–114.

*(60) SN, E. (1951) 'Galathea' måler lyset i havet (in Danish; 'The Galathea expedition'; we measure the light in the sea). *Information* 2 February. (Newspaper article).

(61) SN, E. (1951) Measurement of the production of organic matter in the sea by means of carbon-14. *Nature* **167**, 684–5.

(62) SN, E. (1951) Passive and active transport during photosynthesis in water plants. *Physiologia Plantarum* **4**, 189–98.

(63) SN, E. (1952) Experimental carbon dioxide curves in photosynthesis. *Physiologia Plantarum* **5**, 145–59.

(64) SN, E. (1952) The persistence of aquatic plants to extreme pH values. *Physiologia Plantarum* **5**, 211–17.

(65) SN, E. (1952) On detrimental effects of high light intensities on the photosynthetic mechanism. *Physiologia Plantarum* **5**, 334–44.

(66) SN, E. (1952) The use of radioactive carbon (C^{14}) for measuring organic production in the sea. *Journal du Conseil International pour l'Exploration de la Mer* **18** (2), 117–40.

(67) SN, E. (1952) Måling af havets produktion af organisk stof ved hjælp af radioaktivt kulstof (C^{14}) (in Danish; Measurements of the organic production in the sea by radioactive carbon (C^{14})). *Institut for Vandanalyse*, Vol. **23**, p. 5. (Technical report.)

*(68) SN, E. (1952) Oceaner af mad. På 'Galathea' har man målt planktonproduktionen

ved hjælp af kulstof 14 (in Danish; Oceans of food. During the Galathea expedition the plankton production has been measured by the carbon 14 method). *Vor Viden* **70** 629–34.

*(69)   Spärck, R, SN, E. & Bruun, F. (1952) Galathea-ekspedition 1950–52 (in Danish; The Galathea expedition 1950–52). *Grundris ved folkelig universitetsundervisning*, No. **488**, p. 15, Copenhagen.

(70)   SN, E. (1952) Production of organic matter in the sea. *Nature* **169**, 956–9.

*(71)   SN, E. (1952) Den røde død. Hvor fiskene dør til jul (in Danish; The red death. Where fishes die at Christmas time). *Vor Viden* **77**, 769–73.

*(72)   SN, E. (1953) Der høstes i havet. Havalgerne giver os vigtige råstoffer (in Danish; Harvest in the sea. Important raw material is provided by the macroalgae in the sea). *Vor Viden* **94**, 502–508.

*(73)   SN, E. (1953) En helt ny slags mad. Bliver 'vandbruget' fremtidens næringsvej? (in Danish; A new kind of food. Is aquaculture a future business?). *Vor Viden* **106**, 38–43.

*(74)   SN, E. (1953) Fotosyntesens biokemi (in Danish; The biochemistry of photosynthesis). *Farmaceuten* **16**, 45–9.

*(75)   SN, E. (1953) Måling af havets produktion af organisk stof ved hjælp af radioaktivt kulstof (C^{14}) (in Danish; Measuring the production of the sea by radioactive carbon (C^{14})). *Ingeniøren* **62**, 389–90.

(76)   SN, E. & Jensen, E.A. (1953) A water-sampler for biological purposes. *Journal du Conseil International pour l'Exploration de la Mer* **18**, 296–99.

*(77)   SN, E. (1953) Fra 'Galathea-ekspeditionen': Målinger af oceanernes stofproduktion ved hjælp af radioaktivt kulstof (in Danish; From the Galathea expedition: Measuring the production of the oceans by radioactive carbon). *Naturhistorisk Tidskrift* **17**, 12–13.

(78)   SN, E. (1953) Carbon dioxide concentration and maximum quantum yield in photosynthesis. *Nature* **171**, 1106–1108.

(79)   SN, E. (1953) Carbon dioxide concentration, respiration during photosynthesis, and maximum quantum yield of photosynthesis. *Physiologia Plantarum* **6**, 316–32.

*(80)   SN, E. (1953) Havet og ernæringen af jordens befolkning (in Danish; The sea and the nutrition of the global population). *Naturens Verden* **37**, 53–62.

(81)   SN, E. (1953) Måling af havets stofproduktion (in Danish; Measuring the production of the sea). 'Galatheas Jordomsejling' Vol. **5**, pp. 65–76. Danish Science Press Ltd, Copenhagen.

(82)   SN, E. (1953) Planteproduktionen i de grønlandske farvande (in Danish; Plant production in the waters off Greenland). *Grønland* **1953**, 401–406.

(83)   SN, E. (1953) Om bakterievækst i kondensatorrør (in Danish; About the growth of bacteria in condensator pipes). *Akademiet for de Tekniske Videnskaber Beretning* no. **25**, 41–58.

(84)   SN, E. (1954) On organic production in the oceans. *Journal du Conseil International pour l'Exploration de la Mer* **19**, 309–28.

(85)   SN, E. (1954) On the preference of some freshwater plants in Finland for brackish water. *Botanisk Tidskrift* **51**, 242–7.

*(86)   SN, E. (1954) Måling af plankton produktion (in Danish; Measuring plankton production). *Skolefilm katalog. Statens filmcentral.* (Catalogue for educational movies.)

(87)   SN, E. (1955) An effect of antibiotics produced by plankton algae. *Nature* **176**, 553.

(88)   SN, E. (1955) The production of organic matter by the phytoplankton in a Danish lake receiving extraordinarily great amounts of nutrient salts. *Hydrobiologia* **7**, 68–74.

(89)   SN, E. (1955) The production of antibiotics by plankton algae and its effect upon bacterial activities in the sea. *Deep-Sea Research* (*Supplement on Marine Biology & Oceanography*) **3**, 281–6.

(90)   SN, E. (1955) Influence of pH on the respiration in *Chlorella pyrenoidosa*. *Physiologia Plantarum* **8**, 106–15.

(91)   SN, E. (1955) *Plankton in the Food Cycle of the Sea*. Training Centre for Fishery Administrators, Denmark. (Handout paper.)

(92)   SN, E. (1955) Carbon dioxide as carbon and narcotic in photosynthesis and growth of *Chlorella pyrenoidosa*. *Physiologia Plantarum* **8**, 317–35.

(93)   SN, E. (1955) The interaction of photosynthesis and respiration and its importance for the determination of 14-C discrimination in photosynthesis. *Physiologia Plantarum* **8**, 945–53.

*(94)   SN, E. (1955) Det videnskabelige arbejde på botanisk afdeling, Danmarks Farmaceutiske Højskole (in Danish; The scientific work at the Botany Department, The Royal Danish School of Pharmacy). *Farmaceuten* **18**, 43–5.

*(95)   SN, E. (1955) Er havet fremtidens spisekammer? (in Danish; Is the ocean the main provider of food in the future?). *Kontakt* **6**, 82–5.

(96)   SN, E. (1955) Production of organic matter in the oceans. *Pears Foundation Journal of Marine Research* **14**, 374–86.

(97)   SN, E. & Al Kholy, A.A. (1956) Use of ^{14}C-technique in measuring photosynthesis of phosphorus or nitrogen deficient algae. *Physiologia Plantarum* **9**, 144–53.

(98)   SN, E. & Jensen, Å.E. (1957) Primary oceanic production. The autotrophic production of organic matter in the oceans. *Galathea Report*, Vol. **1**, pp. 49–136.

(99)   SN, E. & Grøntved, J. (1957) Investigations on the phytoplankton in the sheltered Danish marine localities. *Meddelelser fra Kommisionen for Danmarks Fiskeri- og Havundersøgelser, Serie: Plankton* **10** (6), 1–52.

(100) SN, E. (1957) The chlorophyll content and the light utilization in communities of plankton algae and terrestrial higher plants. *Physiologia Plantarum* **10**, 1009–21.

(101) SN, E. & Jensen, P.K. (1958) Concentration of carbon dioxide and rate of photosynthesis in *Chlorella pyrinoidosa*. *Physiologia Plantarum* **11**, 170–80.

(102) SN, E. (1958) The balance between phytoplankton and zooplankton in the sea. *Journal du Conseil International pour l'Exploration de la Mer* **23**, 178–88.

(103) SN, E. (1958) Experimental methods for measuring organic production in the sea. *Rapports et Procès-Verbaux des Réunions* **144**, 38–46.

(104) SN, E. (1958) A survey of recent Danish measurements of the organic production in the sea. *Rapports et Procès-Verbaux des Réunions* **144**, 92–5.

(105) SN, E. (1958) Light and organic production in the sea. *Rapports et Procès-Verbaux des Réunions* **144**, 141–8.

(106) SN, E. (1958) Planteplanktonets årlige produktion af organisk stof i Furesøen (in Danish; The annual production of organic matter by the phytoplankton in Lake Fure). *Folia Limnologica Scandinavica* **10**, 104–9.

(107) SN, E. (1959) The primary production in the waters west of Greenland during July 1958. Paper given at *C.M. Special IGL Meeting No. 7*.

*(108) SN, E. (1959) Om uddannelse af videnskabsmænd (in Danish; About the education of scientists). *Farmaceuten* **22**, 5–7.

*(109) SN, E. (1959) Eksperiment contra iagttagelse i naturen. En biologs betragtninger (in Danish; Experiment versus observation of nature. The thoughts of a biologist). *Farmaceuten* **22**, 144–9.

(110) SN, E. & Hansen, V.K. (1959) Measurements with the carbon-14 technique of the respiration rates in natural populations of phytoplankton. *Deep-Sea Research* **5**, 222–33.

(111) SN, E. & Hansen, V.K. (1959) Light adaptation in marine phytoplankton populations and its interrelation with temperature. *Physiologia Plantarum* **12**, 353–70.

(112) SN, E. (1959) Light adaptation in marine phytoplankton. *Proceedings of the IX International Botany Congress* **11**, 379–80 (Abstract).

(113) SN, E. (1959) Untersuchungen über die Primärproduktion des Planktons in einigen Alpenseen Österreichs. *Oikos* **10**, 24–37.

(114) SN, E. (1959) Chlorophyll as a means of estimating potential photosynthesis of marine phytoplankton. Paper given at *International Oceanographic Congress*, 31 August–12 Septempter, 846–7 (Abstract).

(115) SN, E. & Jørgensen, E.G. (1959) Effect of filtrates from cultures of unicellular algae on the growth of *Staphylococcus aureus*. Paper given at *International Oceanographic Congress* 31 August–12 September, 923–4 (Abstract).

(116) SN, E. (1959) Primary production in tropical marine areas. *Journal of the Marine Biological Association of India* **1**, 7–12.

(117) SN, E. (1959) Book review. *Botanisk Tidskrift* **55**, 70.

(118) SN, E. (1960) Uptake of CO_2 by the plant. In: *Handbuch der Pflanzenphysiologie* (eds W. Ruhland & A. Pirson (subeditor of Vol. 5)), Vol. **5**, No. 1, pp. 70–84. Springer-Verlag, Berlin.

*(119) SN, E. (1960) Nyt fra fotosyntese-forskningen (in Danish; News from the research of photosynthesis). *Naturens Verden* July, 193–5.

(120) SN, E. (1960) Productivity of the oceans. *Annual Review of Plant Physiology* **11**, 341–62.

(121) SN, E. (1960) Dark fixation of CO_2 and measurements of organic productivity. With remarks on chemo-synthesis. *Physiologia Plantarum* **13**, 348–57.

(122) SN, E. (1960) Book review. *Oikos* **11**, 168.

(123) SN, E. & Jørgensen, E.G. (1960) Effect of daylight and of artificial illumination on the growth of *Staphylococcus aureus* and some other bacteria. *Physiologia Plantarum* **13**, 534–8.

(124) SN, E. (1961) Chlorophyll concentration and rate of photosynthesis in *Chlorella vulgaris*. *Physiologia Plantarum* **14**, 868–76.

(125) SN, E. & Jørgensen, E.G. (1961) Effect of filtrates from cultures of unicellular algae on the growth of *Staphylococcus aureus*. *Physiologia Plantarum* **14**, 896–908.

*(126) SN, E. (1961) Fotosyntese på en helt anden måde (in Danish; Another type of photosynthesis). *Naturens Verden*, 61–64.

(127) SN, E. (1961) Plantecellens beskyttelse mod for stærkt lys (in Danish; The protection of the plant cell against too strong light). *Meddelelser fra Videnskabernes Selskab Oversigt* 1960/61, 56–7 (Abstract).

(128) SN, E. (1961) The primary production in the waters west of Greenland during July 1958. *Rapports et Procès-Verbaux des Réunions* **149**, 158–9.

(129)   SN, E. & Hansen, V.K. (1961) Undersøgelser over planteplanktonets stofproduktion i de danske farvande. Fiskeriundersøgelser i 1960 ved Danmark, Færøerne og Grønland (in Danish: Investigations on phytoplankton production in Danish waters. Fisheries investigations in 1960 in Denmark, The Faroes and Greenland). *Skrifter udgivet of Kommisionen for Danmarks Fiskeri- og Havundersøgelser* **21**, 27–38.

(130)   SN, E. & Hansen, V.K. (1961) Influence of surface illumination on plankton photosynthesis in Danish waters (56° N) throughout the year. *Physiologia Plantarum* **14**, 595–613.

*(131) SN, E. (1961) Nytt från fotosynteseforskningen (in Swedish: News from the research of photosynthesis). *Naturens Värld* **1961**, 289–91.

*(132) SN, E. (1962) Fotosynte på ett helt annat sätt (in Swedish: Another type of photosynthesis). *Naturens Värld* **1962**, 381–4.

(133)   SN, E. (1962) Anton Frederik Bruun (Obituary). *Universitetets Festskrift*, **November**, 157–62. Copenhagen.

(134)   SN, E. (1962) On the biology of the Indian Ocean. *ICSU Review* **4**, 9–13.

(135)   SN, E. (1962) Inactivation of the photochemical mechanism in photosynthesis as a means to protect the cells against too high light intensities. *Physiologia Plantarum* **15**, 161–71.

(136)   SN, E., Hansen, V.K. & Jørgensen, E.G. (1962) The adaptation to different light intensities in *Chlorella vulgaris* and the time dependence on the transfer to a new light intensity. *Physiologia Plantarum* **15**, 505–7.

(137)   SN, E. (1962) The relationship between phytoplankton and zooplankton in the sea. *Rapports et Procès-Verbaux des Réunions* **153**, 178–82.

(138)   SN, E. (1962) On the maximum quantity of plankton chlorophyll per surface unit of a lake or the sea. *Internationale Revue der gesamten Hydrobiologie* **47**, 333–8.

(139)   SN, E. & Jørgensen, E.G. (1962) The physiological background for using chlorophyll measurements in hydrobiologie and a theory explaining daily variations in chlorophyll concentration. *Archiv für Hydrobiologie* **58**, 349–57.

(140)   SN, E. (1963) On bicarbonate utilization by marine phytoplankton in photosynthesis. With a note on carbamino carboxylic acids as a carbon source. *Physiologia Plantarum* **16**, 466–9.

(141)   SN, E. (1963) II Fertility of the Oceans. 7. Productivity definition and measurement. In: *The Sea*, Vol. 2, pp. 129–64. Wiley Interscience, New York.

(142)   SN, E. (1964) Det videnskabelige arbejde på Botanisk Laboratorium (in Danish: The scientific work at the Botany Department). *Farmaceuten* **27**, 5–7.

(143)   SN, E. (1964) Recent advances in measuring and understanding marine primary production. *Journal of Ecology* **52** (Suppl.), 119–30.

(144)   SN, E. (1964) Investigations on the rate of primary production at two Danish light ships in the transition area between the North Sea and the Baltic. *Meddelelser Fra Kommisionen for Danmarks Fiskeri- og Havandersøgelser* **4**, 31–77.

(145)   SN, E. & Park, T.S. (1964) On the time course in adapting to low light intensities in marine phytoplankton. *Journal du Conseil International pour l'Exploration de la Mer* **29**, 19–24.

(146)   SN, E. (1964) On a complication in marine productivity work due to the influence of ultraviolet light. *Journal du Conseil International pour l'Exploration de la Mer* **29**, 130–35.

(147)   SN, E. (1964) On the determination of the activity in the ^{14}C-ampoules. International

Conference for the Exploration of the Sea. *C.M. Plankton Committee*, Vol. **105**, pp. 1–2 (typed hand-out not to be cited without prior reference to the author).

(148) SN, E. (1965) On the determination of the activity in ^{14}C-ampoules for measuring primary production. *Limnology and Oceanography* **10** (Suppl.), R247–52.

(149) SN, E. (1965) On the terminology concerning production in aquatic ecology with a note about excess production. *Archiv für Hydrogbiologie* **61**, 184–9.

(150) Jørgensen, E.G. & SN, E. (1965) Adaptation in plankton algae. *Memorie dell Instituto Italiano di Idrobiologia* **18** (Suppl.), 37–46.

*(151) SN, E. (1965) Livets begyndelse og udvikling på jorden (in Danish; The beginning and evolution of life). *Kronik i Politiken* 11 December (newspaper article).

*(152) SN, E. (1965) Om livets begyndelse og udvikling på jorden (in Danish: About the beginning and evolution of life). *Farmaceutisk Tidende* **50**, 1193–8.

(153) SN, E. (1966) The uptake of free CO_2 and HCO_3 during photosynthesis of plankton algae with special reference to the coccolithophorid *Coccolithus Huxleyi*. *Physiologia Plantarum* **19**, 232–40.

(154) SN, E. (1966) The influence of CO_2 concentration and pH on two *Chlorella* species grown in continuous light. *Physiologia Plantarum* **19**, 270–93.

(155) SN, E. (1967) Laboratory experiments with plankton algae as a means to understand primary production. *Helgoländer Wissenshaftliges Meeresuntersuchungen* **15**, 135–6 (Abstract).

(156) SN, E. & Jitts, H.R. (1967) Report of a meeting of the joint group of experts on radiocarbon estimation of primary production. Copenhagen 24–26 October 1966. *Unesco Technical Papers in Marine Science* No. **6**, UNESCO, 6 pp.

(157) SN, E. & Jørgensen, E.G. (1968) The adaptation of plankton algae. I. General part. *Physiologia Plantarum* **21**, 401–13.

(158) SN, E. & Jørgensen, E.G. (1968) The adaptation of plankton algae. III. With special consideration of the importance in nature. *Physiologia Plantarum* **21**, 647–54.

*(159) SN, E. (1968) Planteplankton (in Danish; Phytoplankton). *Danmarks Natur* **3**, 312–29. Politikens Forlag, Copenhagen.

(160) SN, E. (1968) Julius Grøntved, 1899–1967 (Obituary). *Skrifter udgivet of Kommisionen for Danmarks Fiskeri- og Havundersøgelser* **28**, 5–6.

(161) SN, E. (1969) Influence of poison on photosynthesis of unicellular algae. Paper given at *IBP/PP Technical meeting. Productivity of photosynthesis systems. Models and Methods*, Trebon, 1969 (Supplement).

(162) SN, E. (1969) Prediction and measurements of photosynthetic productivity. *Proceedings of the IBP/PP Technical Meeting*, Trebon, September 14–21, 555–8.

(163) SN, E. (1969) General remarks. Mediterranean productivity project. *NATO Subcommittee Oceanographic Research Technical Report* **47**, 1–16.

(164) SN, E. (1969) Ecological Sciences in Denmark. *Intecol Bulletin* **1**, 31–3.

(165) SN, E., Kamp-Nielsen, L. & Wium-Andersen, S. (1969) The effect of deleterious concentrations of copper on the photosynthesis of *Chlorella pyrenoidosa*. *Physiologia Plantarum* **22**, 1121–33.

(166) SN, E. (1969) Book review. *Botanisk Tidskrift* **64**, 281.

(167) Goldman, C.R., SN, E., Vollenweider, R.A. & Wetzel, R.G. (1969) The ^{14}C light and dark bottle technique. In: *A Manual on Methods for Measuring Primary Production*

        in Aquatic Environments, (ed. R.A. Vollenweider) pp. 88–91. IBP Handbook No. **12**.
        Blackwell Scientific Publishers, Oxford.
*(168) SN, E. (1970) Produktionsmålinger i havet og vore ferske vande (in Danish;
        Measuring production in the sea and in fresh waters). *Vand* **1**, 17–21.
 (169) SN, E. (1970) Primary production in the sea. *WHO training course on the coastal
        pollution control*, August 2–29 (Handout).
 (170) SN, E. & Kamp-Nielsen, L. (1970) Influence of deleterious concentrations of copper
        on the photosynthesis of *Chlorella pyrenoidosa*. *Physiologia Plantarum* **23**, 828–40.
 (171) SN, E. & Wium-Andersen, S. (1970) Copper ions as poison in the sea and in
        freshwater. *Marine Biology* **6**, 93–7.
 (172) SN, E. (1970) Mass and energy exchange between phytoplankton populations and
        environment. Discussion Section 6. *Proceedings IBP/PP Technical Meeting. Pro-
        ductivity of photosynthesis systems. Models and methods*, Trebon. Pudoc, Wagen-
        ingen, 1970, pp. 555–8.
*(173) SN, E. (1971) Stofproduktionen i havet og i vore ferske vande (in Danish: The
        production of organic matter in the sea and our fresh waters). *Dansk Natur-Dansk
        Skole, Årsskrift* **1970**, 13–19.
 (174) SN, E. (1971) The balance between phytoplankton and zooplankton in the sea. In:
        Reading in Marine Ecology (ed. J. Nybakken), pp. 116–26. Harper & Row, New York.
 (175) SN, E. & Wium-Andersen, S. (1971) The influence of Cu on photosynthesis and
        growth in diatoms. *Physiologia Plantarum* **24**, 480–84.
 (176) SN, E & Willemoes, M. (1971) How to measure the illumination rate when inves-
        tigating the rate of photosynthesis in unicellular algae under various light conditions.
        Internationale Revue der gesamten Hydrobiologie **56**, 541–66.
*(177) SN, E. (1971) De danske vande (in Danish; The Danish waters). *Kronik i Politiken*
        26 April (Newspaper article).
 (178) SN, E. (1971) Production in coastal areas of the sea. *Tholassia Jugoslavia* **7**, 383–91.
*(179) SN, E. (1971) Measurement of the fertility of the sea. *Nautilus* **11**, 2–3.
 (180) SN, E. (1971) Reciepientgruppens rapport (in Danish: Report from the group to
        evaluate recipients). *Vand* **2**, 53–5.
 (181) SN, E. and Wium-Andersen, S. (1972) Influence of copper on photosynthesis of
        diatoms, with special reference to an afternoon depression. *Verhandlungen Inter-
        nationale Vereinigung für Theoretische und Angewante Limnologie* **18**, 78–83.
 (182) SN, E. (1972) The rate of primary production and the standing stock of zooplankton
        in the oceans. *Internationale Revue der gesamten Hydrobiologie* **57**, 513–16.
 (183) SN, E. (1972) Mekanisk rensning contra biologisk rensning (in Danish; Mechanical
        versus biological waste water treatment). *Stads- og Havneingeniøren* **72**, 239–41.
 (184) SN, E. (1973) *Hydrobiologi. En introduktion til belysning af dens for-
        udsætninger* (in Danish; Hydrobiology. An introduction), pp. 204. Polyteknisk
        Forlag, Copenhagen.
 (185) SN, E. (1973) Fritz Gessner, 27 July 1905–20 December 1972 (Obituary) *Inter-
        nationale Revue der gesemten Hydrobiologie* **58**, 303–305.
 (186) SN, E. (1973) Kaj Berg. 13 April 1899–14 March 1972 (Obituary). *The Royal Danish
        Academy of Sciences and Letters Annual Report for 1972–1973*, 6 pp.
 (187) Jerlov, N.G. & SN, E. (1973) *Optical aspects of Oceanography*. Academic Press, New
        York.

(188)   SN, E. (1974) Store P-koncentrationers indflydelse ved varierende pH på encellede algers fotosyntese og vækst (in Danish; The effect of high P concentrations on the photosynthesis and growth of unicellular algae at variable pH). *Aquanalen*, 16–20.

*(189) SN, E. (1974) Velkomst til nordiske vandforskere (in Danish; A welcome to the aquatic ecologists from the Nordic countries). *Vand* **2**, 23–24.

(190)   SN, E. (1975) *Marine Photosynthesis with Special Emphasis on the Ecological Aspects*. Elsevier Oceanography Series, Vol. **13**, pp. 1–141.

(191)   SN, E., Wium-Andersen, S. & Rochon, T. (1975) On problems in G.M. countings in the C^{14}-technique. *Verhandlungen Internationale Vereinigung für Theoretische und Angewante Limnologie* **19**, 26–31.

(192)   SN, E. (1976) The Carlsberg Laboratory and the exploration of the sea. In: *The Carlsberg Laboratory 1876–1976* (eds H. Holter & K. Max Møller), pp. 331–45. Reitzel, Copenhagen.

(193)   SN, E. & Rochon, T. (1976) The influence of extremely high concentrations of inorganic P at varying pH on the growth and photosynthesis of unicellular algae. *Internationale Revue der gesamten Hydrobiologie* **61**, 407–415.

(194)   SN, E. & Bruun Larsen, H. (1976) Effect of $CuSO_4$ on the photosynthetic rate of phytoplankton in four Danish lakes. *Oikos* **27**, 239–42.

(195)   SN, E. (1976) Johannes Krey 1912–1975 (Obituary). *Journal du Conseil International pour l'Exploration de la Mer* **37**, 1–2.

(196)   SN, E. (1976) Hvad betyder P-tilførslen fra land til Øresund (in Danish; What is the effect of the P-load to the Oeresund). *Øresunds Vand-Kvalitet* **1971–1974**, 111–14.

(197)   SN, E. (1976) The carbon flow in marine ecosystems. *Coastal Pollution Control* **1**, 193–200.

(198)   SN, E. (1977) The carbon-14 technique for measuring organic production by plankton algae. A report on the present knowledge. *Folia Limnologica Scandinavica* **17**, 45–48.

(199)   SN, E. (1977) Poul Laurits Larsen, 30 October 1909–8 March 1976 (Obituary). *Oversigt over Videnskabernes Selskabs Virksomhed*, June 1976–May 1977, 130–35.

(200)   SN, E. (1978) Principal aspects concerning the batch technique in algal assays. *Mitteilungen, Internationale Vereinigung für Theoretische und Angewante Limnologie* **21**, 81–7.

(201)   SN, E. (1978) Growth of the unicellular alga *Selenastrum capricornutum* as a function of P. With some information also on N. *Verhandlungen Internationale Vereinigung für Theoretische und Angewante Limnologie*. **20**, 38–42.

(202)   SN, E. (1978) Growth of plankton algae as a function of the N-concentration, measured by means of a batch technique. *Marine Biology* **46**, 185–9.

Unpublished manuscripts and preparative notes on European churches.

SN, E. *Oldkirkerne i Ravenna og i det øvrige laguneområde* (Primitive (primeval) churches in Ravenna and in the laguna area). Accepted, but never printed.

SN, E. *Nogle gamle engelske katedraler* (Some old English cathedrals).

SN, E. *Nogle romanske og gotiske kirker i Nord-Italien* (Some Roman and Gothic churches in northern Italy).

SN, E. *Gotiske katedraler i Paris-området* (Gothic cathedrals in the area of Paris).

SN, E. *Serbiske og makedonske klosterkirker* (Serbian and Macedonian monastery churches).

SN, E. *Tyske kirker fra renaissancen* (German churches from the renaissance).

Index

Aaby Jensen, E. 4, 5
abiotic carbon cycle 320–24, 345
acclimation 63–4
adaptation 63, 229
ammonium 326–7, 332
assimilation numbers 238—40
Atkins, W.R.G. 26–7
atmosphere 304–5, 320

bicarbonate 2, 20
biological reactions, evolution of carbon cycle and
    324–5
biomass *see* plankton biomass
biomass-limiting nutrients 111
Brandt, Karl 23, 359, 362–3

C⁴ photosynthesis 60–61, 125
Calvin cycle 44, 48–55, 325
carbon 14 (^{14}C) method 4–8, 9–12, 16, 31–2, 33, 36,
    37, 110, 222, 228–9, 366
  calibration 7
  measurement of plankton production 90–98,
    104–5
  interpretation of measurement 87–98
carbon concentrating mechanism 60–61, 124–5
carbon cycle, evolution of 318–46
  abiotic carbon cycle 320–24, 345
  biological reactions 324–5
  carbon burial and role of sedimentation on
    continental margins 335–7
  contemporary view of 344
  energy 326
  glaciation and 322, 334–5, 343–4
  nitrogen crisis of the early Proterozoic 332
  nitrogen fixation and origins of coupled redox
    reactions 326–7
  oceanic mixing and 341–3
  origin of the elements 318–20
  oxygen evolution 327–31
    rise and stabilisation of oxygen 337
    when first appeared 331–2
  Paleozoic oceans 338–41

phosphorus and role of glaciation in
    supercharging the ocean 334–5
carbon dioxide limitation 61, 69–70
carbon isotope fractionation 128–30
carbon reduction (Calvin) cycle 44, 48–55, 325
carbonic anhydrase 127–8
carboxylation
  beta-carboxylation pathway 62
  origins of 325
  photorespiration and 58˙
cells 299–300
  cell quota models 131–3
  diffusion limitation and effect of cell size 119–22
chemical efficiency 294
chlorin synthesis 329
chlorophyll(s) 16, 47, 82, 85, 100, 147, 229–33
climate 156–7
  regional-scale influences on long-term dynamics
    of lake plankton 265–86
    case studies 268–78
    discussion 278–86
    impact of year-to-year weather changes 266–8
    proxy indicators of long-term weather change
      268
climax community 277
co-limitation 134–5
communities
  climax community 277
  patterns of community responses to variable
    pelagic forcing 197–203
  phototrophic microbial community structure
    240–45
  sustained and aperiodic variability in production
    and community structure 222–58
    case studies 245–57
    data set 228–45
compensation point 25
competition 197–9
concentrating mechanisms 60–61, 124–5
Conseil Permanent International pour
    l'Exploration de la Mer (ICES) 12, 25, 34,
    35, 37

386   *Index*

snow 266
soil 306–8
standing waters 309–13
see also freshwaters
Steemann Nielsen, Einer 1–14, 17, 30–31, 34, 109,
350–51, 366
academic career 1
controversy with Riley 8–10, 30, 142, 225
curriculum vitae 2
experimental studies 1–2, 12–13, 27
carbon$_{14}$ method and 4–8, 9–12, 16, 31, 37, 228,
366
Galathea expedition 6, 33, 141, 223
portrait 3
retirement 14
writings 1, 35, 37
sustainability 291, 304

temperature 20, 24, 26, 67
thermal efficiency 294

United Nations
Food and Agricultural Organisation (FAO) 36
UNESCO 12, 34, 37

variability, environmental 63–70
variability of plankton and plankton processes
141–53
directions for future work 152–3
global scale production and 166–71
interannual variability in freshwaters 187–216
patterns of community responses to variable
pelagic forcing 197–203
response thresholds of phytoplankton to
variable pelagic forcing 203–14
understanding interannual variability in
phytoplankton 214–16

variability and constancy in phytoplankton
seasonality 187–94
variability in environmental forcing 194–7
large-scale estimation in the open ocean 142–3,
150–51
parameterisation of mesoscale variability 147–52
parametrisation of mesoscale processes for
large-scale estimation of productivity
150–51
relationship to historical context 151–2
spectral estimation 147–9
transfer functions 149–50
physical mechanisms and scales of mesoscale
variability 143–7
consequences for scales of biological
variability 147
consequences for transport of nutrients 144
eddy pumping 144–6
shear-induced mixing, fronts, jets and
horizontal transport 146
sustained and aperiodic variability in production
and community structure 222–58
case studies 245–57
data set 228–45
vegetation 305–6
vulcanism 321

waste 13-14
water 295–7, 318, 320, 326, 330
weather *see* climate
wind 196, 266–8
World Ocean Circulation Experiment 87

yield measurements of plankton production 79-80,
81, 85-7

zeaxanthin 257

Fig. 1.1   Einer Steemann Nielsen, professor at the Freshwater Biological Laboratory, 1975

Steemann revisited this research area, but now working with phytoplankton species (see appendix references 140, 153, 154).

It has not been possible to find out exactly when and how the idea of the ^{14}C method came to life. Steemann never precisely told any of the many persons I have spoken to and he did not give any clues during the lectures in hydrobiology that I listened to in 1974. My impression is that the idea matured over time and most probably emerged from a combination of factors. One source of inspiration and perhaps the most important one came from his reading about the use of ^{14}C-dating invented by Libby in 1947 and the personal contacts Steemann had with the dating-laboratory of the Danish National Museum in Copenhagen. His personal reprint collection shows that he developed a great interest for the use of radioactive tracers to study photosynthesis; the reprints cover the work with ^{18}O and ^{11}C by Ruben and Kamen from 1939 and they even included a hand-typed reprint on the natural ^{12}C/^{13}C ratio by Nier and Gulbransen from 1939. Unfortunately, he did not date a hand-written request to the University library, which I found stuck into one of these old reprints. Hence, it is not absolutely clear if these reprints were ordered before or after the war. However, most of his reprints on the use of isotopes are with original covers, so I guess they were requested from the authors soon after the war. A note on one of the library orders says 'not yet arrived' for a 1940 issue of the *Journal of the American Chemical Society*. I have the impression that most of his reprints were collected over a short period between 1947 and 1949 and that Steemann was extremely interested in the development of plant physiology in the United States during and shortly after World War II – a time when he had no access to American journals. His reprint collection also covers the partly quantitative use of the carbon isotope to study the chemistry of photosynthesis by Calvin and co-workers in the mid and late 1940s, among these a paper by Weigl and Calvin (1949) comparing ^{14}CO$_2$ and ^{12}CO$_2$ uptake in barley.

The practical and theoretical skills required to use ^{14}C in biological studies were brought to Denmark by Lise Schou and Hilde Levi. In the late 1940s the former had used radiocarbon researching on the chemistry of photosynthesis, publishing together with Bassham and Calvin in 1950, while Hilde Levi had worked in Libbys' laboratory in 1947. Dr Hilde Levi from the Zoophysiological Laboratory, University of Copenhagen, taught Steemann to work with tracers and helped him, according to several acknowledgements, to work out the method (see appendix references 66, 98). The civil engineer E. Aaby Jensen also helped Steemann with the method, but neither of them (Steemann or Aaby) had worked with radioactive tracers before the radiocarbon were available in Denmark, so the help offered by Hilde Levi was appreciated (and needed).

A few hand-written notes on four pieces of paper left in one of the reprints by Ruben show that Steemann trained himself in the use of isotopes and the new discoveries within the biochemistry of photosynthesis. One of the notes has the photosynthetic equation labelled with respect to ^{18}O. With his mind working mostly with the quantitative and ecological aspects of photosynthesis, it seems an obvious step to replace the ^{18}O with ^{14}C, when he learned about this isotope shortly after the war and from the work of American plant physiologists and Libby (the ^{14}C-dating method). Steemann knew the carbon uptake mechanisms in photosynthesis: '... it is of no importance whether this process [photosynthesis] is measured through the assimilation of CO_2, ... the production of O_2, or ... the formation of organic matter' (see appendix reference 66). He also was very familiar with the chemistry of inorganic carbon from his close co-operation with the chemist Professor Faurholt at the School of Pharmacy. From 1944 to 1969 Steemann headed the Botany Department of the Royal Danish School of Pharmacy. Steemann's previous work with the oxygen light–dark bottle technique made him fully aware that this method was not sensitive enough to be used in oligotrophic oceans (see appendix references 2, 11). He stated that results from such areas based on the oxygen method and long-term incubations (the high values published by Riley (1939) and co-workers):

'... seem incredible [and] seemed to be in absolute disagreement both with measurements of the standing crop of phytoplankton in the ocean and with theoretical considerations concerning the background of oceanic productivity' (see appendix references 66, 98).

Steemann's previous work with plankton production made him in 1949 the first choice to be invited to join the Danish *Galathea* expedition round the world:

'The senior author [Steemann] was excited by the idea of attempting to measure the organic productivity in the sea ... but a new method therefore [due to the low sensitivity of the oxygen method] had to be developed' (see appendix reference 98).

This is as close as we get to an answer in any of his articles. The first quantity of radiocarbon from Oak Ridge, USA had arrived in Denmark during the summer of 1949, paid for by the Carlsberg Foundation and 'the availability of this tracer made it possible to develop a new technique. Although the time was rather short for such work, there was no choice' (see appendix reference 98).

In co-operation with E. Aaby Jensen, the more technical aspects of the method were developed over the next half-year at the School of Pharmacy. According to the Acknowledgements in the 1952 paper (see appendix reference 66) Hilde Levi was also involved. Unfortunately, she had no recollection of that period when I talked with her; however, she is now at the age of 90 (October 2000). It is my understanding that the idea itself most likely was present before the radiocarbon arrived in Denmark and that Steemann, as usual, had no doubt about his idea and did not bother to involve other scientists too much in the work despite the fact that he and Aaby Jensen had never worked with radioactive isotopes before.

All filters from the initial experiments were counted at the Zoophysiological Laboratory, where Hilde Levi participated in the construction of the first Geiger counter in Denmark. During this period Steemann also used Calvin's book on isotopic carbon as 'the laboratory bible' (Calvin *et al.*, 1949). All samples from the *Galathea* expedition were counted on board using a specially constructed thin-window Geiger-Müller counter made by Brüel and Kjær. The current for the counter was supplied by a special generator isolated from other circuits on the ship.

The beauty of the method can be presented by the very simple equation:

$$^{12}CO_2 \text{ uptake}/\,^{12}CO_2 \text{ present} = \,^{14}CO_2 \text{ uptake}/\,^{14}CO_2 \text{ present} \qquad (1.1)$$

Thus, the $^{12}CO_2$ uptake can easily be calculated, if the other three variables are known.

The method was ready for use when the *Galathea* expedition started in October 1950. Steemann joined the ship in Lisbon (Fig. 1.2) and worked on the ship for three months. Aaby Jensen did all the measurements after Steemann returned to Denmark. A short note of the results from the first leg in the southern Atlantic along the African coast was prepared on the ship and published in *Nature* in 1951 (see appendix reference 61). The method was described in more detail in 1952 (see appendix reference 66) together with the results from the Indian Ocean. This article was also prepared on the ship. It is impossible to carry out an experiment with the ^{14}C method from the description in the *Nature* article, thus, the 1952 paper (see appendix reference 66) is identified as the proper publication of the method.

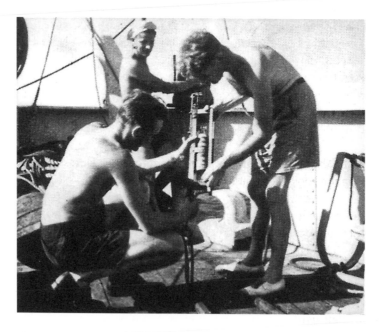

Fig. 1.2 Working on *Galathea*. Emptying the all-glass water sampler mounted in a teak carrier. Steemann Nielsen to the right with the fancy white shoes

Although Steemann over the next years continued to improve the method and the interpretation of its results, especially concerning the interference from and interpretation of algal respiration (see appendix references 93, 97, 110) and calibration of ampoules (see appendix references 147, 148), the original method has basically survived intact. He was very sceptical about changes and, over a number of years, refused to change to liquid-scintillation counting at the Carbon-14-Agency and at his home base from 1969, the Freshwater Biological Laboratory in Hillerød. One major change occurred when Steemann in the late 1960s had to accept that his original calibration of the specific activity in the ampoules using $BaCO_3$ were approximately 31% in error and that previous ^{14}C measurements should be increased by a factor 1.45. The calibration values found by Charles Goldman (1968) using gas-phase counting were, after a long period with an extremely sceptical attitude, finally accepted by Steemann. The argument with Goldman was settled with a bottle of Danish Aquavit. However, Steemann only once quoted the work by Goldman (see appendix reference 191).

The canonical status of the method and a certain unwillingness to read the original papers has meant that many subsequent workers have relied on received information, which has resulted in some misuse and misinterpretation of results compared with the original assumptions and warnings given by Steemann. Two of these are that the method cannot be used at low irradiances and when the heterotrophic bacterial activity is high compared with photosynthesis (high dark uptake). The method has often been challenged, but most aspects (e.g. contamination of ampoules and water samples by metals, impurities of the isotope, organic excretion, use at low light intensities, use with high bacterial activity and the question of whether net or gross photosynthesis is measured) were all treated by Steemann. The review of the method from 1957 (see appendix reference 98) might even be of use in student classes today to demonstrate experimental planning and how to care for details. Furthermore, the paper is a good indicator of the status of biological oceanography in the 1950s, where Steemann viewed plankton production in its ecological context related to both the chemical and the physical conditions of the sea.

In relation to the question of whether the method measures gross or net production, Steemann took the stand:

'In incubations lasting 3–4 hours, the method measures something between gross and net production as about 50 to 70% of the respiratory loss is recently assimilated $^{14}CO_2$. In experiments of very long duration the technique measures net production' (see appendix reference 93).

To correct for the respiratory loss in short incubations he used the following approach: at high light intensities microalgal respiration is about 10% of photosynthesis (in his terminology photosynthesis = gross production) and with an average 60% respiratory loss in the form of $^{14}CO_2$, the ^{14}C incorporation has to be corrected with 6%, because 60% of the 10% algal respiration is lost as $^{14}CO_2$. Together with an isotopic discrimination of 5% and a normal dark fixation of 1%, he therefore suggested to correct the incorporated ^{14}C by 10% (+5 +6 −1) to calculate gross production (see appendix references 66, 93). Steemann was fully aware that measurements at low light intensities had to be corrected with higher values. However, he explicitly said: 'The ^{14}C method is not suitable for the measurement of the intensity of photosynthesis at very low light intensities' (see appendix reference 66). His conclusion was also that for the calculation of photosynthesis per area, such corrections were of minor quantitative importance and could be ignored.

Steemann did measure low primary production in oligotrophic oceans, e.g. 43–58 mg C m^{-2} day^{-1} in the Sargasso Sea (these values should be some 45% higher due to the problem with calibration of the specific activity in the ampoules). In his own opinion he proved the ocean primary production data by Riley to be wrong and that previous estimates (reviewed by Rabinowitch, 1945) of ocean primary production had to be lowered by a factor of 10.

Steemann had a temper and was not very diplomatic in his critique of other people's work, if he considered it erroneously or falsely interpreted. He never strove for a diplomatic tone in his judgements. His controversy with Gordon A. Riley from Bingham Oceanographic Laboratory was never settled at the personal level. As far as I know, Riley did not want to speak to Steemann during his visits to the United States. Steemann has been quoted:

> 'He [Riley] has never understood productivity measurements [and] the results of these investigations [the oxygen technique with three to seven days incubation and subtraction of the light and dark bottles used by Riley] can hardly be considered realistic. The measured production seems to be the reduction of the bacterial respiration in illuminated experimental bottles, which was measured instead of the photosynthesis of the plankton algae' (see appendix reference 98).

The reduction was hypothesised to be caused by the detrimental effect of ultraviolet (UV) radiation.

The production controversy with Riley and Ryther started with Steemann's 1952 paper, when he had measured the production in the Indian Ocean. In this deep blue ocean he measured a surface and area photosynthesis some 20-fold lower than the values Riley (1939) had published for the Sargasso Sea and Rabinowitch (1945) had used to estimate a tenfold higher global production in the sea than on land. Steemann claimed that the long incubation times increased the bacterial respiration in the dark bottles due to wall growth and that an equivalent bacterial respiration in the light bottles was suppressed by UV radiation. Thus, he posited, the difference between the light and dark bottles, which Riley used as the measure of production, was not photosynthesis. In 1954 and 1955 Steemann (see appendix references 84, 87) changed the argument and now suggested that antibiotics produced by the algae were the respiration suppressing agents. One experiment with *Chlorella* was carried out to prove the case (see appendix references 87, 89) and was much later followed by a few further experiments (see appendix reference 125).

In 1953 Riley defended his high values in a lengthy letter to the editor:

'Without denying the possibility that the bactericidal effect may produce a significant error, I shall demonstrate by the following examples that the extremity of his [Steemann's] position is untenable.' (Riley, 1953)

Riley then forwarded four examples in defence of the oxygen method. These examples included a comparison of oxygen data from different sites in 'an ecological analysis', which apparently differed little more than 10%. To me the argument is incomprehensible and Steemann has pencilled a big question mark in his reprint of the 'letter'. Further, Riley, with good arguments, advocated that some of his results off the Dry Tortugas in the Gulf of Mexico (Riley, 1938) were contradictory to a bactericidal effect of light and that Steemann did not take account of the effects of temperature and nutrients on the relationship between photosynthesis and respiration. However, Riley neither could nor wanted to reject the high values from the Sargasso Sea ($1.5 \, \text{g C m}^{-2} \, \text{d}^{-1}$), but mentioned that in August 1947 photosynthesis was undetectable by oxygen experiments. He continued the defence by pointing out: (1) that Steemann's numbers might be invalidated by the loss of ^{14}C by respiration, which was not accounted for at low light intensities. Riley commented: 'One might wish he [Steemann] had spent more time experimenting with the method before using it on a long and important expedition'; (2) that the observed low and high values in the Sargasso Sea could be due to natural seasonal and geographical variability and that the Sargasso Sea might be different from the Indian Ocean. Riley added: 'Whether this difference indicates a different order of magnitude of productivity in the two areas or a serious error in the ^{14}C method remains to be seen'; and (3) what Riley considered his ultimate argument, namely that the biomass of the zooplankton in the Sargasso Sea had a daily carbon requirement of about $300 \, \text{mg C m}^{-2} \, \text{d}^{-1}$ and therefore had to be supported by an even higher photosynthetic production. During the defence Riley claimed: 'Although the attack on the oxygen method seems as empty as it is violent', he did make an opening by admitting 'it is immediately apparent that the ^{14}C technique is superior in sensitivity and accuracy'.

In the paper from 1954 (see appendix reference 84) Steemann presented ^{14}C values from the Sargasso Sea and also had Aaby Jensen doing a three-day oxygen experiment. In an appendix, Steemann presents and discusses the ^{14}C values from the surface waters, which were about 200 times lower than Riley's oxygen values. 'It would appear therefore that the "O_2-production" [Steemann's quotation marks] obtained by the oxygen technique cannot be due to photosynthesis'. In addition, Steemann with some arrogance dismisses Riley's zooplankton argument by stating '...derived from estimates of zooplankton food requirements, but details of the computation are, however, not given'. The strongest argument in the 1954 paper, however, came from a very detailed analysis of the light absorption in the Sargasso Sea. The conclusion was that with a photic zone of about 120 m, the water itself and dissolved organic matter were the main contributors to light attenuation and that an algal density and light absorption leading to a production as claimed by Riley would have resulted in a narrow photic zone. 'The waters of the Sargasso Sea are transparent because of the insignificant rate of organic production that occurs there as a result of the supply of nutrient salts'. As far as I know, Riley never replied, but in his unpublished autobiography, *Reminiscences of an Oceanographer*, he wrote a very short section on the controversy with the final statement: 'We parted on quite friendly terms. (Of course, all this is still far from settled.)' (Riley, 1984).

Ryther from Wood Hole Oceanographic Institution might also have been slightly disturbed by Steemann's choice of words: 'In these American investigations it was often not the actual photosynthesis by phytoplankton which was measured, but something else' (see appendix reference 66). Ryther (1954) said: '[Steemann's] ...preliminary data for mid-ocean photosynthesis ... was at once both stimulating and disturbing.' Ryther and Vaccaro (1954), in a comparison between the oxygen and the ^{14}C methods, mostly observed good agreement in exponentially growing algal cultures, but in apparently starved cultures where the ratio of respiration and photosynthesis approached unity, 'the ^{14}C experiments give values which are lower by an order of magnitude'. However, they had problems with the oxygen method in long-term incubations (96 h) of diluted cultures. A direct comparison between the methods in oligotrophic oceans, including the Sargasso Sea, and with the use of three-day oxygen incubations showed ^{14}C values from 10 to 100 times lower than the oxygen values (Ryther, 1954). Furthermore, Vaccaro and Ryther (1954) could not experimentally verify the suggestion of a bactericidal effect of UV radiation. Ryther took the stand that net photosynthesis was measured with the ^{14}C method and that the difference between the methods could be explained if algal respiration was equal to gross photosynthesis. By claiming that the loss of ^{14}C by respiration could not be accounted for in experiments of more than 24 h duration, Ryther and Vaccaro tried to 'reverse' the discussion by the phrase 'the ^{14}C method is not recommended for use in long-term experiments'. These results were presented with a reasonable amount of diplomacy and the 'honour of the American school' was saved. Ryther also used the argument that the Indian Ocean and the Sargasso Sea might not be comparable; however, he soon had to abandon this argument. Ryther did use some diplomacy to tell both Riley and Steemann that they should not use their most extreme values to support an average oceanic production value.

Steemann never doubted that prolonged incubations in small bottles could not be used to measure photosynthesis and specifically so in deep blue oligotrophic waters and with the oxygen method. However, he was probably slightly disturbed by the argument that nutrient-deficient algae with very high respiration:photosynthesis ratios would result in erroneous ^{14}C measurements. So, with the use of nutrient-deficient *Chlorella* cultures he showed experimentally that high respiration at the level of 86% of gross photosynthesis did not violate his interpretation of the ^{14}C method (see appendix reference 97). Thus, he 'was unable to confirm Ryther's results' and further, with the so familiar attitude 'it was claimed [by Ryther] without giving any details that the experiments were made under conditions of nutrient deficiency'. Steemann continues: 'There is not, however, much reason to consider this possibility [i.e. that Steemann had used a different species], as Ryther and Vaccaro (1954) have given an experimental proof of the real cause, which seems to be inadequate experimental technique.' Steemann then evaluates the experiments and shows that the culture used by Ryther and Vacarro must have been dying and losing a large amount of biomass to the dissolved phase. He also criticised that Ryther and Vaccaro did not use stirring for such long incubations. And Steemann kept the hard tone: 'The experiments published by Ryther (1954) were made with the same inadequate technique. Thus, there is not much reason to discuss the results which differ so much from those published in the present article.' Steemann also presented an analysis of the ecological consequences if the phytoplankton in the Sargasso Sea had such a high respiration: 'The algae respiration must consequently according to Ryther always be at least 180–198% of the photosynthesis rate there. It is self-evident that such conditions cannot exist constantly in nature.'

At a jointly sponsored ASLO and ESA symposium on primary production in Michigan in 1955, Ryther (1956) once more advocated that in his experience the oxygen method could be used for 24 to 48 h, but 'as a general practice should probably not exceed the daylight portion of a day'. Further, he presented new evidence that the ^{14}C method measures net photosynthesis, although he realised that the reason for this is not clear. Concerning the ^{14}C method, Ryther's conclusion was: 'It does not provide, as currently employed, a quantitative measurement of either gross or net daily production.' It might, at this point, be emphasised that a direct comparison between oxygen and ^{14}C is invalid if not corrected by 1.45 due to the error of the specific activity determination with the $BaCO_3$ technique. A comparison also includes an unknown photosynthetic quotient. The problem with respiratory loss of recently assimilated $^{14}CO_2$ has to my knowledge not been solved. One experimental attempt was done by Williams *et al.* (1996). In a direct comparison between ^{12}C and ^{14}C uptake, their results could best be explained by 100% recycling of respiratory losses and that the ^{14}C values were very close to net photosynthesis.

At an ICES meeting in 1957 (see appendix reference 103), Steemann once more addressed his American 'friends', explaining the shortcomings of the oxygen method and that he did not accept the explanation by Ryther (1954) that the rate of photosynthesis nearly equals the optimal rate of photosynthesis in the surface layers of deep blue ocean waters. Steemann argued that the algal biomass present in the Sargasso Sea, and assuming a growth rate of about 1 day^{-1}, could support 'a production of 0.7 mg C m^{-3} day^{-1}, as measured by Ryther (1954) with the carbon-14 technique, [and] fits excellently', but could not support the 100 times higher oxygen value. In the book on Danish Limnology prepared for the SIL congress in Copenhagen in 1977 (see appendix reference 198), Steemann did not take the opportunity to summarise or comment on the controversy. It should here be mentioned that Ryther and Steemann were on friendly terms, had personal communication and Ryther's paper from 1954 was forwarded 'To Einer. With best regards. John.'

Although some questions were left behind, the ^{14}C method became internationally accepted at an ICES meeting in Bergen in 1957 (see Barber and Hilting, Chapter 2). Many researchers found their way to Copenhagen to learn the method, among these Wilhelm Rodhe from Uppsala. After the Bergen meeting, Steemann approached UNESCO with a view to sponsorship of an independent and international Agency to carry out teaching and services in connection with the method. In 1958, UNESCO granted $1000 and Steemann started the International Carbon-14-Agency with ten pieces of equipment (the inventory purchase list to UNESCO) in a very small room at the Fisheries Research, Charlottenlund Castle, Copenhagen. Vagn Hansen was the first practical leader of the laboratory and Steemann became its chief supervisor until retirement.

Steemann was a man with a lot of humour and a good (noisy) laugh. He very much enjoyed being the centre of scientific discussions. He used his humour during such times and would tease to stimulate discussions and to have fun. He was also very supportive to people he wanted to promote professionally. A few misjudgements can, however, be identified. He always greeted the many guests arriving in his laboratory with a positive attitude and was prepared to discuss and share his theoretical and practical experience. For lunch he always had a strange type of bitter malt beer (ship beer). Testing guests with a serving of this low-alcoholic speciality was an amusing experiment. In contrast, he was extremely direct and sharp, if he disagreed scientifically with someone. The controversy with Riley is a good example and some of his book reviews are testimonies to his sharp pen.

As might be expected, Steemann also became the first to use the ^{14}C method in freshwater. The measurements were carried out between 1953 and 1954 in Furesø and agreed with measurements with the oxygen technique (see appendix reference 106). Actually, the ^{14}C method was tested with water from Furesø before the *Galathea* expedition and some of the samples were filtered in Steemann's kitchen at home (no safety officer was around in those days). The results from Furesø were published in 1958, the same year as the results from Lake Erken done by Wilhelm Rodhe in Uppsala. Some of the leftover ampoules prepared for the *Galathea* expedition were in 1953 generously handed over to Pétur Jónasson, who succeeded Steemann as professor at the Freshwater Biological Laboratory in 1977. The ampoules were wrapped in toilet paper of different colours, according to their isotopic contents. The five-year primary production survey in Lake Esrum was published in 1959. Since then countless measurements with ^{14}C have been carried out both at sea and in lakes. An enduring standard method had arrived.

The laboratory experiments by Steemann were planned and carried out with an eye to every detail and most results had a very short time span from idea to experiment and manuscript. The combination of detailed planning, the demand for high-quality results and temper, did not make it an easy task helping Steemann in the laboratory. When he said to his technician and assistants 'tomorrow is laboratory day' it was not exactly the best cue for a good night's sleep. Mistakes or departure from his detailed protocols were not allowed: 'don't do that again' or in grave cases 'damned, never do that again'. Diplomacy was not the strong side of Steemann and a closed door was not enough to dampen an outspoken critique. He could be shaking in temper, for example when the technical staff, by law, were allowed every third Saturday off. If he considered a fundamental mistake to be present in an experiment or he was presented with a view showing signs of lack of knowledge and logic, he gave the direct and personal advice never to do ecological work in the future.

Although he never left the studies on primary production and its methodological aspects, most of his work in the 1960s and 1970s was on phytoplankton physiology, including a series of seminal papers on light adaptation via chlorophyll and pigment changes. These papers were almost exclusively published in *Physiologia Plantarum*. Very early in his career he also became aware of the toxicity of metals for algae. The construction of an all-glass water sampler for the *Galathea* expedition was apparently his idea (see appendix reference 76). However, an experimental approach to metal toxicity was first started in the mid-1960s together with his PhD student Lars Kamp-Nielsen. Lars was the only PhD student formally trained by Steemann, although he advised the doctoral thesis presented by E. G. Jørgensen. It was Steemann's hypothesis that trace amounts of copper ions in the upwelling from deep oceans can affect photosynthesis and kill some of the algae at the beginning of a bloom and can explain years with low recruitment of fish larvae. Low recruitment was a national disaster in Iceland. The experiments resulted in a series of relevant ecotoxicological papers and showed that organic excretion from diatoms can bind copper and remove the toxic effect. However, the original hypothesis was never proven.

The many experiments by Steemann on microalgal photosynthesis and growth are not famous for their number of replicates, but for their clear results. 'An impeccable experiment is more worth than a lousy experiment with lots of replicates', was in short Steemann's message to students and assistants. His attitude was that it should be possible to reproduce the general outcome of an experiment and not to waste energy doing a lot of replications within a single experiment. If the results did not show what he expected, something had gone wrong with the experiment.

Steemann was devoted to science and during his whole academic life he very willingly shared his experience and knowledge with others. He was more interested than his peers in getting the results presented to a broader public and was always ready to present a talk. He published a substantial number of small popular articles in newspapers, magazines and local journals.

In the 1970s Steemann very actively participated in the growing debate on environmental questions and how the increasing load of waste water from towns and nutrients from agriculture would affect the marine environment. He had a sharp debate with Danish environmentalists when he presented the view that the nutrient load from Denmark to its coastal waters would be of no harm and probably a benefit for fisheries. Major oxygen depletion events in the 1980s proved him wrong.

Steemann retired in 1977 and his last scientific paper was published in 1978. From retirement day he entirely withdrew from science: 'As an experimentalist I have nothing left to offer science.' Retirement gave him extra time for his great passion and hobby – the study of European churches, their paintings and stained-glass windows. Every spring he spent several weeks travelling around Europe. He had visited almost all churches of any importance in southern Germany, Switzerland and both northern and southern Italy; he was in Rome 13 times. Most Danish churches were of course known to him in great detail. Within this area too he wanted to share his knowledge with others. One of his manuscripts was accepted by the publisher Branner and Korch. Unfortunately, the publisher had economic problems and merged with another publisher. Steemann died before the situation with the new publisher was resolved. None of his works on churches was ever published. At the end of the publication list (see appendix at the back of the book), I present the titles of these books-to-be. For a lay person it is difficult to understand that he, as an amateur, wanted to compete with the professionals within this major area of cultural history. As a scientist he was very hard on people presenting what he considered wrong statements and misunderstandings; 'amateurs' he called them.

(Halfdan) Einer Steemann Nielsen was in all aspects a remarkable and singular man and an outstanding scientist, who with a very creative mind offered marine ecology and limnology an insight to the world of plankton production.

Acknowledgements

I appreciate the great help offered by Lise Steemann Nielsen, Eivind Gargas, Ann-Mari Bresta, Gunni Ærtebjerg, Lars Kamp-Nielsen, Hilde Levi, Daniel Conley and by Hans Fastig from the Botanical Library. A special thank you for the enthusiasm shown by Pétur M. Jónasson, who helped me to contact some of the people who worked with Steemann at the time when the ^{14}C method was developed. Finally, I acknowledge the many good suggestions made by Peter Williams and Colin Reynolds.

References

Calvin, M., Heidelberger, C., Reid, J.C., Tolbert, B.M. & Yankwich, P.F. (1949) *Isotopic Carbon. Techniques in its Measurements and Chemical Manipulation.* Wiley, New York.
Goldman, C.R. (1968) The use of absolute activity for eliminating serious errors in the measurements of primary productivity with C^{14}. *Journal du Conseil International pour l'Exploration de la Mer* **32**, 172–179.

Rabinowitch, E.I. (1945) *Photosynthesis and Related Processes*, Vol. II, Part 1. Interscience Publications Inc., New York.

Riley, G.A. (1938) Plankton studies. I. A preliminary investigation of the plankton of the Tortugas region. *Journal of Marine Research* **1**, 335–352.

Riley, G.A. (1939) Plankton studies II. The western North Atlantic, May–June, 1939. *Journal of Marine Research* **2**, 145–162.

Riley, G.A. (1953) Letter to the editor. *Journal du Conseil International pour l'Exploration de la Mer* **19**, 85–89.

Ryther, J.H. (1954) The ratio of photosynthesis to respiration in marine plankton algae and its effect upon the measurements of productivity. *Deep-Sea Research* **2**, 134–139.

Ryther, J.H. (1956) The measurement of primary production. *Limnology and Oceanography* **1**, 72–84.

Ryther, J.H. & Vaccaro, R.F. (1954) A comparison of the oxygen and the ^{14}C methods of measuring marine photosynthesis. *Journal du Conseil International pour l'Exploration de la Mer* **20**, 25–34.

Vaccaro, R.F. & Ryther, J.H. (1954) The bactericidal effects of sunlight in relation to 'light' and 'dark' bottle photosynthesis experiments. *Journal du Conseil International pour l'Exploration de la Mer* **20**, 18–24.

Weigl, J.W. & Calvin, M. (1949) An isotope effect in photosynthesis. *Journal of Chemistry and Physics* **17**, 210.

Williams, P.J.le B., Robinson, C., Søndergaard, M., *et al.* (1996) Algal ^{14}C and total carbon metabolism. 2. Experimental observations with the diatom *Skeletonema costatum*. *Journal of Plankton Research* **18**, 1961–1974.

Chapter 2:
History of the Study of Plankton Productivity

Richard T. Barber and Anna K. Hilting

2.1 Introduction

1952 is a milestone in the advance of biological oceanography; the introduction of the ^{14}C method in that year marked the end of a century-long struggle to develop a method to determine oceanic primary productivity with precision, accuracy and efficiency. Armed with Steemann Nielsen's new method, oceanographers could make more and better determinations and greatly increase resolution of temporal and spatial variability. More importantly, however, the primary productivity procedure was in principle standardised, so determinations from research groups around the world were comparable and could be synthesised to obtain an improved holistic estimate of ocean primary productivity. 1952 was, therefore, the end of an era of subjective and individualistic estimates of primary productivity and the beginning of an era of objective, team-driven international collaboration. This entirely new style of research exploited the increased technological and human resources that gave oceanography so much more analytical power in the second half of the twentieth century.

Determination of time-varying plankton productivity in the world ocean has been a major, if not *the* major, goal of biological oceanography from its beginning in the mid-nineteenth century. Soon after Steemann Nielsen introduced the ^{14}C method in 1952, it was adopted by several leading research groups to estimate global primary productivity. Combined with the second great breakthrough of the twentieth century – the remote sensing of phytoplankton biomass by optical means – the ^{14}C method has accomplished the goal that eluded biological oceanography for nearly 150 years. Steemann Nielsen's method provides productivity per unit chlorophyll and remote sensing provides a highly resolved chlorophyll field. Taken together they make possible the objective determination of time-varying global plankton productivity, which in itself is one of the leading accomplishments of the natural sciences in the twentieth century.

The early history of biological oceanography was treated comprehensively by Eric L. Mills (1989) in his scholarly volume: *Biological Oceanography: An Early History, 1870–1960*. In this chapter we focus on the history of the study of plankton productivity. Because recent and current developments in plankton productivity are the basis of many of the chapters in this volume honouring the jubilee of Steemann Nielsen's contribution, we concentrate here on the early and middle history of the study of plankton productivity – to the 1970s.

This history is an intellectual drama involving a large cast of interesting and strong-willed characters from several countries, but the role of Steemann Nielsen in the play is unique. He was, at the beginning of his career, a classical marine biologist following in the track of Gran and other prominent Scandinavian marine biologists. He shifted from taxonomy to productivity when his eyes gave him some trouble (see Chapter 1), but we suspect that the eye trouble provided a welcome excuse. As with Gran, it is hard to imagine that taxonomic studies would satisfy Steemann Nielsen's enormous drive, originality and fascination with powerful ideas. His pre-war career, which included the publication of several key papers on plankton production, was a success by any standard. As the war ended he set out to solve an emerging problem in plankton productivity studies.

2.2   Prehistory (to 1850)

It is difficult to study something without a concept of what is to be studied. While the concept of plankton primary productivity is central to an understanding of the ocean, it is a twentieth century construct. The use of the metaphor 'All flesh is grass' by the Old Testament prophet Isaiah indicates that his audience understood something of trophodynamics and such an understanding is not the least surprising in a nomadic culture dependent on grazing. In contrast to Isaiah's forceful use of food chain dynamics to make his point, the New Testament apostle Mark said: 'If someone would scatter seed on the ground, the seed would sprout and grow, he does not know how.' Mark's words emphasise that plant growth or production was a mystery. That mystery was so profound that there wasn't a name for the process until about 200 years ago. The central features of the primary productivity process – light adsorption, uptake of gaseous carbon dioxide and production of gaseous oxygen – could not be sensed by humans until the necessary technology was developed in the eighteenth century.

The concept of primary production waited in the wings until an accurate description of photosynthesis entered the scene late in the eighteenth century. Experiments and rapid publication of results (Table 2.1) fuelled a progression of discoveries from 1770 to 1804 by Lavosier, Priestley, Ingenhousz, Senebier and de Saussure that culminated in the understanding that in green plants:

$$CO_2 + H_2O + light \rightarrow new\ organic\ matter + O_2 \qquad (2.1)$$

and, further, that CO_2, H_2O and O_2 react in apparently fixed ratios (stoichiometrically).

Table 2.1 Milestones in the discovery of the nature of photosynthesis as presented by Rabinowitch (1945).

Name (nationality and profession)	Date	Milestone
Priestley (English theologian)	1772	Plants produce O_2
Ingenhousz (Dutch physician)	1779	Plants need light to produce O_2
Senebier (Swiss clergyman)	1782	Plants need CO_2 to produce O_2
Ingenhousz (Dutch physician)	1796	Plants need light and CO_2 to produce O_2 and organic matter
de Saussure (Swiss scientist)	1804	Participation of H_2O and light in O_2 production and stoichiometry of photosynthesis

The review of photosynthesis by Rabinowitch (1945; Table 2.1) evokes the heady intellectual competition that unfolded in Europe as these individuals, almost all amateur scientists, competed with each other to gain priority for their discoveries. In the end, however, it was their composite contribution that described the nature of photosynthesis, solved the ancient mystery of where the grass comes from and provided one of the crowning achievements of this era.

2.3 Further Discovery (1850–1912)

By 1800, when a reductionist, laboratory-based understanding of terrestrial photosynthesis was available, the stage was set for applying this new concept to aquatic and marine environments. Following the discovery of photosynthesis, however, it took about a half-century – until about 1850 – for curiosity about the nature of photosynthetic production to surface in aquatic science and it was another half-century before the process was accurately described (Chambers, 1912).

In the last decades of the nineteenth century, an understanding of aquatic primary production as an environmental chemical process was slowly taking shape in Europe and North America (Chambers, 1912). The initial advances to understand marine plankton productivity necessitated work with macroalgae, cyanobacterial mats or aquatic macrophytes to get the requisite photosynthetic biomass. Most of this work was done in freshwater in which the total carbon dioxide content either varied naturally or could easily be manipulated. Chambers (1912) reviewed the works of over 30 researchers who between 1841 and 1912 investigated the environmental chemical changes that characterise aquatic photosynthesis (Table 2.2). About half of the studies observed *in situ* changes in carbon dioxide and oxygen concentrations as plants grew in ponds or small lakes, and about half determined those changes with plants growing in glass bottles. These pioneering studies convincingly showed that an increase in aquatic plant biomass necessarily involves uptake of carbon dioxide

Table 2.2   Chronology of the use of O_2 production to quantify marine planktonic primary productivity. *In situ* ΔO_2 refers to the change in O_2 concentration in a body of unconfined water based on separate determinations. All other methods are based on the change in O_2 concentration in light and dark bottle incubations (L–D) except the O_2 bubble method and the O_2 production versus spectrum method. References can be found in Gaarder and Gran (1927), Mills (1989), Riley (1941, 1944 or 1963), Steemann Nielsen (1952), Steemann Nielsen and Jensen (1957) or Vinberg (1960).

Author	Date	Region	Accomplishment
Pütter	1908	Bay of Naples	1st ambient L–D O_2 production
Gran	1918	Christiania Fjord	1st immersed L–D O_2 production
Pütter	1924	Kiel Bay	L–D O_2 production
Pütter	1925	Kiel Bay	L–D O_2 production
Gran and Ruud	1926	Oslo Fjord	*In situ* ΔO_2
Pütter	1926	Tenerife Island	L–D O_2 production
Gaarder and Gran	1927	Christiania Fjord	1st immersed L–D O_2 production
Gran	1929	Norwegian coast	L–D O_2 production
Gran and Thompson	1930	Puget Sound, USA	L–D O_2 production
Marshall and Orr	1930	Scottish Fjord	*In situ* ΔO_2
Ruttner	1931	Java, Sumatra and Bali	O_2 bubble method
Gaarder	1932	Norwegian sea	L–D O_2 production
Steemann Nielsen	1932	Danish waters	L–D O_2 production
Cooper	1933	English Channel	*In situ* ΔO_2
Clarke and Oster	1934	US Atlantic coastal waters	1st O_2 production v. spectrum
Cooper	1934	English Channel	*In situ* ΔO_2
Gran and Braarud	1935	Bay of Fundy, Gulf of Maine	L–D O_2 production
Seiwell	1935	Western N. Atlantic	*In situ* ΔO_2
Seiwell	1935	Tropical N. Atlantic (3–13° N)	*In situ* ΔO_2
Steemann Nielsen	1935	Færøe Isl./Iceland/E. Greenland	L–D O_2 production
Steemann Nielsen	1937	Danish waters	1st annual cycle L–D O_2 production
Riley	1938	Gulf of Mexico	Ambient L–D O_2 production
Riley	1939	Western N. Atlantic (23–41° N)	1st on-deck L–D O_2 production
Steemann Nielsen	1940	North Sea/Baltic Sea	L–D O_2 production

Continued

Table 2.2   *Continued*

Author	Date	Region	Accomplishment
Riley	1941	Long Island Sound	*In situ* ΔO_2
Riley	1941	Long Island Sound	L–D O_2 production
Riley	1941	Georges Bank	On-deck L–D O_2 production
Riley	1946	Georges Bank	On-deck L–D O_2 production
Riley	1947	US Atlantic coastal waters	On-deck L–D O_2 production
Riley and Gorgy	1948	Subtropical N. Atlantic	*In situ* ΔO_2
Riley, Stommel and Bumpus	1949	Western N. Atlantic	On-deck L–D O_2 production
Steemann Nielsen	1951	Danish waters	L–D O_2 production

and production of oxygen. The source of inorganic carbon for aquatic plants was much more problematic than that for terrestrial plants, but as early as 1833 Raspail advanced the idea that the bicarbonate ion provides carbon for aquatic photosynthesis (Steemann Nielsen, 1947). Details of the carbon source were not worked out until 1907 (Gran, 1912), but the debate about the carbon source did not prevent experimentalists (Table 2.2) from measuring the uptake of inorganic carbon (Chambers, 1912).

Morren in 1841, working with a red-tide organism, was apparently the first to measure experimentally this inverse carbon dioxide uptake–oxygen production relationship (Chambers, 1912). After Morren, Hassak in 1888 showed the light dependence of inorganic carbon uptake using pH sensitive indicator dyes, an experimental method that was conceived independently by both Hassak and Chambers. In addition, many of the investigators in Chambers' (1912) review used the new Winkler method, introduced in 1888, for measuring oxygen concentration and gained for the first time a quantitative estimate of the magnitude of new organic material synthesised during the duration of the observation.

During the late nineteenth century there was a lively debate by naturalists of whether plant growth in lakes and in the ocean required enhanced (supersaturated) concentrations of oxygen or indeed produced them (Chambers, 1912). In hindsight this debate seems misguided because the first principles of the photosynthesis equation had been known for almost a hundred years (Table 2.1). Naturalists argued strongly that the association of cool water and increased plant growth was best accounted for by the greater oxygen solubility of cool waters. Knowledge of photosynthesis as carried out by terrestrial plants did not convince naturalists that aquatic photosynthesis was a parallel process, but experimental results both *in situ* and *in vitro* (Table 2.2) clearly settled the issue by the first decade of the twentieth century.

Investigators cited by Chambers (1912) represent a cross-section of the biological scientific community from the latter half of the nineteenth century, including individuals from several countries and walks of life. Talented amateurs, especially physicians and clergymen, are represented as well as professional biologists associated with museums or, in the United States, the numerous state or national biological surveys. In contrast to this widespread 'basic' research, research on plant productivity at sea was done for pragmatic reasons by a very small group of professional fisheries oceanographers.

The study of plankton production, the centrepiece of early oceanography, had one purpose: to enhance fishery resources. The end of the nineteenth century had seen a general depression in Germany with widespread economic suffering and a famine in Norway. In both the UK and the US, overfishing had started to reduce yields, particularly catch per unit effort. Governments in both Europe and North America were concerned about these decreases, and it was generally assumed by enlightened governments that a better understanding of the biological economy of the oceans was required to increase, or even maintain, the societal benefits provided by ocean fisheries.

2.4 Forging the Paradigms (1850–1912)

In the mid-nineteenth century, two individuals, Hooker and Ørsted, independently made the remarkable observation that tiny unicellular planktonic plants support large oceanic animals, but a wide recognition that these plants were the primary producers supporting impressive oceanic food webs took another half-century. Hooker had participated in the Ross Antarctic Expedition (1839–1843), whose scientific instructions reflected an interest the Royal Society had developed in plankton distribution. This interest was based on Ehrenberg's observations of 'the microscopic composition of [sedimentary] rocks and the discovery of the great share which microscopic organisms take in the formation of the crust of the earth at the present day' (Huxley, 1915). Hooker's 1847 observations are impressive and his excitement is tangible.

'It is, therefore, with no little satisfaction that I now class the *Diatomaceæ* with plants, probably maintaining in the south Polar Ocean that balance between the vegetable and the animal kingdom which prevails over the surface of our globe. Nor is the sustenance and nutrition of the animal kingdom the only function these minute productions may perform; they may also be the purifiers of the vitiated atmosphere, and thus execute in the Antarctic latitudes the office of our trees and grass turf in the temperate regions, and the broad leaves of the palm ... in the tropics.' (Huxley, 1915)

Hooker's observations were based on the abundant diatom blooms that provide food for the rich animal life in the Antarctic. The biomass poverty of the Sargasso Sea provided a similar inspiration in 1844 to a Danish marine ecologist named Ørsted. After a cruise through the Sargasso Sea, he proposed that in the vast reaches of the open ocean microscopic algae provide the primary food of oceanic food webs. Understandably, this iconoclastic concept of the tiny feeding the large was not accepted by most contemporary marine biologists, and Ørsted did not again refer to this idea during his long career in botany (Wolff and Peterson, 1991).

About 25 years after Hooker and Ørsted first made their suggestion, the Norwegian oceanographer G. O. Sars, on the basis of observations during an 1876–1878 North Atlantic expedition, also proposed that small ocean plants are the nutritional basis of animals that live in the open sea (Gran, 1912). This remarkable idea of Hooker, Ørsted and Sars was reintroduced, and more forcefully advocated in 1887 by Victor Hensen, the great German marine biologist who gave us the word *plankton* to describe the concept of a drifting existence in the ocean. Hensen was an imposing individual with great creativity and intellect, but with flaws of mythic proportions (Mills, 1989). He engaged in a long intellectual dispute with the other German marine biology intellectual giant, Ernst Haeckel (see Chapter 13) and, since Haeckel is a common name in biology and Hensen is not, it appears that Haeckel won in the court of popular scientific opinion. Haeckel argued that oceanic food webs are based on plant material from attached macroalgae and aquatic plants as well as terrestrial plant debris washed into the open sea from shores and rivers. It was preposterous, he said, to believe that tiny, invisible drifting plants could provide enough food to support vast schools of fish, birds and whales. Hensen, vehemently promoting the idea that Hooker, Ørsted and Sars had introduced, insisted that microscopic planktonic plants were the major producers and it was his lifelong goal to measure the planktonic production of the ocean. An indication of Hensen's vision was his conviction that 'if we should succeed in determining the consumption or reproduction of food plants, especially certain phytoplankton, then the amount of animal material in the sea may be determined or at least delimited' (Mills, 1989), a prediction that was realised some 60 years later by Ryther (1969). The irony was that the important plant producers were not only small, but even smaller than Hensen could have appreciated with the nets designed specifically for plankton. One of his errors was that his nets simply missed the most abundant phytoplankton: Lohmann's nanoplankton were discovered between 1896 and 1909 (Mills, 1989). For most of the twentieth century, Hensen's error of underestimating small phytoplankton continued to plague biological oceanography (Barber and Hilting, 2000).

Hensen's great 1911 monograph on life in the sea was an attempt to rebut Haeckel. It makes clear both Hensen's accurate vision of the importance of plankton and his erroneous thesis that plankton abundance was, because of assumed universal horizontal mixing, the same everywhere in the ocean (Mills, 1989). How wrong he was is revealed by ocean colour imagery. What would Hensen, the stubborn genius, say about his uniformity thesis if he could see the rich spatial variability in a global ocean colour image?

In the same year (1912) as Chambers' review on the relationship of algae to dissolved oxygen and carbon dioxide, Gran (1912) published a chapter, 'Pelagic plant life', in a book edited by Hjort and Murray, *The Depths of the Ocean*. Many issues of modern biological oceanography are discussed in Gran's chapter; his goal was to determine 'the laws of production in the sea', but he lamented 'our investigations are as yet too incomplete to justify us in framing laws for plant production in the ocean.' With that caveat, Gran then proceeded to set down a series of hypotheses to account for the regional, latitudinal, vertical and seasonal variations in plant production. Comparison of Gran's many observations, ideas and hypotheses on regulation of biomass, productivity, species composition, species succession and variability with those in a textbook by Johnstone (1908), *Conditions of Life in the Sea*, shows how innovative and original Gran's synthesis was.

Assuming that Johnstone's (1908) textbook represented the state of knowledge about plankton productivity, one sees that Gran's 1912 chapter was revolutionary. Using Whipple's (1899) pioneering observations of phytoplankton growth in bottles suspended at various depths, Gran described the concept of a photic zone limited to the upper 200 m in tropical waters. He cautioned, however, that light and temperature cannot be the major regulators of the latitudinal variability of oceanic plant production as evidenced by temperate abundance versus tropical poverty. Gran evoked the importance of the nutrient elements, N, P and Si, as initially proposed by Brandt and Nathansohn, but rejected Brandt's erroneous hypothesis that denitrification was the major process controlling horizontal nutrient variations and hence the varying spatial abundance of plant plankton. Gran cited the work of Nathansohn, which showed that upwelling and mixing have a great influence on plankton production through the supply of subsurface nutrients. Around 1900, knowledge of the vertical distribution of nutrients was sparse, but because of the light dependence of plant production established by Whipple, it was accepted that deep-living animals lived on food that rained down from the photic zone (Johnstone, 1908). It was known from experiments that animals produce inorganic nutrients, so it was logical to hypothesise that deep waters, with their exclusively animal communities, accumulated nutrients. That chain of reasoning led Gran to suggest that divergent, cyclonic oceanic systems, with their ascending water motion which supplies subsurface nutrients to the photic zone, are richer in plant production than the convergent, anti-cyclonic gyres that dominate the tropics. Gran provided a complete causal sequence involving gyre physics and chemistry to account for the poverty of plant production of the great subtropical gyres.

Nathansohn and Gran worked together in Christiana (now Oslo) Fjord on the seasonal pattern of plant production using counting chambers to assess weekly changes in phytoplankton cells. This pioneering work documented the spring bloom and the summer decrease in phytoplankton abundance. Experiments done in August after the phytoplankton decline showed that addition of ammonium would cause the ambient phytoplankton to bloom again. This work established the importance of nutrients in regulating seasonal plant production, although Gran acknowledged that circulation and mixing in the fjord also had a large, but undetermined, effect on plankton abundance. Nevertheless, he and Nathansohn were convinced that 'plant production in the sea is mainly regulated by the amount of dissolved nutritive substances.' As further evidence, Gran cited Whipple's (1899) thesis that the cold water–increased nutrient content relationship is the causal factor in the well-known cold water–phytoplankton abundance relationship observed seasonally in lakes as well as in temperate and polar seas, and that the opposite relationship (warm water–low abundance) characterises tropical waters.

Gran believed the diatom–copepod food link is a dominant process in the ocean and disagreed with Hensen's idea that diatoms are not a major food source for oceanic food webs. Gran accepted and promoted Lohmann's ideas on the abundance and importance of nanoplankton, especially in relatively poor tropical waters, and suggested that heterotrophic plankton protozoa need a nanoplankton food supply equal to about half of their own volume each day – an idea that was somehow lost by biological oceanographers until it was revived in modern form in the micrograzer limitation concept of the microbial loop (Landry *et al.*, 1997). Noting that the open ocean has considerably less plankton abundance and plant production than coastal regions, Gran allowed that terrestrial runoff may play some role in this spatial variability, but he hypothesised that increased mixing and upwelling are most responsible for coastal richness. Gran's synthesis lays out a series of hypotheses concerning the seasonal and regional patterns and variability of plant production in the ocean. These ideas were shaped by and in some cases originated with the productive German oceanographers, such as Nathansohn, Lohmann and Brandt. Gran had an exceptional ability to take specific ideas that arose in the German school and refine them, on the basis of his experience at sea, to produce an integrated description of the idea, calling on physical, chemical and biological observations from many sources.

Three things stand out in Gran's 1912 synthesis: his speculations, many intuitive, about the 'laws of plant production' were remarkably accurate; he mentioned most of the processes that we now think affect primary production in the ocean and he was able to use the results and observations from Hensen's powerful Kiel oceanography group without becoming embroiled in their unfortunate polemics regarding, for example, uniform abundance or denitrification. Gran, as a sea-going Scandinavian scientist, seemed to have a more intuitive understanding of the ocean than Hensen or Brandt, and he had the advantage of a personality that enabled him to incorporate the ideas of other oceanographers with his own keen observations to form a remarkably prescient series of hypotheses.

Surprisingly, the progress made by Gran was accomplished without having a method that actually measured plant production. Gran must have felt keenly the need for such a method. In 1887, Victor Hensen made it his life's goal to measure ocean productivity, but he failed to do so because appropriate productivity concepts and requisite technology were not available in the 1880s and 1890s. Hensen depended on nets for his sampling, but as Lohmann and Gran learned, the plankton nets of the day were woefully inadequate for collecting nanoplankton, or even quantitative samples of net phytoplankton, because of net clogging (Mills, 1989). Another stumbling block in the 1800s was that Hensen could apparently conceive only of 'growth per year', a hold-over from land productivity work; the concept of rapid turnover rates and the idea of 'growth per day' emerged surprisingly slowly.

2.5 New Methods (1910–1940)

During the period 1910–1940, all of the national fisheries oceanography groups devoted much effort and many resources to developing quantitative methods to measure plant productivity. By 1912, a cohort of fisheries oceanographers was closing in on what they needed: a method to measure the bulk production of a community of planktonic plants over a short time interval. Chambers' (1912) review shows that basic knowledge was in hand to do that in either of two ways: measuring oxygen production with the Winkler method, or measuring carbon dioxide uptake by determining pH changes with indicator dyes. A method for measuring the uptake of phosphate and using a C to P ratio to calculate carbon production soon provided a third method (Mills, 1989). The stage was set for a saltatory advance in the study of plankton production.

   To assess the seasonal pattern of plant productivity, Apstein had used *in situ* changes in phytoplankton counts as early as 1910 in the Baltic Sea and Gran did so in Christiania Fjord about the same time. However, Gran (1912) reported that the coastal circulation and mixing in the fjord made it impossible to assess growth quantitatively by counting the change in phytoplankton numbers on the time-scale of weeks. The inability to use *in situ* phytoplankton counts (a procedure that works well in lakes) stimulated Gran to look for an *in vitro* method that could assess productivity in a sample isolated from local advection and turbulence. The work described in Chambers (1912) established the possibility of using quantitative oxygen production or carbon dioxide uptake. Both the Winkler method for oxygen and the pH indicating dyes method for carbon dioxide were developed in the late 1800s, but the pH method was not successfully adapted to seawater until the 1910s (Harvey, 1955). Gran pursued a number of questions regarding plant productivity; one was to determine the depth at which there was just enough light to balance the respiratory demand, the *compensation point*. To pursue this interest, Gran and Gaarder suspended bottles at various depths down to 40 m with the deepest depth also having a 'dark' bottle (wrapped in black cloth). This investigation evolved into the procedure known today as the *light and dark bottle method* for determining plant productivity and in the early applications of the method, both oxygen and carbon dioxide changes were followed. In hindsight it is clear that this method was an important advance. The work was done in 1916 and reported at the 1917 Conseil Permanent International pour l'Exploration de la Mer (ICES) meeting, but neither German nor British scientists attended that meeting because of World War I;

furthermore, the subsequent publication by Gran (1918) was in Norwegian and went unnoticed by both English- and German-speaking oceanographers. Eventually, the method was published in English (Gaarder and Gran, 1927), and within a year it was in use in Great Britain by Marshall and Orr; in the following years it was used widely (Table 2.2).

Simultaneously with advances in using oxygen production led by Gran, Gaarder, Braarud, Marshall and Orr between 1915 and 1939, efforts were made to assess plant productivity by the uptake of inorganic nutrients, nitrate, silicic acid or phosphate (Mills, 1989). For example, early measurements in the Baltic Sea made by Raben showed that phosphate disappeared first and reached the lowest values in the summer. As nutrient observations began to accumulate, it became apparent that phosphate is an inorganic nutrient ion that is atom-ratio deficient in seawater relative to the composition of phytoplankton (Mills, 1989). With hindsight it now appears that the limiting-phosphate assumption was driven by methodological artefacts: at low values the nitrate and silicate methods gave false positive readings. In 1915, Matthews, using a new colorimetric phosphate analysis developed in 1909, found that the phosphate concentration in the English Channel was much lower than the concentration Raben had reported for the Baltic Sea (Mills, 1989). In 1921, Atkins began a study of the seasonal nutrient, mixing and light variability in the English Channel which showed that the timing of phosphate depletion was coincident with both the spring diatom bloom and the isolation of the surface layer by a newly formed thermocline. He interpreted this sequence as being caused by local solar heating that stabilised the nutrient-rich surface layer, creating an isolated layer where phytoplankton could deplete the phosphate, take up carbon dioxide, increase the pH, take up nitrate and silicic acid and produce oxygen (Atkins, 1926).

Atkins, between 1922 and 1933, wrote a remarkable series of 26 publications on the seasonal cycle of physical, chemical and biological properties in the English Channel. Several of these publications describe the necessity for convective winter mixing to restore nutrients to the surface layer and the subsequent need for thermal stabilisation to maintain phytoplankton in the well-lighted and initially nutrient-rich surface layer (Harvey, 1955). Atkins' narratives in straightforward language contain much the same information that was given by Gran (1912 and 1932) and later described with quantitative rigour by Sverdrup (1953) as the *critical depth* concept.

The following quotes from Atkins and Gran illustrate how well these men understood the need for both mixing and stabilisation in the regulation of regional and seasonal intensity of primary productivity.

'As we proceed northwards the surface cooling becomes greater and the deep water temperatures are also lower, so the vertical circulation will proceed to progressively greater depths. Conversely near the equator the temperature changes throughout the year are small, so that vertical circulation there must be due in the main to wave motion – which cannot be effective to any very great depth – and not to density changes. Thus no considerable seasonal change in phosphate content is to be expected.' Atkins (1926)

'The richness of northern waters may accordingly be considered as due to the better vertical circulation as regards the temperate zone; also to the fact, which becomes more marked in high latitudes, that since light is the limiting factor for plant growth for a considerable portion of the year, the water accumulates a greater store of phosphate. Thus when daylight returns, or lengthens its duration, a great outburst of the phytoplankton takes place.' (Atkins, 1926)

'These examples show the importance of the stabilization of the surface layers. A marked stratification excludes circulation of the nutrient salts with the result that the surface layers become depleted of plant nourishment and the diatoms with their high requirements sink and disappear. On the other hand, continuous vertical mixing prevents them from accumulating in the lighted zone and from utilizing the nutrient salts present to such an extent as might have been expected.' (Gran, 1932)

At the Plymouth Laboratory, working in the English Channel using C, N, P and Si uptake and oxygen production methods, Atkins, Cooper, Harvey and their many coworkers measured an annual productivity of 60 to 80 mg C m^{-2} y^{-1} for the total phytoplankton assemblage with about 9% by diatoms (estimated from silicate uptake) and 6% by coccolithophorids (estimated from Ca uptake) (Harvey, 1955). Steemann Nielsen (1937) was the first to use light and dark bottle oxygen production methods for determining the seasonal cycle of primary production in the North Sea off Denmark (Table 2.2). In 1938–1939 Riley began a similar intensive study of the seasonal cycle of productivity in Long Island Sound off the US coast (Riley, 1941; Table 2.2). All of these intensive investigations of the seasonal productivity cycle were exceptionally successful in documenting the successive changes in stratification, nutrients, phytoplankton production and zooplankton production. While the methods available just prior to World War II were relatively labour-intensive, with poor precision and unknown accuracy, they were adequate to document reproducibly the annual cycle in temperate-latitude, coastal ocean environments. By the end of this period (1940), however, it was apparent that for determining primary productivity in the open ocean, all of these methods had flaws.

In the spring of 1939, Riley participated in a cruise in the western North Atlantic. These observations were, we believe, the first use of the light and dark bottle oxygen production method in an open ocean setting and the first use of on-deck incubation (Table 2.2). Each day Riley started a new three-day incubation, with the long incubation making it possible to obtain oxygen changes large enough to be detected with the Winkler method. The productivity values Riley estimated for these oligotrophic tropical waters with a very deep euphotic zone were as high, on a square metre basis, as values obtained in eutrophic coastal waters with much shallower euphotic zones. At the time they were published, Riley's high estimates of open ocean primary productivity produced little controversy. They disagreed, however, with the conventional wisdom of Brandt, Lohmann and Gran, who knew that tropical open ocean phytoplankton biomass was low and speculated that productivity also was low because the year-round concentration of nutrients was low. Riley's estimates were state-of-the-art quantitative determinations done with the most modern technology, while the conventional wisdom was just a naturalist's intuition. Riley (1944) used the high rates of about $1.0 \, \mathrm{g} \, \mathrm{C} \, \mathrm{m}^{-2} \, \mathrm{d}^{-1}$ to calculate high rates of global ocean productivity (Table 2.3). Others, for example Rabinowitch (1945), used these estimates to argue that ocean productivity was considerably higher than terrestrial productivity, a view that persisted as late as 1961 (Perutz and Kendrew, 1964).

Table 2.3 Global oceanic primary productivity estimates from 1919 to 2002. Following the practice of Platt and Subba Rao (1975), original estimates of 'gross productivity' are cited when both net and gross estimates are available. Citations not included in the references can be found in Cobb and Harlin (1976), Cushing (1975), Lieth (1975), Platt and Subba Rao (1975), Berger (1989), Longhurst *et al.* (1995), Vinogradov *et al.* (1996), Barber and Hilting (2000) or Behrenfeld (see Chapter 7). The Laevastu (1957) map was presented at the 1957 Bergen meeting but is unpublished.

Date	Author	PP (Pg C y^{-1})	Method
1919	Schroeder	22	Speculation
1934	Zernov	60	(Not available)
1937	Noddack and Komor	29	One prod. determination
1944	Riley	126 ± 82	O$_2$ method (long incubations)
1950	Skopintsev	50	Based on sediments
1952	Steemann Nielsen	20	Few ^{14}C determinations
1955	Fleming and Laevastu	Map only	FAO prod. data (O$_2$, ^{14}C, etc.)
1955	Sverdrup	Map only	Based on physical processes (1955)
1957	FAO (Laevastu)	Map only	Based on Fleming and Laevastu (1955)
1957	Fleming	Map only	Based on Fleming and Laevastu (1955)
1957	Steemann Nielsen and Jensen	20 to 25	Few ^{14}C determinations
1958	Fogg	32	FAO prod. data (Fogg, 1958)
1958	Kesteven and Laevastu	20 + map	Based on Fleming and Laevastu (1955)
1959	Gessner	20 + map	Based on Fleming and Laevastu (1955)

Continued

Table 2.3 *Continued*

Date	Author	PP (Pg C y^{-1})	Method
1968	Koblentz-Mishke *et al.*	23 + map	Synthesis of many ^{14}C prod. stations
1969	Bogorov	25	Synthesis of many ^{14}C prod. stations
1969	Ryther	20	^{14}C and spatial model
1970	Koblentz-Mishke *et al.*	25 to 30 + map	Slight revision of 1968 manuscript
1972	FAO	Map only	Based on Koblentz-Mishke *et al.* (1970)
1975	Cushing	Map only	Based on Koblentz-Mishke *et al.* (1970)
1975	Lieth	20 + map	Based on Gessner (1959) (see Berger, 1989)
1975	Bunt	19 + map	Based on Gessner (1959) (see Berger, 1989)
1975	Rodin *et al.*	Map only	Based on Koblentz-Mishke *et al.* (1970)
1975	Platt and Subba Rao	31	New ^{14}C synthesis
1976	Cobb and Yarlin	Map only	Based on Rodin *et al.* (1975)
1980	Fogg	Map only	Based on Cushing (1975)/FAO (1972)
1984	Parsons *et al.*	Map only	Based on Koblentz-Mishke *et al.* (1970)
1984	Romankevich	Map only	Based on Koblentz-Mishke *et al.* (1970)
1985	Shushkina	56	New ^{14}C and biomass data (Shushkina, 1985)
1987	Berger *et al.*	27 + map	New ^{14}C synthesis
1987	Martin *et al.*	51	Based on Koblentz-Mishke *et al.* (1970)
1995	Longhurst	45 to 50	CZCS and ^{14}C calibration
1996	Antoine *et al.*	33	CZCS and ^{14}C calibration
1996	Vinogradov *et al.*	100	CZCS and ^{14}C calibration
1997	Behrenfeld and Falkowski	46	CZCS, ^{14}C calibration and irradiance
2001	Behrenfeld *et al.*	54 to 59	SeaWiFS, ^{14}C calibration and irradiance
2002	Behrenfeld *et al.*	41	SeaWiFS, ^{14}C calibration and irradiance

Riley, after this one pre-war investigation of the tropical open ocean, turned after World War II to an intensive study of Georges Bank. There, using the light and dark bottle oxygen method together with physical, nutrient and zooplankton observations, he worked out the fundamental relationships that control productivity with enough accuracy to develop a predictive model of the spring bloom on Georges Bank (Riley, 1946).

Riley's Georges Bank study and his tropical open ocean measurements have had vastly different trajectories in the development of biological oceanography. The former created the field of *ecosystem modelling* and earned Riley the title of 'father of modern biological oceanography' (Barber and Hilting, 2000). His open ocean measurements gave rise to a long dispute with Steemann Nielsen (see Chapter 1), a dispute that Steemann Nielsen won decisively by showing that three-day incubations give very misleading light versus dark oxygen values. There are two important points about this dispute. First, the light and dark method in a rich coastal environment, unlike open ocean incubations, gave Riley reproducible results because the high biomass allowed short incubations to give oxygen changes large enough to measure reliably. The patterns, relative magnitude and timing of light, stability and nutrients as regulators of primary productivity were well-determined. Like Atkins, Cooper, Harvey and Steemann Nielsen, Riley could determine much about spring bloom dynamics and seasonal cycles of productivity, but with the oxygen method he was using, he could not reliably quantify open ocean productivity.

The second point is that Steemann Nielsen recognised immediately that Riley's open ocean values were wrong. Steemann Nielsen was, like Gran, a shrewd naturalist who had excellent intuition about ocean ecology. Steemann Nielsen was also a methodical laboratory worker who checked and rechecked the details of his procedure. In his various studies of the annual primary productivity cycle in the North Sea, Steemann Nielsen checked the consequences of long incubations and showed that the three-day procedure produced large artefacts due to bacterial growth in the dark bottle (Steemann Nielsen and Jensen, 1957).

The contrast between the methodical, but innovative experimentalist, Steemann Nielsen, and Riley, the brilliant analytical thinker whose casual lab manner was legendary, could not have been greater. The message from this vignette is that the productivity methods of the pre-World War II period were good enough to reveal the temporal pattern but not good enough to answer quantitative questions involving primary productivity. Steemann Nielsen recognised this better than any other biological oceanographer.

2.6   Breakthrough (1940–1950)

In the late 1930s and the 1940s Steemann Nielsen had the good fortune to work in Copenhagen, a city that was an international centre of nuclear science by virtue of being home to the Institute of Theoretical Physics (now Niels Bohr Institute) and also a centre of biological radiotracer studies by virtue of being the home of von Hevesy. Von Hevesy was the pioneer of the use of isotopes to trace physiological processes, work for which he received a Nobel Prize in 1943.

Development of Steemann Nielsen's ^{14}C method had roots at the Radiation Laboratory at the Berkeley campus of the University of California. In 1936 Lawrence built a new cyclotron and in 1937 put a brand new PhD graduate named Kamen in charge of making radioisotopes for research use. Kamen made and supplied ^{32}P radioisotope to a number of leading scientists, including von Hevesy in Copenhagen for use in biological tracer experiments (Kamen, 1986). Kamen also teamed up with Ruben to use ^{11}C, the isotope with a 21-minute half-life, as a radiotracer to determine the path of carbon in photosynthesis. In 1940 Kamen was given control of the cyclotron for three days and nights to bombard graphite in an attempt to produce ^{14}C, an isotope which had been predicted to exist by Kurie in 1934 (Kamen, 1986). The attempt was successful. Ruben and Kamen were now poised to start definitive experiments on the path of carbon in photosynthesis using the long half-life ^{14}C isotope. However, the start of World War II, the switch of resources to the atomic bomb project, the untimely death of Ruben in a laboratory accident and security clearance disputes with the US government kept Kamen from doing definitive photosynthesis experiments with ^{14}C. Immediately after the war, Calvin and Benson began the photosynthesis work made possible by Kamen's production of ^{14}C (Kamen, 1986).

After the war, a number of developments converged quickly. By 1947 Calvin and Benson had begun producing papers on the path of carbon in photosynthesis (work for which Calvin received a Nobel Prize in 1961). In the same year that Steemann Nielsen was developing his method in Copenhagen, his former student, Lise Schou, spent a year at Calvin's photosynthesis lab in Berkeley. Returning to Copenhagen in 1950 to finish her dissertation in plant physiology, Schou brought with her firsthand experience with the use of ^{14}C uptake to label the products of photosynthesis. Although Schou and Steemann Nielsen never discussed Schou's work directly (Wilkinson [née Schou], pers. comm.), after her return to Copenhagen she gave a talk on the method of using radioactive carbon isotopes in the study of photosynthesis. Well-worn reprints of her talk were found among Steemann Nielsen's papers together with a similarly well-worn copy of Calvin's 1949 valuable isotope cookbook entitled *Isotopic Carbon* (see Chapter 1).

In 1947, a colleague of von Hevesy at Copenhagen, Hilde Levi, a Berlin-trained PhD physicist experienced with radioisotopes and Geiger-Müller counting technology, received a grant to work in Libby's laboratory at the University of Chicago for a year. In 1948 she returned to Copenhagen and in that year began running the monthly 'Isotop Kollokvier' series led by von Hevesy. Steemann Nielsen later gave talks in the colloquium. Furthermore, in the winter of 1949 Levi began teaching a course in tracer technology that was so popular among her colleagues that classes were held at night so they could attend (Niels Bohr Archive staff, pers. comm.). Levi was familiar with low-level counting from her work with Libby, and in Copenhagen, together with the engineer Jensen, built a Geiger-Müller counter that was capable of counting ^{14}C efficiently (see Chapter 1).

Levi and Schou tie Steemann Nielsen to the work of the three Nobel Prize recipients – von Hevesy, Calvin and Libby – whose accomplishments, respectively, were (1) the use of biological radiotracers, (2) the use of ^{14}C as a radiotracer to determine the path of carbon in photosynthesis, and (3) the development of counting methods sensitive enough to make possible ^{14}C dating. Steemann Nielsen had the insight that the radiotracer methodology could be used to quantify a bulk ecological process such as primary production. Many independent events had to come together for Steemann Nielsen's idea to become reality. He was the right person in the right place to bring together separate ideas from Chicago, Berkeley and most importantly, from von Hevesy's pioneering radiotracer work.

2.7   Expansion (1950–1970)

Timing is everything. In the early 1950s professional demand was stronger than ever for answers to the fundamental productivity questions laid out by Hensen, Brandt and Gran a half-century earlier. The pre-World War II effort had sharpened the focus on questions about (1) the absolute magnitude of ocean primary productivity, (2) the cause of regional productivity variations and (3) the regulation of season-ality, especially the mid- and high-latitude spring diatom bloom dynamics; but in essence the questions remained unanswered. After World War I, German ocea-nography had decreased in relative influence. Japan, the US, the USSR and other countries made considerable effort in fisheries exploration and biological collecting cruises (Table 2.4), but there was little biological oceanography *per se* on a par with the work of the countries around the North Sea.

For the Scandinavian and British researchers, 1920–1940 was a golden age of biological oceanography because ample support and international political stability enabled this small group of about 50 scientists to communicate well and make excellent progress year after year. New analytical methods were developed and applied to the measurement of light, nutrients and plant production, especially at the Plymouth Laboratory (Harvey, 1955), and good qualitative descriptions were worked out, but for the most part the productivity work was neither analytically nor quantitatively rigorous.

After World War II the individuals, countries and manner of doing biological oceanography all changed dramatically. Scandinavian and German laboratories never regained their pre-war hegemony; Great Britain was a de facto leader in biological oceanography in spite of a large post-war migration of young British oceanographers to the US and Canada where there was a plethora of job oppor-tunities. The greatest change was the explosive growth of ocean sciences in the US and the USSR. The Cold War was in part a cultural rivalry, with science being one of the major areas of intense competition between the US and the USSR. Coupled

Table 2.4 List of expeditions where some biological collecting or other scientific biological studies were carried out from a list compiled by Tjärnö Marine Biological Laboratory, Strömstad, Sweden. The year 1766 is significant because it was the first year in which ship-based expeditions named marine organisms according to Linnaeus's binomial nomenclature.

Nationality	Expeditions	
	1766–1913	1915–1952
British	55	4
Swedish	22	1
French	19	1
US	17	4
German	11	1
Danish	9	6
Russian/USSR	7	5
Norwegian	5	1
Austrian/Hungarian	2	—
Irish	2	—
Italian	2	—
Prussian	2	—
Austrian	1	—
Belgian	1	—
Dutch	1	1
Hungarian	1	—
Indian	1	—
Monacan	1	—
Scottish	1	—
Spanish	1	—
Dutch/Danish	—	1
Canadian	—	1
Total	161	26

with this Cold War East–West expansion of oceanography was a flowering of oceanography in many other countries, including what were then called 'third world' countries. Biological oceanography became a regular component of government-supported science in several Mediterranean and Eastern European countries as well as in South America and Asia.

In this rapidly changing milieu, Steemann Nielsen's new method made its debut on the 1950 *Galathea* Expedition. The association of the ^{14}C method with the *Galathea* Expedition marks both the ending of the old era and the beginning of the era of expansion (1950–1970). Since the time of the great cruises of Cook or La Perouse it had been a custom for European countries to send out discovery cruises that more or less wandered across the world's oceans exploring and collecting (Table 2.4). The *Galathea* Expedition was one of the last, if not *the* last, of this illustrious series of national discovery expeditions. In contrast the International Geophysical Year (IGY), which was carried out in 1957, was emblematic of the new way to study the ocean. The new effort was international rather than national, and all countries were invited to contribute to the overall plan. The IGY cruise tracks were organised with assigned stations to obtain a desired level of spatial resolution. Oceanography was responding to the need to better sample the space and time variability of the ocean by deploying numerous ships simultaneously in a co-ordinated pattern. Although the IGY supported little biological oceanography, this international endeavour was so successful that the International Indian Ocean Expedition (IIOE) was set in motion soon after the IGY. The era of organised multinational cruises began just as Steemann Nielsen's method to measure primary productivity became available. The IIOE employed this powerful new, standardised method on 17 cruises throughout the Indian Ocean (Table 2.5). The international organisations – ICES, the Intergovernmental Oceanographic Commission (IOC) and the United Nations Educational, Scientific and Cultural Organisation (UNESCO) – provided forums for the planning of international expeditions and international meetings such as the Oceanographic Congresses in New York (1959), Moscow (1966) and Edinburgh (1976).

Table 2.5   Characteristics of 221 primary research papers published in the 20-year period, 1954 to 1973, involving the ^{14}C uptake method to measure primary productivity. These papers either describe primary productivity in a given region or habitat or discuss a process or the ^{14}C method *per se*. Only individuals with three or more papers are listed. No 1953 papers using the ^{14}C method were found. Information was obtained from references cited in this chapter.

Individual	No.	Country	No.	Province	No.
Steemann Nielsen	20	US	57	Process studies	59
Ryther	15	USSR	41	Atlantic	47
Doty	13	Denmark	26	Pacific	42
Sorokin	11	Great Britain	21	Global	25
El-Sayed	9	Japan	17	Antarctic	17
Koblentz-Mishke	8	France	12	Indian	17
Steele	8	Canada	10	Mediterranean	9
Saijo	7	Germany	8	Arctic	4
Kavanov	6	India	5	Baltic	1
Bunt	4	Australia	4		
Cushing	4	Argentina	2		
Angot	3	Norway	2		
Parsons	3	Italy	1		
Takahashi	3	Greece	1		
Vinberg	3	Peru	1		
		Sweden	1		

The post World War II expansion of biological oceanography research was also accompanied by a turnover in the individuals doing productivity research. Few pre-war leaders continued as post-war leaders; Harvey, Ketchum, Krey, Riley and, of course, Steemann Nielsen come to mind. Table 2.5 shows individuals publishing results obtained with the ^{14}C uptake method between 1954 and 1973; of these individuals, only Steemann Nielsen published equally actively before and after the Second World War as well as in both oxygen production and ^{14}C production (Tables

2.3 and 2.5). Among 221 papers reporting ^{14}C observations between 1954 and 1973 there was good geographic coverage with a surprisingly large number of papers reporting results from the Indian Ocean and Antarctic waters. Thus, by the 1970s the ^{14}C method had been used in most oceanic regions of the world. This effort at data collecting provided a rich resource for scientists who wanted to estimate the global ocean primary productivity or characterise the productivity of the various regions of the oceans.

2.8 Milestones (1950–1970)

Four milestone events in the 1950–1970 period document the vigorous advance of plankton productivity research made possible by the ^{14}C method. They are:

(1) the 1957 Bergen meeting of ICES,
(2) the 1963 publication of *The Sea* with chapters by Steemann Nielsen, Riley, Ryther and Provasoli (Hill, 1963),
(3) the publication of Koblentz-Mishke, Volkovinsky and Kabanova's (1968) global productivity estimate, and
(4) Ryther's 1969 *magnum opus*, 'Photosynthesis and fish production in the sea'.

The symposium, 'Measurements of primary production in the sea', held in Bergen in September 1957, was chaired by Steemann Nielsen and Cushing. This event, occurring a scant five years after first publication of the ^{14}C method, is testimony to the speed and breadth of adoption of the new ^{14}C method. Ten countries were represented by the 31 papers given, and of those papers, 16 either presented results obtained by the ^{14}C method or discussed the method *per se*. The proceedings, discussions and recommendations were published in Volume 144 of the *Rapports et Procès-Verbaux* of ICES (Steemann Nielsen and Cushing, 1958). This volume is, in our opinion, the most influential single volume of papers in the history of primary productivity studies. It has four themes: methods of estimating standing crop in phytoplankton, methods of measurements of production, measurements of primary production, and environmental conditions for primary production.

One paper, 'Productivity in relation to nutrients', by the Woods Hole group of Ketchum, Ryther, Yentsch and Corwin, elicited 'an exhaustive discussion', according to Steemann Nielsen and Cushing (1958). There was consensus among participants that nutrient supply is the factor 'which in most cases' determines the productivity of a given ocean region. Gran and Nathansohn were well vindicated, but according to the published proceedings there was 'disagreement about the way in which lack of nutrients limits productivity'. The discussion included a persistent disagreement between the Woods Hole group and Steemann Nielsen himself. Ketchum, Ryther, Yentsch and Corwin argued that nutrient depletion reduces net productivity because the ratio of photosynthesis to respiration approaches 1, producing low or zero net photosynthesis. Steemann Nielsen countered by pointing out that in oligotrophic gyre regions the photosynthesis to respiration ratios are, in fact, not low, but normal. In retrospect, it appears that the Woods Hole group was describing what happens at the termination of the spring bloom, while Steemann Nielsen was describing photosynthetic performance in permanently low-nutrient situations where the ambient phytoplankton are well-adapted to low nutrient concentrations.

In contrast to the enthusiasm expressed at this meeting for the ^{14}C method, Cooper expressed serious reservations that *in situ* consumption of nutrients could ever be a successful means of measuring primary production. British investigators, especially the famous team at the Plymouth Laboratory, had been using the nutrient uptake method since 1923. Now after 25 years of diligent work in the English Channel, the Plymouth scientist Cooper argued that this approach was flawed because recycling and complex hydrography precluded the rigorous nutrient budgeting necessary to estimate productivity (Steemann Nielsen and Cushing, 1958). This represented a passing of the torch in two ways, from nutrient uptake methods to the ^{14}C uptake method and from the Plymouth group to a new generation of young British scientists such as Cushing, Steele and Fogg, who went on to distinguish themselves in biological oceanography.

Similarly, the symposium signalled the obsolescence of the oxygen production method for measuring total primary productivity. Table 2.2 indicates how widely the oxygen method was used in the period before 1952. Two oxygen method papers, both by Soviet scientists, were presented in the methods session at the Bergen meeting, but there was no presentation of primary productivity measurements obtained with the oxygen method. Riley, the one scientist who would have been motivated to defend the oxygen method, did not attend the meeting.

A mystery associated with the 1957 Bergen meeting involves a global primary productivity map that was presented at the meeting (Berger, 1989), but not mentioned in either the formal papers or written proceedings. The map, by the Food and Agricultural Organisation (FAO) was the first objective mapping of oceanic productivity (Berger, 1989). It is surprising that the Bergen proceedings omitted discussion of its presentation, as it surely must have drawn considerable attention. Karl Banse of the University of Washington (pers. comm.) believes that the original map was compiled by Laevastu from scattered ^{14}C and oxygen productivity observations and observations on the colour of the sea. The map first appeared in an unpublished manuscript by Fleming and Laevastu (1955) and was published two years later (Fleming, 1957). Fleming (1957) refers to the map as 'semi-quantitative', acknowledging the temporal and spatial limitations of the dataset. The map was reproduced frequently and global primary production estimates (Table 2.3) were calculated using the map as late as 1975. In 1959, Gessner published the map with a drafting error which was subsequently reproduced in 1975 by both Lieth and Bunt (Berger, 1989). The original FAO estimate of global ocean productivity (20 Pg y^{-1}) is still cited with little discussion about how it was compiled (Table 2.3).

At the meeting the ICES Committee on Methods for the Measurement of Primary Production made, among others, the following recommendations:

'It was agreed that the carbon-14 technique, at the present moment is a most useful technique for measuring production of organic matter in the sea, and that every effort should be made to facilitate workers in the field to utilize the technique.

It is suggested that a central agency be established, for example at Charlottenlund under the direction of Prof. E. Steemann Nielsen, which would provide standardized ampoules of ^{14}C, counting of ^{14}C samples, and/or calculation of carbon assimilation rates... It is proposed that UNESCO be approached concerning the possibility of their providing the necessary funds for establishing the agency and guaranteeing the economy during the first years.

In most smaller laboratories and research groups in the initiation of such studies, field kits for sample collecting and processing, including water samplers, filtration apparatus, filters, etc. will be provided on request and at cost by Prof. E. Steemann Nielsen, Prof. Maxwell S. Doty and Dr John H. Ryther.'

This innovative central agency was created and it made state-of-the-art primary productivity methods widely available. When one of us (RTB) entered oceanography in 1964 it was standard practice in US laboratories to obtain ^{14}C from the agency. By agreeing to direct this new agency, Steemann Nielsen demonstrated his recognition that the ^{14}C method needed a new infrastructure to realise its full potential.

A second milestone in primary productivity studies in the period from 1950 to 1970 was the publication in 1963 of *The Sea: Ideas and Observations on Progress in the Study of the Seas*, Vol. 2, edited by Hill, which contained major articles by Steemann Nielsen, 'Productivity, definition and measurement'; Ryther, 'Geographic variations in productivity'; Riley, 'Theory of food chain relations in the ocean'; and Provasoli, 'Organic regulation of phytoplankton fertility' (Hill, 1963). All of these chapters were written in 1960, so they represent a decadal celebration of the introduction of the ^{14}C method. Each comprises a comprehensive review, a clear synthesis and an evaluation of work needed. Steemann Nielsen focused on the ^{14}C method *per se* and the regulation of photosynthesis by light. The results of his many studies of the ^{14}C method, his basic research on photosynthesis and his actual observations from around the world are distilled in his chapter.

Riley focused on the theory of oceanic food webs. With 40 equations – impressive even by the standards of physical oceanography – Riley's chapter served as a primer in ecosystem modelling for a generation of students before John Steele's monograph became available in 1974 (Barber and Hilting, 2000). Citation indices must be used with caution, but one of them indicates that Riley's chapter in Hill (1963) has been cited three times as often as either Steemann Nielsen's or Ryther's chapter. We suggest that Riley's chapter in *The Sea* was his *magnum opus* because it is the most readable and accessible account of his ideas and demonstrates better than his longer reports his originality and analytical power. Unlike Ryther, Riley was not given to succinct *Science* articles, so for him *The Sea* was an important avenue to a broader oceanographic audience.

Ryther's chapter was a masterful analysis of the regulation of productivity in different geographic locations. Not surprisingly, as it was written before large numbers of ^{14}C productivity measurements were available, the chapter has no single synthetic global map. One can see, however, the analytical pattern that led Ryther in a few years to his seminal paper, which was published in *Science* in 1969 (Ryther, 1969). More than Steemann Nielsen and much more than Riley, Ryther is the intellectual descendant of Gran. Like Gran, Ryther looked at all aspects of a given region – water colour, temperature, biomass of higher trophic organisms, wind and seasonality – along with carefully evaluating the ^{14}C uptake measurements. Like Gran, he is an intuitive and consummate ocean naturalist.

Provasoli, in his chapter, comprehensively reviewed the subject of organic growth factors, concentrating on both organic chelators and trace metal availability (Hill, 1963). This was the first detailed treatment of this topic in a volume intended for a general oceanographic audience. While Provasoli's chapter did not directly refer to primary productivity measurements, the issues he raised and discussed regarding trace metal effects on phytoplankton growth became central primary productivity issues in the 1980s and 1990s. This chapter provides foreshadowing of the trace metal issues that were reborn in John Martin's iron hypothesis and became, as the importance of iron was resolved, one of the major achievements of biological oceanography in the post World War II era (Barber and Hilting, 2000). Provasoli's chapter, like Riley's, has also been cited several times as often as the Steemann Nielsen and Ryther chapters. Just as Steemann Nielsen dominated the 1957 Bergen ICES meeting, Riley dominated *The Sea* (Hill, 1963) with his complex, but tightly reasoned analysis of how oceanic food webs should work. The Steemann Nielsen and Ryther chapters in Hill (1963) are solid and scholarly, but lack the panache of Riley's contribution.

The third milestone of this period was the most widely used global primary productivity map, by Koblentz-Mishke *et al.*, published in Russian in 1968 and in English in 1970 (Koblentz-Mishke *et al.*, 1968, 1970). This map differs from all previous efforts because it is based on a very large suite of data, much of which came from the far-flung USSR cruises of the late 1950s and 1960s. Koblentz-Mishke *et al.* (1970) cited as the source of these data the international expeditions of the preceding 15 years; a generous attribution in that most of the data came from Soviet cruises. The authors refer to data from more than 7000 stations; in addition to the 16 countries listed in Table 2.5, Brazil, Indonesia, Nigeria, Republic of Congo and South Africa are acknowledged for providing data. It appears that the ^{14}C central agency at Charlottenlund was operating successfully around the world. For the most part, however, Koblentz-Mishke *et al.* (1970) used the huge Soviet collection of daily surface production rates (g C m^{-3} d^{-1}), which they transformed to annual water column rates (g C m^{-2} y^{-1}) and used to derive a global ocean productivity estimate of 25–30 Pg C y^{-1} (Table 2.3). Their map is still widely reproduced in textbooks and it is the most frequently cited article dealing with global productivity.

The Koblentz-Mishke *et al.* (1970) paper represents in several ways the culmination of the effort started by Hensen and Gran in the nineteenth century. Just as the nineteenth century oceanographers' sole purpose was to increase, or at least maintain, the yield of fish from the ocean, Koblentz-Mishke *et al.* (1970) also justify their work in terms of its benefit to improving and managing fisheries. Furthermore, this grand synthesis was based on single stations occupied by ships, as were the analyses of Hensen, Brandt and Gran. The Soviet authors, like the nineteenth century oceanographers, believed that with enough ships and enough stations an accurate description of the oceans could be obtained. When satellite remote sensing provided an accurate picture of oceanic variability, this belief in the efficacy of ship stations to resolve ocean variability ebbed. In 1968 and 1970, however, the optimism of these authors carried them through the arduous and tedious work involved in this synthesis. From 1950 to 1987 estimates of global productivity, based mostly on the work by Koblentz-Mishke *et al.* and the FAO, hovered around a mean of about 24 Pg C y^{-1}. It appears that for a period, biological oceanographers assumed that this work was satisfactorily completed. Of course, a mean annual productivity map necessarily fails to capture the intense seasonal variability of the productivity process in regions such as the Arabian Sea and the Southern Ocean. Nevertheless, Koblentz-Mishke *et al.* (1970) introduced objective regional productivity for the entire ocean to a new generation. This figure prepared oceanographers for the global information that started to become available from satellites in the 1980s (Table 2.3). Publication of Koblentz-Mishke *et al.*'s productivity map is a milestone in the history of the study of plankton productivity that biological oceanographers remember as the culmination of a simpler time, a time without remote sensing and before anyone had the ability to process massive amounts of data.

A fourth milestone was the publication of John Ryther's *magnum opus*, 'Photosynthesis and fish production in the sea' (Ryther, 1969). This contribution published in *Science* in 1969 has found a wide audience and is discussed in almost all oceanography and marine ecology textbooks. Ryther made a powerful simplification, positing that there are only three basic kinds of productivity regimes in the ocean: oceanic, coastal and upwelling. He then used his own ^{14}C estimates of primary productivity in each of these regimes together with physical estimates of the area of each regime to calculate the global productivity of the ocean. The estimate was 20 Pg C y^{-1}, remarkably similar to earlier estimates obtained with less sophisticated logic. What makes Ryther's 1969 work so revolutionary is that he went on to provide a quantitative estimate of fish yields that primary productivity can support in each of the three productivity regimes.

The basis of Ryther's simple partitioning of the ocean was, of course, the original suggestion by Gran (1912) that convergent gyres are nutrient-poor because of decreased exchange between deep waters and the euphotic layer, whereas upwelling and coastal regimes are richer because of enhanced vertical exchange. Starting with an estimate of annual primary productivity, Ryther then determined the number of ecological transfers between phytoplankton and harvestable fish as well as the efficiency of the biomass transfer at each step. Using the primary productivity, number of transfers and efficiency of each transfer, he calculated the fish yield for each regime. This simple, but revolutionary, logic enabled Ryther to make a bottom-up estimate of the global fish yield, thus realising Hensen's elusive objective.

Ryther's bold contribution sparked a lot of controversy, particularly among fisheries biologists who argued that his powerful simplifications invalidated his estimates (Alverson *et al.*, 1970). Despite these criticisms Ryther (1969) continues today to be widely read and cited and the subsequent history of fisheries bore out his contentions.

2.9   Conclusion

This chapter recounts the history of the effort to measure oceanic primary productivity, a process that has obvious importance to the management of natural resources, but also has far-reaching significance at this time when human activity is changing the carbon dioxide concentration of Earth's atmosphere. Understanding the 'laws of plant production in the ocean', the goal sought by the great pioneers Hensen, Gran and, of course, Steemann Nielsen, now has more significance than they could have envisioned.

This history has involved, albeit peripherally, some of the world's great scientists (Priestley, for example, and the von Hevesy, Calvin and Libby group as well), but the first half of this history is mainly the story of an elite group of sea-going oceanographers from countries around the North Sea. Technical limitations prevented accurate measurement of plankton productivity before World War II, although good qualitative descriptions of the dynamics of plankton productivity were achieved.

After World War II a greatly expanded international cohort of biological oceanographers, newly armed with Steemann Nielsen's ^{14}C uptake method, enthusiastically set out to answer three key questions: How productive is the ocean? What is the regional pattern of productivity? And what is the seasonal pattern of productivity? As it turned out, answers did not come easily because temporal and spatial variability of the ocean was more important than had been suspected and also because the sensitive ^{14}C method revealed other problems that had to be addressed before the original questions could be answered satisfactorily. Nevertheless, in the period of 1950 to 1970, biological oceanography marked a series of milestones as answers to the questions originally posed by Hensen, Gran and Steemann Nielsen became available. This history is, therefore, a success story – a story in which Steemann Nielsen plays a pivotal role: he bridged the pre-war and post-war periods and armed the struggling discipline of biological oceanography with the technology needed to succeed in the new era.

Acknowledgements

We gratefully acknowledge Felicity Pors of the Niels Bohr Archive; David Talbert, Duke University Marine Laboratory librarian; Dr Annette Vogt, Scholar in Residence at the Max Planck Institute for the History of Science; and Lady Lise Schou Wilkinson for their help. They provided essential information or directed us to appropriate sources.

References

Alverson, D.L., Longhurst, A.R. & Gulland, J.A. (1970) How much food from the sea? *Science* **168**, 503.

Atkins, W.R.G. (1926) A quantitative consideration of some factors concerned in plant growth in water. Part II. Some chemical factors. *Journal du Conseil International pour l'Exploration de la Mer* **1**, 99–126.

Barber, R.T. & Hilting, A.K. (2000) Achievements in biological oceanography. In: *Fifty Years of Ocean Discovery*, pp. 11–21. National Academy Press, Washington, DC.

Berger, W.H. (1989) Global maps of ocean productivity. In: *Productivity of the ocean: past and present*, Vol. 44 (eds W.H. Berger, V.S. Smetacek & G. Wefer), pp. 429–445. Dahlem Workshop Reports.

Chambers, C.O. (1912) *The relation of algae to dissolved oxygen and carbon-dioxide. With special reference to carbonates.* Missouri Botanical Garden 23rd Annual Report. The Board of Trustees, St Louis, Missouri.

Cobb, J.S. & Harlin, M.M. (1976) Introduction. In: *Marine Ecology: Selected Readings* (eds J.S. Cobb & M.M. Harlin), pp. 5–16. University Park Press, Baltimore.

Cushing, D.H. (1975) *Marine Ecology and Fisheries.* Cambridge University Press.

Fleming, R.H. (1957) Features of the oceans. In: *Treatise on Marine Ecology and Paleoecology* (ed. J.W. Hedgpeth), Vol. 1, pp. 87–107. The Geological Society of America Memoir 67. Waverly Press, Baltimore.

Fleming, R.H. and Laevastu, T. (1955) *The influence of hydrographic conditions on the behaviour of fish: A preliminary literature survey.* Unpublished manuscript.

Fogg, G.E. (1958) Actual and potential yields in photosynthesis. *Advances in Science* **14**, 395–400.

Gaarder, T. & Gran, H.H. (1927) Investigations of the production of plankton in the Oslo Fjord. *Rapports et Procès-Verbaux des Réunions, Conseil International pour l'Exploration de la Mer* **42**, 1–48.

Gran, H.H. (1912) Pelagic plant life. In: *The Depths of the Ocean* (eds J. Murray & J. Hjort), pp. 307–387. MacMillan and Co., London.

Gran, H.H. (1918) Kulturforsök med planktonalger. *Forhandlinger Skand. Naturforskeres 16 de möte Kristiania (1916)*, 391.

Gran, H.H. (1932) Phytoplankton. Methods and problems. *Journal du Conseil International pour l'Exploration de la Mer* **7**, 343–358.

Harvey, H.W. (1955) *The Chemistry and Fertility of Sea Waters.* University Press, Cambridge.

Hill, M.N. (1963) *The Sea, Ideas and Observations on Progress in the Study of the Seas.* John Wiley and Sons, London.

Huxley, T.H. (1915) On some of the results of the expedition of H.M.S. Challenger. In: *Discourses Biological and Geological: Essays*, Vol. VIII. D. Appleton, New York.

Johnstone, J. (1908) *Conditions of Life in the Sea: A Short Account of Quantitative Marine Biological Research.* University Press, Cambridge.

Kamen, M.D. (1986) A cupful of luck, a pinch of sagacity. *Annual Review of Biochemistry* **55**, 1–34.

Koblentz-Mishke, O.J., Volkovinsky, V.V. & Kabanova, J.G. (1968) Noviie dannie o veli-chine pervichnoi produktsii mirovogo okeana. *Doklady Akad. Nauk SSSR* **183**, 1189–1192.

Koblentz-Mishke, O.J., Volkovinsky, V.V. & Kabanova, J.G. (1970) Plankton primary production of the world ocean. In: *Scientific Exploration of the South Pacific* (ed. W.S. Wooster), pp. 183–193. National Academy of Sciences, Washington, D.C.

Landry, M.R., Barber, R.T., Bidigare, R.R. *et al.* (1997) Iron and grazing constraints on primary production in the central equatorial Pacific: an EqPac synthesis. *Limnology and Oceanography* **42**, 405–418.

Lieth, H. (1975) Historical survey of primary productivity research. In: *Primary Productivity of the Biosphere* (eds H. Lieth & R.H. Whittaker), pp. 7–16. Springer Verlag, New York.

Longhurst, A., Sathyendranath, S., Platt, T. & Caverhill, C. (1995) An estimate of global primary production in the ocean from satellite radiometer data. *Journal of Plankton Research* **17**, 1245–1271.

Mills, E.L. (1989) *Biological Oceanography: An early history, 1870–1960.* Cornell University Press, Ithaca.

Perutz, M.F. & Kendrew, J.C. (1964) *Nobel Lectures, Chemistry 1942–1962.* Elsevier Publishing Company, Amsterdam.

Platt, T. & Subba Rao, D.V. (1975) Primary production of marine microphytes. In: *Photosynthesis and Productivity in Different Environments* (ed. J.P. Cooper), pp. 249–280. Cambridge University Press.

Rabinowitch, E.I. (1945) *Photosynthesis and Related Processes, Volume 1. Chemistry of Photosynthesis, Chemosynthesis and Related Processes in Vitro and in Vivo.* Interscience Publishers, Inc., New York.

Riley, G.A. (1941) Plankton Studies IV. Georges Bank. *Bulletin of the Bingham Oceanographic College* **1**, 1.

Riley, G.A. (1944) Carbon metabolism and photosynthetic efficiency. *American Scientist* **32**, 132–134.

Riley, G.A. (1946) Factors controlling phytoplankton populations on Georges Bank. *Journal of Marine Research* **5**, 54–73.

Riley, G.A. (1963) Theory of food-chain relations in the ocean. In: *The Sea*, Vol. 2 (ed. M.N. Hill), pp. 438–463. John Wiley & Sons, New York.

Ryther, J.H. (1969) Photosynthesis and fish production in the sea. *Science* **166**, 72–76.

Shushkina, E.A. (1985) Production of principal ecological groups of plankton in the epipelagic zone of the ocean. *Oceanology* **25**, 653–658.

Steemann Nielsen, E. (1937) The annual amount of organic matter produced by the phytoplankton in the Sound off Helsingor. *Meddelelser fra Kommissionen for Danmarks Fiskeri- og Havundersøgelser, Serie: Plankton* **3**, 1–37.

Steemann Nielsen, E. (1947) *Photosynthesis of Aquatic Plants with Special Reference to Carbon Sources.* H.P. Hansens, Copenhagen.

Steemann Nielsen, E. (1952) The use of radio-active carbon (C^{14}) for measuring organic production in the sea. *Journal du Conseil International pour l'Exploration de la Mer* **16**, 117–140.

Steemann Nielsen, E. & Cushing, D.H. (1958) Measurements of primary production in the sea. *Conseil Permanent International Pour L'Exploration de la Mer: Rapports et Procès-Verbaux* **144**, 1–158.

Steemann Nielsen, E. & Jensen, E. (1957) Primary ocean production. The autotrophic production of organic matter in the oceans. *Galathea Report* **1**, 49–136.

Sverdrup, H.U. (1953) On conditions for the vernal blooming of phytoplankton. *Journal du Conseil International pour l'Exploration de la Mer* **18**, 287–295.

Sverdrup, H.U. (1955) The place of physical oceanography in oceanographic research. *Journal of Marine Research* **14**, 287–294.

Vinberg, G.G. (1960) *Pervichnaya Produktsiia Voedoemov* (The Primary Production of Bodies of Water). Izdatel'stvo Akademii Nauk, Belorusskaya, SSR, Minsk.

Vinogradov, M.E., Shushkina, E.A., Kopelevich, O.V. & Sheberstov, S.V. (1996) Photosynthetic productivity of the world ocean from satellite and expeditionary data. *Oceanology* **36**, 531–540.

Whipple, G.C. (1899) *The Microscopy of Drinking Water.* John Wiley and Sons, New York.

Wolff, T. & Petersen, M.E. (1991) A brief biography of A.S. Ørsted, with notes on his travels in the West Indies and Central America and illustrations of collected polychaetes. *Ophelia Supplement* **5**, 669–685.

Chapter 3
Physiology and Biochemistry of Photosynthesis and Algal Carbon Acquisition

Richard J. Geider and Hugh L. MacIntyre

3.1 Introduction

Photosynthesis involves a series of reactions that start with light absorption, involve synthesis of NADPH and ATP as intermediate energy-conserving compounds, and lead to CO_2 fixation in the Calvin cycle (Falkowski and Raven, 1997). It be represented by the deceivingly simple reaction:

$$CO_2 + H_2O + 8 \text{ photons} \rightarrow O_2 + 1/6 \ C_6H_{12}O_6 \qquad (3.1)$$

Although this equation captures the essence of photosynthesis, it is an incomplete model for several reasons. Oxygen cycling within the photosynthetic electron transfer chain due to the Mehler reaction can occur independently of carbon cycling at up to 50% of the gross O_2 evolution rate (Kana, 1992). Actively growing, nutrient-replete algae allocate about 50% of recent photosynthate to protein synthesis (Li and Platt, 1982), leading to direct or indirect competition for reductant (NADPH) and ATP between CO_2 fixation, transport processes, nitrate reduction and amino acid and protein synthesis (Turpin and Bruce, 1990; Noctor and Foyer, 1998). Furthermore, O_2 competes with CO_2 in a process known as photorespiration (Osmond, 1981) because the enzyme that fixes CO_2, ribulose-1,5-bisphosphate carboxylase, is also an oxygenase (RUBISCO). Although phytoplankton commonly suppress the oxygenase activity of RUBISCO by actively concentrating CO_2 within the chloroplast (Badger *et al.*, 1998), this also requires energy. Finally, CO_2 and HCO_3^- can be incorporated by non-photosynthetic reactions (Raven, 1997).

Primary production can be quantified in terms of four currencies: photons, electrons, oxygen and carbon (equations (3.1) and (3.2)). Primary production is commonly measured by inorganic ^{14}C assimilation, although it can also be measured as O_2 evolution or TCO_2 uptake in light/dark bottle gas exchange experiments. However, under some circumstances, oxygen consumption in the light can differ significantly from oxygen consumption in the dark. The difference is most pro-

44

nounced at irradiances that saturate photosynthesis. For example, in a range of estuarine microflagellates, the maximum oxygen consumption rate in the light was 3–20-fold greater (mean sevenfold) than the dark respiration rate, and light respiration equalled 12–40% (mean 26%) of light-saturated photosynthesis (Lewitus and Kana, 1995). In *Synechococcus* sp. (WH7803), the gross O_2 evolution rate at light saturation was up to 50% greater than net O_2 evolution (Kana, 1992), the effect being most pronounced in cells cultured at low irradiance (45 μmol m^{-2} s^{-1}) and then exposed to photoinhibiting irradiance (> 1000 μmol m^{-2} s^{-1}). The rate of gross oxygen evolution can be obtained directly by measuring evolution of $^{18}O^{16}O$ from $H_2^{18}O$ (Grande *et al.*, 1989) or indirectly by adding net O_2 evolution to $^{18}O_2$ consumption in the light (Kana, 1992; Lewitus and Kana, 1995). At low light, gross O_2 evolution is limited by the rate of photon absorption and the maximum quantum efficiency, providing a direct link to the bio-optical calculation of photosynthesis. At high irradiance, inefficiencies in photosynthesis lead to reductions of the quantum efficiency that can be assessed by fluorescence techniques such as pulse amplitude modulated (Gilbert *et al.*, 2000) or fast-repetition-rate fluorescence (Kolber *et al.*, 1998). Reconciling measurements of primary production obtained in terms of these four currencies is an active area of research (Gilbert *et al.*, 2000; Suggett *et al.*, 2001).

3.2 Light harvesting and ATP/NADPH production

The essential features of the light reactions of photosynthesis (Fig. 3.1) are:

(1) photon absorption in the light-harvesting antennae,
(2) migration of excitation energy of the absorbed photon (i.e. an exciton) to the reaction centres,
(3) electron transfer from H_2O to $NADP^+$, and
(4) generation of ATP by a trans-thylakoid pH gradient that is set up as a consequence of electron transfer (Falkowski and Raven, 1997).

Under ideal conditions, the light reactions can be summarised as:

$$8 \text{ photons} + 2\ H_2O + 2\ NADP^+ + 3\ ADP + 3\ P_i$$
$$\rightarrow O_2 + 2\ H^+ + 2\ NADPH + 3\ ATP \tag{3.2}$$

This equation summarises photosynthetic electron transfer (PET), which links O_2 evolution to NADPH production (Falkowski and Raven, 1997). PET is catalysed by three major, membrane-spanning complexes and small molecules that ferry electrons between them. The large complexes are Photosystem II (PSII), the cytochrome b_6/f complex ($cyt_{b/f}$) and Photosystem I (PSI). PSII and $cyt_{b/f}$ are linked by the mobile carrier plastoquinone, whereas $cyt_{b/f}$ and PSI are linked by plastocyanin or cytochrome c. Finally, ferredoxin links PSI to the reduction of $NADP^+$. In addition to the PET chain, the thylakoid membranes contain the ATP synthase that draws on the proton gradient generated by the electron transfer chain to synthesise ATP.

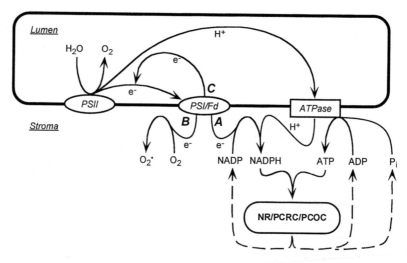

Fig. 3.1   The light reactions of photosynthesis. Light is absorbed by pigments in the antennae of PSII and PSI. Hydrolysis occurs at PSII. The resulting electron is passed to the PSI/Fd complex through a series of electron carriers (not shown). Protons produced by hydrolysis and from a trans-thylakoid pump between the two photosystems (not shown) generate a proton gradient across the thylakoid membrane. This is coupled to ATP synthesis by ATP synthase, a membrane-spanning complex. Electrons can be passed from the PSI/Fd complex to a number of acceptors. These include: (A) $NADP^+$, in its reduction to NADPH; (B) O_2, in formation of the superoxide radical by the Mehler reaction; and (C) the electron transport chain in cyclic flow around PSI. The ATP and NADPH produced by the light reactions are consumed by nitrate reduction, the Calvin (PCRC) cycle and photorespiration (PCOC), regenerating the substrates NADP, ADP and P_i. *Abbreviations*: ATPase, ATP synthase; Fd, ferredoxin; NR, nitrate reduction; PCOC, photosynthetic carbon oxidation cycle; PCRC, photosynthetic carbon reduction cycle; PSI, photosystem I; PSII, photosystem II.

Equation (3.2) describes linear electron flow from water to NADPH. The structural complexity of the PET chain allows considerable versatility in electron flow. In particular, it allows light-driven cycling of electrons around PSI that is linked to ATP synthesis independently of NADPH production. This cyclic electron flow allows the supplies of ATP and NADPH to be closely matched to the varying requirements of nutrient (including CO_2) transport, photosynthesis and biosynthesis.

Photosynthetic unit size

A photosynthetic unit (PSU) is defined operationally as the ratio of chlorophyll *a* (or in some cases total chlorophyll or total pigments) to functional reaction centres. Although methods of measuring photosynthetic unit size were developed in the 1930s for PSII and 1950s for PSI (Falkowski and Raven, 1997), it was not until the late 1970s that these techniques were applied within the context of algal ecophysiology (Prézelin, 1981). Ambiguity in reports of PSU size arises from variability in the relative abundance of the components of the PET chain (Raven, 1990). In chromophytes the PSII:PSI ratio may exceed 2, whereas in cyanobacteria it is typically 0.25–0.5. In addition, this ratio may change with growth irradiance (Neale and Melis, 1986).

Light harvesting

Photosynthesis begins with photon absorption by pigments located in the thylakoid membranes. Chlorophyll *a* (or divyl chlorophyll *a*) is the only pigment that is present in all oxygenic phytoplankton, and it is the essential pigment that is required for photosynthesis to occur. In addition to chlorophyll *a*, phytoplankton contain accessory chlorophylls, carotenoids and/or phycobilins. Major differences in accessory pigment composition amongst higher taxonomic groups can significantly affect the ability of phytoplankton to absorb light (Sathyendranath *et al.*, 1987). Although most light absorption is due to accessory pigments, chlorophyll *a* is still the most common index of phytoplankton abundance, and light absorption is typically expressed as the chlorophyll *a*-specific light absorption coefficient, designated a^{Chl}. Fortunately, in eukaryotic phytoplankton, the ratios of accessory carotenoids to chlorophyll *a* appear to be relatively stable within a species (Goericke and Montoya, 1998; MacIntyre *et al.*, 2002), although this is not the case with the cyanobacterial phycobilins. It should be noted that a^{Chl} is not only affected by accessory pigment complement, but also by cell size and intracellular pigment concentration (Bricaud *et al.*, 1988).

Bio-optical calculation of photosynthesis rate

Photosynthesis can be calculated as the product of the photon efficiency of photosynthesis and the rate of light absorption. On a chlorophyll *a*-specific basis, this is simply:

$$P^{Chl}(E) = a^{Chl}\phi(E)E \tag{3.3}$$

where: $P^{Chl}(E)$ is the chlorophyll *a*-specific photosynthesis rate (μmol O_2 [g chl$a$]$^{-1}$ s^{-1}), $\phi(E)$ is the irradiance-dependent photon efficiency of photosynthesis (mol O_2 [mol photon]$^{-1}$) and E is irradiance (μmol photons m^{-2} s^{-1}). All of the terms in this equation, but especially a^{Chl}, depend on wavelength.

Implicit in equation (3.1) is the maximum value for ϕ of 1 O_2 per 8 photons absorbed. The photon efficiency will be lower if:

(1) photons are absorbed by pigments that are uncoupled from the reaction centres such as zeaxanthin located in the cell walls of cyanobacteria (Bidigare *et al.*, 1989),
(2) excitons are converted to heat (Olaizola *et al.*, 1994),

(3) electrons are diverted from the PET chain, for example by the Mehler reaction (Kana, 1992), and/or

(4) cyclic electron flow is required to synthesise additional ATP.

A correction for light absorption by non-photosynthetic pigments can be made if carotenoid content is known, or if fluorescence excitation spectra are available for comparison with absorption spectra (Sakshaug *et al.*, 1997). The remaining inefficiencies are accounted for by determining the irradiance dependence of ϕ. For example, if the photosynthesis–irradiance (PE) response curve is described by a Poisson function (equation (3.5)), then

$$\phi(E) = \phi_m E_K [1 - \exp(-E/E_K)]/E \tag{3.4}$$

where ϕ_m is the maximum photon efficiency of photosynthesis and E_K is the light saturation parameter of the PE curve. Thus, equation (3.3) can be rewritten as:

$$P^{Chl}(E) = a^{Chl} \phi_m E_K[1 - \exp(-E/E_K)] \tag{3.5}$$

Defining the functional relationship between ϕ and irradiance typically requires surface truth observations of PE curves:

$$P^{Chl}(E) = P_m{}^{Chl}[1 - \exp(-a^{Chl} \phi_m E/P_m{}^{Chl})] \tag{3.6}$$

where $E_K = P_m{}^{Chl}/a^{Chl} \phi_m$. Alternatively, limnologists and oceanographers have turned to fluorescence techniques (Gilbert *et al.*, 2000; Suggett *et al.*, 2001) to establish the light dependence of ϕ (equation (3.4)). Equations (3.5) and (3.6) are equally valid summaries of the light dependence of photosynthesis, and provide identical results. However, equation (3.6) explicitly considers the light-saturated photosynthesis rate.

3.3   The Photosynthetic Carbon Reduction (Calvin) Cycle

Net photosynthetic carbon fixation involves a cycle of reactions named the photosynthetic carbon reduction cycle, but also referred to as the Calvin or Calvin–Benson–Bassham cycle to honour its discoverers. Carbon enters the Calvin cycle as CO_2 and leaves in the form of sugar phosphates (Fig. 3.2). The Calvin cycle requires energy in the form of ATP and reducing equivalents in the form of NADPH. It can be summarised as:

$$CO_2 + 2\ NADPH + 2\ H^+ + 3\ ATP$$
$$\rightarrow 1/6\ C_6H_{12}O_6 + H_2O + 2\ NADP^+ + 3\ ADP + 3\ P_i \tag{3.7}$$

(Note: equations (3.2) and (3.7) sum to yield equation (3.1).)

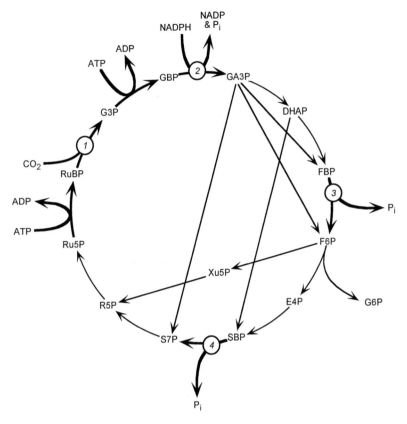

Fig. 3.2 The Calvin cycle. Carboxylation by RUBISCO is fuelled by ATP and NADPH generated by the light reactions and ultimately results in net synthesis of sugars, with regeneration of the substrate RuBP. The rate of CO_2 consumption is determined by either the rate of carboxylation or the rate of RuBP regeneration. Note that P_i enters the cycle with formation of RuBP and is not released for subsequent synthesis of ATP until FBP and SBP are dephosphorylated. This has important consequences for regulation of the enzymes that participate in the cycle. Enzymes whose activities are regulated are numbered: ① ribulose-1,5-bisphosphate carboxylase/oxygenase (RUBISCO); ② 3-phosphoglyceraldehyde dehydrogenase; ③ fructose-1,5-bisphosphatase (FPBase); and ④ sedoheptulose-1,7-bisphosphatase. *Abbreviations*: DHAP, dihydroxyacetone phosphate; E4P, erythrose 4-phosphate; FBP, fructose 1,5-bisphosphate; F6P, fructose 6-phosphate; GA3P, glyceraldehyde 3-phosphate; GBP, glycerate 1,3-bisphosphate; G3P, glycerate 3-phosphate; G6P, glucose 6-phosphate; P_i, inorganic phosphate; RuBP, ribulose 1,5-bisphosphate; Ru5P, ribulose 5-phosphate; R5P, ribose 5-phosphate; SBP, sedoheptulose 1,7-bisphosphate; S7P, sedoheptulose 7-phosphate; Xu5P xylulose 5-phosphate.

At the heart of the Calvin cycle is the carboxylation of ribulose 1,5-bisphosphate (RuBP) to form two molecules of glycerate 3-phosphate (G3P). G3P is reduced to glyceraldehyde 3-phosphate (GA3P). The remaining steps of the Calvin cycle serve to regenerate RuBP. The cycle is autocatalytic, with net incorporation of CO_2 leading to a build-up of intermediates, some of which are withdrawn from the cycle for biosynthesis or storage. Under steady-state photosynthesis, sugar-phosphates can be exported from the Calvin cycle: for every five molecules of GA3P that are recycled to RuBP, one molecule can be exported.

The carboxylation of RuBP is catalysed by RUBISCO, an enzyme that is unique and central to the Calvin cycle (Tabita, 1999). It is a monophyletic enzyme that exists in two forms (Table 3.1). These are Form I with eight large and eight small (L_8S_8) subunits and Form II with two large subunits (L_2). Each L_8S_8 Form I enzyme contains eight active sites. The large subunits alone are capable of fixing CO_2; however, the small subunits of the Form I enzyme increase catalytic efficiency. The Form I enzyme is found in cyanobacteria and all eukaryotes, with the exception of some peridinnin-containing dinoflagellates which contain the Form II enzyme.

Table 3.1   Diversity of forms of RUBISCOs within phytoplankton.

Form (subunit organisation)	Form I (L_8S_8)		Form II (L_2)
Origin	Cyanobacterial	β-Proteobacterium	α-Proteobacterium
Taxa	Form 1B (green type) Cyanobacteria	Form 1D (red type) Rhodophyta Cryptophyta	Dinophyta
	Chlorophyta Chlorophyceaea Prasinophyceae Eugleophyceae	Haptophyta Heterokonta Bacillariophyceae Chrysophyceaea Diophyta	
Location of gene	Chloroplastic *rbc*L Nuclear *rbc*S	Chloroplastic *rbc*L Chloroplastic *rbc*S	Nuclear *rbc*L

RUBISCO can rapidly lose catalytic activity following extraction. Under optimal extraction and assay conditions, *in vitro* RUBISCO activity is often sufficient to support the *in vivo* photosynthetic CO_2 fixation rates in cyanobacteria and chlorophytes. However, *in vitro* RUBISCO activity in some chlorophyll c-containing algae such as the diatom *Thalassiosira weissflogii* is typically $< 40\%$ of the *in vivo* activity, despite efforts to optimise assay conditions (MacIntyre *et al.*, 1997). In the Form II-bearing dinoflagellate *Amphidinium carterae*, RUBISCO activity *in vitro* was only 1% of the rate *in vivo* (Whitney and Yellowlees, 1995). The loss of catalytic activity *in vitro* limits progress in understanding the role of RUBISCO in phytoplankton ecophysiology. This is because accurate measurements of the *in vitro* activity are necessary to determine whether RUBISCO or PET limits light-saturated photosynthesis and to determine the concentration of substrate CO_2 needed to support a specified photosynthesis rate. This has implications for assessing the inhibition of carbon fixation by oxygen (see section 3.4) and the role of a CO_2 concentrating mechanism (see section 3.5).

RUBISCO is a large, catalytically inefficient enzyme. The maximum reported carboxylation rate of 80 mol CO_2 (mol RUBISCO)$^{-1}$ s^{-1}, or 10 mol CO_2 (mol active sites)$^{-1}$ s^{-1} (Table 3.2) is low compared with other carboxylases (Raven, 1997). As a consequence, RUBISCO can account for a sizeable fraction of cell protein and is arguably the most abundant protein on the planet (Ellis, 1979). RUBISCO accounts for 1–10% of cell carbon (about 2–10% of total protein) in microalgae and cyanobacteria (Sukenik *et al.*, 1987; Falkowski *et al.*, 1989; Beardall *et al.*, 1990; Giordano and Bowes, 1997). Under high light conditions RUBISCO can account for up to five times as much cell mass as chlorophyll *a*, the most common benchmark of phytoplankton abundance. However, the cell quotas of RUBISCO and chlorophyll *a* are more equal under low light conditions.

Table 3.2   Kinetic constants for RUBISCO carboxylation (K_{CO_2}) and oxygenation (K_{O_2}) reactions (abstracted from Badger *et al.*, 1998). V_{mCO_2} and V_{mO_2} are the maximum, cell specific carboxylation and oxygenation rates. S_{rel}* is the specificity factor (see equation (3.11)).

Species	K_{CO_2} [1]	K_{O_2} [1]	V_{mCO_2} [2]	V_{mO_2} [2]	S_{rel}*
Green algae (Form 1B)	12–38 ($n=4$)	410–660 ($n=3$)	No data	No data	54–83 ($n=4$)
Cyanobacteria (Form 1B)	105–185 ($n=5$)	530–1300 ($n=4$)	160–170 ($n=2$)	13–17 ($n=2$)	35–55 ($n=5$)
Chromophytes (Form 1D)	31–59 ($n=4$)	568–2074 ($n=4$)	11–80 ($n=4$)	1.4–35 ($n=4$)	80–110 ($n=6$)
Rhodophyta (Form 1D)	6.6–22 ($n=3$)	1574 ($n=1$)	17–23 ($n=4$)	12 ($n=1$)	129–238 ($n=4$)

[1] K_{CO_2} and K_{O_2} are in μM.
[2] V_{mCO_2} and V_{mO_2} are in units of μmol g^{-1} s^{-1} and were calculated assuming eight active sites per molecule of RUBISCO with a molecular weight of 560 000.

The rate of carbon fixation depends on the amount of RUBISCO in the cell, the proportion of RUBISCO that is catalytically competent, the maximum catalytic activity of activated RUBISCO and the intracellular concentration of CO_2 *at the active site* of RUBISCO. On evolutionary time-scales, the maximum catalytic rate, the half saturation constant and the susceptibility to O_2 inhibition of CO_2 fixation have been subject to selective pressure (Table 3.2). The concentration of CO_2 at the active site depends on whether cells possess a carbon concentrating mechanism (see section 3.5). Variations in the cell content of RUBISCO are considered in section 3.7. Here, we consider the mechanisms whereby the activation state of RUBISCO is modulated.

Regulation of RUBISCO activity in vivo

On physiological time-scales, RUBISCO activity is regulated to maintain a balance between the generation of NADPH and ATP from the light reactions and sinks for the triose-phosphates in biosynthesis or storage product (starch or polysaccharide) synthesis. Were RUBISCO activity not regulated, the imbalance could sequester phosphate in phosphorylated sugar intermediates (Fig. 3.2), restricting the regeneration of ATP. There are occasional reports of RUBISCO that is little if at all regulated across the light gradient, for example maize and *Pavlova lutheri*. Regulation of RUBISCO may occur through limitation of CO_2 supply at subsaturating irradiance in the former (Sage and Seeman, 1993) and may be an artifact of the *in vitro* assay in the latter (MacIntyre *et al.*, 1997).

Most of the research on regulation of RUBISCO's activity has been done on vascular plants and a few chlorophytes and cyanobacteria. The mechanisms include:

(1)   a reversible activation by modification of the active site,
(2)   the binding of a non-competitive inhibitor in the active site, and
(3)   the binding of effectors at sites other than the active site.

Relatively little is known about regulation in chromophytes and rhodophytes, but the highly conserved structure of the active site and the responses of those species studied to date suggest that the same mechanisms may operate in these taxa.

The first means of regulating RUBISCO's activity is by modification of the active site, which must bind both a non-substrate CO_2 (carbamylation) and a magnesium ion in order to become catalytically competent (Miziorko and Lorimer, 1983). Catalysis is facilitated by conformational changes in the large subunit, on which the active site resides. Critical differences in the structures of the RUBISCOs confer differences in CO_2 versus O_2 specificity and catalytic turn-over rates (Table 3.2). In both the vascular plants and the chromophyta/rhodophyta, elongation of the C-terminus parallels the increased specificity for CO_2 at the expense of reduced maximum catalytic rate (Spreitzer, 1993).

The second level of regulation involves binding of sugar phosphates as non-competitive inhibitors in the active site. This can occur in three ways (Fig. 3.3). First, substrate RuBP can be bound in non-carbamylated active sites (Jordan and Chollet, 1983), preventing activation. Chromophytic and rhodophytic RUBISCOs are inhibited by RuBP to different degrees (Reid and Tabita, 1994; although see Newman *et al.*, 1989), whereas the cyanobacterial RUBISCO is not (Tabita and Coletti, 1979). Second, 2-carboxyarabinitol 1-phosphate (CA1P), which is synthesised in low light or darkness, can bind to carbamylated active sites (Seemann *et al.*, 1990; see Fig. 3.3). The presence of CA1P has not been demonstrated in organisms other than some vascular plants, but responses characteristic of CA1P regulation have been observed in *Dunaliella tertiolecta* and *Thalassiosira weissflogii* (MacIntyre *et al.*, 1997). Third, there can be a back-reaction of catalysis that produces the sugar phosphates xylulose 1,5-bisphosphate (XuBP) and 3-keto arabinitol 1,5-bisphosphate (KABP). The former is produced by the carboxylase activity of RUBISCO and the latter by the oxygenase activity (Zhu *et al.*, 1998). These are tight-binding inhibitors that reduce RUBISCO activity (Edmondson *et al.*, 1990). This process (Fig. 3.3) is termed fallover and is one cause of declining activity *in*

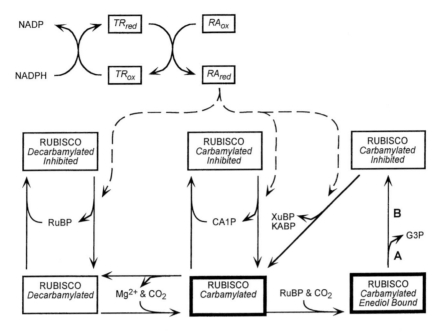

Fig. 3.3   Regulation of RUBISCO. The primary step in activation of the enzyme is carbamylation, binding of a magnesium ion and non-substrate CO_2, which proceeds spontaneously. The decarbamylated enzyme can be inhibited by binding RuBP in the active site and the carbamylated enzyme can be inhibited by binding CA1P in the active site. Substrates, RuBP and CO_2, bind in the active site of the carbamylated enzyme to form an enediol, following deprotonation of RuBP. This has two possible fates: (1) decomposition to two molecules of G3P in carboxylation (indicated by A in the diagram) or one G3P and one PG in oxygenation (not shown); or (2) reprotonation to form the inhibitors XuBP and KABP in fallover (indicated by B in the diagram). Removal of the bound sugar phosphate inhibitors in all three cases is facilitated by RUBISCO activase. This is in turn regulated by ATP and NADPH. The activity of RUBISCO is therefore regulated by the availability of its substrates. RUBISCO activase activity is regulated by NADPH in a redox cascade via thioredoxin. It is also regulated by the relative abundance of ATP and ADP. *Abbreviations*: CA1P, carboxyarabinitol 1-phosphate; GA3P, glyceraldehyde 3-phosphate; KABP, 3-keto arabinitol 1,5-bisphosphate; P_i, inorganic phosphate; PCRC, photosynthetic carbon reduction cycle; RA, RUBISCO activase; RuBP, ribulose 1,5-bisphosphate; TR, thioredoxin; XuBP, xylulose 1,5-bisphosphate.

vitro. It may also modulate RUBISCO's activity *in vivo*. Fallover has been observed in chromophytes (Uemura *et al.*, 1998) but not in cyanobacteria (Andrews and Ballment, 1984) nor chlorophytic microalgae (Uemura *et al.*, 1998). The inhibition of RUBISCO activity by all three of these modes of sugar phosphate binding can be reversed by the enzyme RUBISCO activase (see below).

The third level of regulation is by binding of effectors at sites other than the active site. The effectors include sulphate (Parry and Gutteridge, 1984), phosphate (Marcus and Gurevitz, 2000) and sugar phosphates, including RuBP (Uemura *et al.*, 1998; Marcus and Gurevitz, 2000). RuBP can act as a substrate, inhibitor or effector for carboxylation, depending on the nature of its binding. Its multiple roles and the stimulatory effect of phosphate on catalysis are critical in regulating RUBISCO's activity to match the rate at which reductant and ATP are produced in the light reactions and the rate of sugar synthesis and RuBP regeneration in the Calvin cycle.

RUBISCO activase

RUBISCO activase regulates RUBISCO activity by catalysing the removal of sugar phosphates from RUBISCO's active site (see above). It exists in two isoforms, both of which are sensitive to ADP:ATP (Zhang and Portis, 1999). The larger of the two isoforms is also regulated by the redox state of PSI (Zhang and Portis, 1999) in a redox cascade that involves thioredoxins (Ruuska *et al.*, 2000). These are enzymes that are reduced by the passage of electrons from ferredoxin via ferredoxin-thio-redoxin reductase (Fig. 3.3) and which in turn reduce disulphide bonds in target enzymes, increasing their activity (Buchanan, 1992). In species that have both iso-forms, the redox-regulated larger isoform imposes a level of control on the smaller isoform (Zhang and Portis, 1999). Information on RUBISCO activase comes almost exclusively from chlorophytes (both microalgae and vascular plants). A RUBISCO activase-like protein has been found in the cytosol (but interestingly, not in the carboxysomes) of *Synechococcus* sp. (Friedberg *et al.*, 1993) and a gene for activase has been isolated from *Nostoc* and *Anabaena* (Li *et al.*, 1993). Of the few obser-vations for microalgae, it appears that *Chlamydomonas* has only the smaller iso-form, but the enzyme is sensitive to proteases and the lack of the more highly regulated larger isoform may be an artefact of the assay conditions. The protein has not been documented in the chromophyte/rhodophyte lineage.

  The sensitivity of RUBISCO activase to ADP:ATP and redox conditions allows it to modify RUBISCO activity in response to changes in irradiance or substrate availability. A transient rise in ADP:ATP can be caused either by a reduction in irradiance or through sink limitation of photosynthesis. This will inhibit activase activity with a corresponding decline of RUBISCO activity through accumulation of sugar phosphate inhibitors in the active sites (Fig. 3.3). Conversely, a rise in irra-diance will cause a transient decrease in ADP:ATP, followed by enhanced activase activity promoting an increased rate of removal of the sugar phosphate inhibitors, a higher activation state of RUBISCO and a higher rate of catalysis. This will cause an increase in the rate of ATP consumption so that the enhanced activity of activase will decline when a new steady state has been achieved.

Sink limitation

CO_2 fixation can be limited by the RuBP-saturated rate of carboxylation or the rate of RuBP regeneration (Ruuska *et al.*, 1998). Under low light and/or low CO_2, RuBP is saturating and the rate of CO_2 fixation is limited by RUBISCO activity. However, under conditions of high light and high CO_2, CO_2 fixation is limited by regeneration of RuBP (Ruuska *et al.*, 1998), a process known as sink limitation. This occurs when the rate of dephosphorylation of fructose bisphosphate by FBPase (Fig. 3.2) is slow (Sharkey, 1990). Low RuBP levels favour formation of inhibited, decarbamylated RUBISCO (Portis, 1995), down-regulating RUBISCO activity to match the rate of RuBP regeneration. Accumulation of FBP sequesters phosphate (Fig. 3.2). As a consequence, RUBISCO activity declines through loss of phosphate from non-catalytic regulatory sites (Marcus and Gurevitz, 2000) and reduction of the phosphate-stimulation of RUBISCO carbamylation (Belknap and Portis, 1986). Last, the sequestration of phosphate also limits the rate of ATP generation (Sharkey, 1990; Fig. 3.2), causing an increase in the ADP:ATP ratio and a decrease in RUBISCO activase activity and so down-regulation of RUBISCO activity. The reduction in the rate of ATP synthesis also causes a build-up in the trans-thylakoid proton gradient, inducing non-photochemical energy dissipation (Sharkey, 1990; Fig. 3.2).

3.4   Photorespiration

Photorespiration (Osmond, 1981) is a term used by vascular plant physiologists to refer to a specific sequence of reactions that starts with the formation of phosphoglycolate via the oxygenation of RuBP at the active site of RUBISCO (Fig. 3.4). We will use the term in the restricted sense of oxygenation of RuBP by RUBISCO and refer to the subsequent steps as photorespiratory glycolate metabolism. Competition between O_2 and CO_2 for RuBP reduces the rate of CO_2 assimilation into triose-P, reduces the energetic efficiency of photosynthesis, and may reduce the photosynthetic quotient. Thus, photorespiration is an important process that can limit photosynthetic CO_2 fixation and growth at low CO_2. Less well documented is the possibility that photorespiration can reduce oxidative stress at high irradiance (Kozaki and Takeba, 1996).

Photorespiratory glycolate metabolism

The phosphoglycolate formed by photorespiration is subsequently metabolised to glycolate, which may be excreted. Alternatively, glycolate may be oxidised to glyoxylate, and then converted to glycerate 3-phosphate by a series of reactions including condensation and decarboxylation (Fig. 3.4). The chlorophytes, prasinophytes and eugleophytes contain at least some of this sequence of reactions, known from vascular plants as the photosynthetic carbon oxidation cycle (PCOC) (Raven, 1997), involving the formation of serine from glycine. Diatoms, and perhaps other microalgae and cyanobacteria may employ a pathway in which glyoxylate is converted to tartronic semialdehyde (Fig. 3.4). Diatoms may also employ a pathway of glyoxylate metabolism that involves synthesis and subsequent metabolism of malate (Fig. 3.4), which in turn is completely oxidised to CO_2 (Raven *et al.*, 2000).

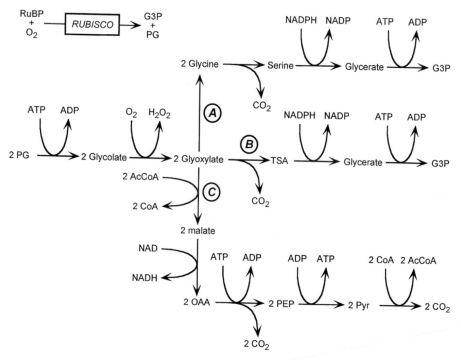

Fig. 3.4 Photorespiration. The oxygenase activity of RUBISCO (upper left) produces one G3P and one PG, rather than the two G3P produced by carboxylation. PG may be recycled through the PCOC Ⓐ or tartronic semialdehyde pathway Ⓑ, with a net loss of carbon as two PG (C₂) yield one G3P (C₃). Alternatively it may be oxidised to CO_2 Ⓒ. *Abbreviations*: AcCoA, acetyl CoA; CoA, co-factor A; G3P, glycerate 3-phosphate; OAA, oxaloacetic acid; PEP, phosphoenol pyruvate; PG, phosphoglycolate; Pyr, pyruvate; RuBP, ribulose 1,5-bisphosphate; TSA, tartronic semialdehyde.

Returning now to the initial step of photorespiration, the production of glycolate can be summarised as:

$$RuBP + O_2 \rightarrow G3P + glycolate \qquad (3.8)$$

If excreted, each molecule of glycolate removes two carbons from the Calvin cycle, which requires two carboxylations to regenerate intermediates. To this must be added the energy cost of regenerating the RuBP that acts as a substrate (Table 3.3; Raven *et al.*, 2000).

The PCOC and tartronic semialdehyde cycles dissipate less energy than does glycolate excretion by recovering 1 G3P for every 2 glycolates processed. The PCOC is essentially neutral in terms of consumption of reductant and may generate ATP (Table 3.3). In contrast, the tartronic semialdehyde cycle consumes one NADPH, but may yield some ATP in regenerating G3P (Table 3.3). Thus, these two pathways will tend to minimise the reduction of net CO_2 fixation into triose-P that could accompany inorganic carbon limitation of photosynthesis by reducing the loss

Table 3.3 Energetics of glycolate metabolism (see Raven *et al.*, 2000 for more details).

	Reactions
Glycolate synthesis	6 RUBP + 4 CO_2 + 2 O_2 + 12 NADPH + 15 ATP → 10 G3P + 2 Glycolate
Photosynthetic carbon oxidation cycle	2 Glycolate + O_2 + NADPH + ATP → G3P + CO_2 + NADH (+4 ATP)
Tartronic semialdehyde cycle	2 Glycolate + O_2 + NADPH + ATP → G3P + CO_2 (+4 ATP)
Malate cycle	Glyoxylate + 2 acetyl CoA → 2 acetyl CoA + 4 NADH + 4 CO_2

	Energetics				
	O_2/glycolate	CO_2/glycolate	G3P/glycolate	NADPH[1]/glycolate	ATP/glycolate
Glycolate excretion	−1	−2	0	−6	−7.5
Glycolate synthesis plus PCOC	−1.5	−1.5	1	−6	−8 to −6
Glycolate synthesis plus tartronic semialdehyde pathway	−1.5	−1.5	1	−6.5	−8 to −6
Glycolate synthesis plus malate pathway	−1	0	0	−4	−7.5

(1) NADPH and NADH are assumed to be equivalent.

of intermediates from the Calvin cycle. They would not be as effective as glycolate excretion in dissipating excitation energy. The pathway from glyoxylate through malate generates 2 NADH (Table 3.3) for each glycolate processed (Raven *et al.*, 2000). Although this pathway does not regenerate G3P, the reductant produced can partially offset the energetic inefficiency of photorespiration.

Carboxylation and oxygenation rates

Knowledge of the cell quota for RUBISCO, the kinetic constants for oxygenation and carboxylation, and the O_2 and CO_2 concentrations at the active site allows calculation of the carbon fixation rate from the following pair of equations:

$$V_{\text{grossCO}_2} = AS \ Q_{\text{RUBISCO}} \ V_{\text{mCO}_2} \ [CO_2]_i / \{[CO_2]_i + K_{\text{mCO}_2}(1 + [O_2]_i / K_{\text{mO}_2})\} \quad (3.9)$$

$$V_{\text{grossO}_2} = AS \ Q_{\text{RUBISCO}} \ V_{\text{mO}_2} [O_2]_i / \{[O_2]_i + K_{\text{mO}_2}(1 + [CO_2]_i / K_{\text{mCO}_2})\} \quad (3.10)$$

where V_{grossCO_2} is the cell-specific gross RUBISCO carboxylation rate (mol CO_2 cell^{-1} s^{-1}), V_{grossO_2} is the RUBISCO oxygenation rate (mol O_2 cell^{-1} s^{-1}), AS is the activation state of the enzyme (dimensionless), Q_{RUBISCO} is the cell RUBISCO content (mol RUBISCO cell^{-1}), V_{mCO_2} and V_{mO_2} are the maximum, carboxylation and oxygenation rates (mol CO_2 or O_2 mol RUBISCO^{-1} s^{-1}), $[CO_2]_i$ and $[O_2]_i$ are the CO_2 and O_2 concentrations at the active site (M), and K_{mCO_2} and K_{mO_2} are the half saturation constants for carboxylation and oxygenation (M). The kinetic constants are related through the identity:

$$S_{\text{rel}}{}^* = [V_{\text{mCO}_2} / V_{\text{mO}_2}][K_{\text{mO}_2} / (K_{\text{mCO}_2})] \quad (3.11)$$

where $S_{\text{rel}}{}^*$ is the specificity factor.

The net carboxylation rate (V_{netCO_2}) will be lower than V_{grossCO_2} due to photorespiration and subsequent metabolism of glycolate. If glycolate is excreted or fully oxidised, then the maximum rate of particulate organic carbon production (neglecting mitochondrial respiration) will be 2 V_{grossO_2} less than V_{grossCO_2}. If photorespiration occurs at a significant rate, then care needs to be taken in the definition of gross and net CO_2 fixation and its relation to ^{14}C assimilation or light–dark CO_2 exchange.

Photorespiration and the photosynthetic quotient

The ratio of O_2 to CO_2 availability can have a pronounced impact on the photosynthetic quotient (PQ = O_2 evolved/CO_2 assimilated) for net photosynthesis. PQs in *Glenodinium* sp. and *Pavlova lutheri* ranged from 1.2 to 1.8 at low O_2 partial pressures (<0.1 atmosphere) and high net photosynthesis rates (130 mmol O_2 [g Chl*a*]$^{-1}$ h^{-1}) to under 0.3 at high O_2 partial pressures (>0.3 atmospheres) and low net photosynthesis rates (3 mmol O_2 [g Chl$a$]$^{-1}$ h^{-1}) (Burris, 1981). Values of PQ of 1.2–1.8 are consistent with protein and lipid as the major products of photosynthesis (Laws, 1991). Values of the PQ as low as 0.75 O_2/CO_2 can be obtained from photorespiration and glycolate excretion at the CO_2 compensation point. Values of the PQ below 0.75 cannot be explained in terms of photosynthesis alone. They may reflect an unbalanced growth condition where respiration contributes to net CO_2 and O_2 exchange. For example, if respiration of carbohydrate with a respiratory quotient of 1.0 occurred at 50% of the rate of gross oxygen evolution, then the PQ would be 0.33 at the CO_2 compensation point.

Evidence for an effect of photorespiration on net photosynthesis

The effect of oxygen concentration on the rate of ^{14}C assimilation has been used to assess photorespiratory metabolism in phytoplankton. Carbon assimilation by *Dunaliella tertiolecta* and *Synechococcus* sp. was reduced by 17–19% in air and 38% in O_2, relative to uptake in N_2 (Beardall *et al.*, 1976; Morris and Glover, 1981). Oxygen inhibition of ^{14}C fixation was less pronounced in *Skeletonema costatum* and *Gonyaulax tamarensis* (Beardall *et al.*, 1976), consistent with the relatively higher specificity factors in chromophytes than chlorophytes and cyanobacteria (Table 3.2). Beardall *et al.* (1976) reported 4–20% lower rates of ^{14}C assimilation in air than in N_2 and 30–60% lower rates in O_2 than in N_2 in coastal phytoplankton. If the observed reductions of particulate ^{14}C assimilation in air and O_2, relative to N_2, are due to photorespiration with excretion of glycolate, this implies that oxygenation is occurring at 7.5–10% of the rate of carboxylation in air and 15–30% in O_2. The available kinetic data for RUBISCO oxygenase and carboxylase activities (Table 3.2) are consistent with this suggestion. Significantly, the decline in particulate ^{14}C assimilation may be accompanied by increased dissolved organic ^{14}C production (Fogg, 1977).

Photorespiration and photoinhibition

Under high excitation pressure, regulated glycolate excretion may provide a safety valve that reduces photo-oxidative stress. The upper limit on the operation of this hypothetical safety valve will depend on the attainable oxygenase activity of RUBISCO relative to its maximum carboxylase activity. Given that V_{mO_2} is typically only 10–25% of V_{mCO_2} (Table 3.2), and that oxygenation is likely to be inhibited by operation of a CO_2 concentrating mechanism, it appears that there is little potential for photorespiration to increase markedly the rate of energy dissipation. However, if there are two pools of RUBISCO in the cell, one operating in a high CO_2 microenvironment to maximise carboxylase activity and a second operating in a low CO_2 high O_2 microenvironment to maximise oxygenase activity, then glycolate excretion might become a viable safety valve. If, as hypothesised, the pyrenoid provides a high CO_2 environment, then changes in the proportion of

RUBISCO localised in the pyrenoid might provide a mechanism for regulating the relative rates of carboxylation and oxygenation.

3.5   Carbon-concentrating mechanisms

The biophysical CO_2-concentrating mechanism (CCM) of microalgae and cyano-bacteria has been the subject of a number of recent reviews (Raven, 1997; Badger *et al.*, 1998; Moroney and Chen, 1998). The kinetic characteristics of RUBISCO (high K_m and low V_{max} for CO_2 and inhibition of carboxylation by O_2) require high concentrations of CO_2 at the active site of RUBISCO to maintain carboxylase and inhibit oxygenase activity. Indirect evidence for a CCM comes from the low rates of photorespiration observed in phytoplankton relative to values expected for air-equilibrated sea water. Direct evidence for a CCM includes intracellular TCO_2 levels that exceed TCO_2 in the suspending medium (Burns and Beardall, 1987; Bloye *et al.*, 1992; Tortell *et al.*, 1997), and net CO_2 efflux from actively photo-synthesising cells due to leakage through the cell membrane (Tchernov *et al.*, 1998). Admittedly, much of this physiological evidence has been accumulated for laboratory 'weeds', although coastal diatom assemblages also possess a CCM (Tortell *et al.*, 1997, 2000).

Most of our understanding of the CCM is based on studies of cyanobacteria and chlorophytes. The CCM may involve both light-dependent and light-independent signal transduction pathways (Matsunda *et al.*, 1998). CO_2 and/or HCO_3^- transport at the plasma membrane in cyanobacteria and at the chloroplast and/or thylakoid membranes in eukaryotes is essential to the CCM, as is localised catalysis of CO_2 and HCO_3^- by carbonic anhydrase. In cyanobacteria, the CCM is associated with the carboxysomes, whereas in chlorophytes and other eukaryotes it is associated with pyrenoids. These bodies contain crystaline arrays of RUBISCO, RUBISCO activase and carbonic anhydrase. As much as 99% or as little as 1% of RUBISCO may be associated with the pyrenoid, the amount depending on growth conditions and experimental technique (Süss *et al.*, 1995; Moroney and Chen, 1998).

C_4 photosynthesis

An alternative to the biophysical CCM based on membrane transport of CO_2 or HCO_3^- is the biochemical elevation of CO_2 at the active site of RUBISCO via C_4 photosynthesis. C_4 photosynthesis refers to the initial carboxylation of bicarbonate to form malate followed by the subsequent decarboxylation to liberate CO_2 near RUBISCO. The initial carboxylation into a C_4 acid does not involve reduction of CO_2, nor does it result in net fixation of CO_2. C_4 photosynthesis is known in vascular plants, where CO_2 acquisition is separated either spatially (Hatch-Slack metabo-lism) or temporally (Crassulacean acid metabolism) from RUBISCO carboxylation. Early reports of C_4 photosynthetic carbon fixation in microalgae (Beardall *et al.*,

1976) have failed to stand the test of time. However, Reinfelder *et al.* (2000) recently concluded that C_4 photosynthesis may occur in Zn-limited diatoms. They suggest that β-carboxylation can operate in lieu of a carbonic anhydrase-mediated CCM, for which Zn is a necessary co-factor. In contrast to vascular plants, where the C_4 carboxylation is 'widely' separated in space or time, carboxylation of malate appears to occur in the cytoplasm of the diatom *Thalassiosira weissflogii*, with transport of malate to the chloroplast prior to decarboxylation. This is a controversial hypothesis. In particular, the lack of early analysis (1–2 s) fails to show whether G3P or a C_4 acid is the first product (Johnston *et al.*, 2001) and even though high percentages of ^{14}C go through malate, this is not sufficient to demonstrate true C_4 labelling. In fact, Raven *et al.* (2000) proposed a photorespiratory pathway in diatoms that involves cycling through malate (Fig. 3.4).

CO_2 limitation of phytoplankton growth and photosynthesis in the sea

The potential for inorganic carbon limitation of phytoplankton growth in the sea has been a contentious issue since Riebesell *et al.* (1993) concluded that some large marine diatoms rely exclusively on diffusive supply of CO_2 to the cell surface to support growth. Reduction of growth rates and cell-specific TCO_2 assimilation rates were observed when pCO_2 was reduced below about 10–20 μM in otherwise nutrient-replete cultures growing under optimum irradiance (Riebesell *et al.*, 1993). Severe inhibition of growth rate was observed when pCO_2 was reduced to about 5 μM. However pH would have been >8.5 in the very low pCO_2 treatments (versus 7.8 at air-equilibrium CO_2). Hein and Sand-Jensen (1997) showed that increasing sea water pH by adding NaOH tended to reduce ^{14}C assimilation whereas decreasing pH by adding HCl tended to increase ^{14}C assimilation in oceanic phytoplankton in the North Atlantic. The effect of pH was presumed to be indirect by altering the CO_2 concentration from 10 μM in sea water to 3 μM in the +NaOH treatment and 36 μM in the +HCl treatment. Stimulation of ^{14}C assimilation ranged from 0 to 100% but varied regionally, presumably due to the differences in the taxonomic composition of the phytoplankton. In contrast, other species such as *Skeletonema costatum* and *Thalassiosira weissflogii* do not show this dependence of growth rate on pCO_2 as long as TCO_2 is available, even at pH as high as 8.8 (Burkhardt *et al.*, 1999). Similarly, the growth rate of three large oceanic diatoms was unaffected by pH (and by implication pCO_2) in closed bottles during pH drift experiments until pH rose to >8.5 and pCO_2 dropped to about 4 μM (Goldman, 1999). Finally, rapid growth ($1 \, d^{-1}$) of coastal *Thalassiosira* assemblages was unaffected by a reduction of pCO_2 to 100 ppm, a concentration that would only support a diffusion-limited growth rate of $0.2 \, d^{-1}$ (Tortell *et al.*, 1997, 2000).

3.6   Other pathways for carbon acquisition

The dominant means of carbon incorporation in photosynthetic organisms is through carboxylation in the Calvin cycle. Two other pathways exist, β carboxylation and mixotrophy. The former is responsible for 'dark' uptake of either HCO_3^- or CO_2 by a number of carboxylases including phosphoenol pyruvate carboxylase, phosphoenol pyruvate carboxykinase, pyruvate carboxylase and acetyl CoA carboxylase (Raven, 1997). Anapleurotic carboxylation is a critical component of several metabolic pathways, replacing intermediates that are siphoned off for lipid, amino acid, porphyrin and pyrimidine synthesis (Turpin *et al.*, 1988; Falkowski and Raven, 1997; Raven, 1997). These reactions do not generally result in a net increase in carbon fixation. The exception occurs in C_4 photosynthesis where the uptake and subsequent release of CO_2 may enhance the supply of CO_2 to the active site of RUBISCO (section 3.5).

Anapleurotic carboxylation usually occurs at a small fraction of the rate of RUBISCO caraboxylation except in C_4 and CAM plants. For phytoplankton in balanced growth with C_3 photosynthesis, about 95% of carboxylation will be by RUBISCO, with the remaining 5% occuring via β carboxylation (Raven, 1997). An even greater percentage of carboxylation would be catalysed by RUBISCO in cells growing on a light–dark cycle with continued protein, nucleic acid, pigment and lipid synthesis in the dark period. However, under transient conditions, β carboxylation can temporarily occur at rates that equal or exceed the rate of RUBISCO carboxylation. For example, β carboxylation is enhanced and RUBISCO carboxylation is inhibited immediately following the addition of nitrate or ammonium to nitrogen-limited cells (Turpin and Bruce, 1990). This transient response involves the diversion of triose phosphate from the Calvin cycle thus reducing the substrate for RUBISCO and hence RUBISCO activity. It also involves enhancement of β carboxylation associated with mobilisation of carbohydrate energy reserves for amino acid synthesis, and the diversion of reductant and ATP for nitrate reduction or ATP for ammonium assimilation.

The second alternative pathway for carbon acquisition is mixotrophy. Some dinoflagellates are voraciously heterotrophic, ingesting other protists phago-trophically and/or utilising dissolved organic compounds osmotrophically (Stoecker, 1999) and many other groups have been shown to be facultative osmotrophs (Lewitus and Kana, 1994, 1995; Gervais, 1997). Mixotrophy in facultative species is not necessarily a substitute for autotrophy, although it can provide an energetic subsidy that may be stimulated under conditions of reduced light or nutrient (N or P) supply (Li *et al.*, 2000). Facultative heterotrophs may have higher levels of pigmentation and enhanced rates of light-saturated and light-limited photosynthesis (Lewitus and Kana, 1994; Li *et al.*, 1999) relative to their performance under strictly autotrophic conditions. The reverse pattern also holds, where pigmentation is reduced under mixotrophic conditions (Hansen *et al.*, 2000) and where photosynthetic rates decline above a threshold prey density. The regulation of mixotrophy is not well understood but it may well be a critical component of carbon metabolism, particularly in estuarine species.

3.7   Response of phytoplankton photosynthesis to environmental variability

Phytoplankton grow, compete and evolve within a physical/chemical environment that fluctuates on time-scales from seconds to millennia. Adjustment to these fluctuations involves biophysical, biochemical, physiological, ecological and evolutionary components that operate with characteristic response times. Here we differentiate between the terms *adaptation, acclimation* and *regulation*. Adaptation describes an outcome of selection that involves changes in the genetic composition of a species or population. Acclimation describes a change of the macromolecular composition of an organism that occurs via synthesis or breakdown of specific components, operating within the limitations of the genetic make-up of the population. Regulation describes the adjustments of catalytic or energetic efficiency that occur without net synthesis or breakdown of macromolecules, but which involve slight structural modifications. We are concerned with acclimation and regulation in this chapter, taking the gene pool as given. For a recent review of the time-scales of photosynthetic regulation, see MacIntyre *et al.* (2000).

An understanding of how environmental factors influence a^{Chl}, the chlorophyll *a*-to-carbon ratio and the parameters of the PE curve is central to evaluating and modelling primary productivity. Knowledge of these variables is particularly relevant to mapping phytoplankton production from ocean colour data (Platt and Sathyendranath, 1988) and assimilating observations of chlorophyll *a* distributions into biogeochemical models. Physiology models developed by Bannister and Laws (1979), Kiefer and Mitchell (1983), and Geider *et al.* (1996, 1997, 1998) capture much of the systematic variability in the PE curve and Chl*a*:C ratio under conditions of balanced growth. Although these models differ in structure and notation, they exhibit many fundamental similarities (MacIntyre *et al.*, 2002). Coupling such regulatory/bio-optical models with allometric rules (Finkel and Irwin, 2000) should lead to more general formulations of phytoplankton growth that would apply to interspecific variability as well.

Dynamic balance models and photoacclimation

A key to understanding the physiological response to irradiance, temperature and nutrient-limitation is the imbalance between excitation energy supplied by light absorption, and electron demands for biosynthesis and maintenance (Kana *et al.*, 1997). Photon supply and electron demand can be compared through a regulatory ratio (ω) that can be defined as follows:

$$\omega = \text{electron demand/excitation energy supply} = 2\,A_{PET}/A_{hv} \qquad (3.12)$$

where A_{PET} equals the activity of the photosynthetic electron transfer chain (μmol electrons [g C]$^{-1}$ s^{-1}), A_{hv} is the carbon-specific rate of light absorption (μmol photons [g C]$^{-1}$ s^{-1}), and 2 photons electron^{-1} is a factor that accounts for the ratio of photons absorbed to electrons transferred during linear photosynthetic electron flow. This formulation assumes that 100% of excitation energy is transferred from the pigment bed to the reaction centres. However, the equation can be generalised to account for non-photochemical quenching. A_{PET} can be evaluated from the gross oxygen evolution estimated from ^{18}O exchange (Kana, 1992):

$$A_{PET} = 4\,P_g{}^C \qquad (3.13)$$

where $P_g{}^C$ is the carbon-specific gross O_2 evolution rate (μmol O_2 [g C]$^{-1}$ s^{-1}) at growth irradiance. The regulatory ratio can be cast in terms of the PE curve, the relative quantum efficiency of photosynthesis or a 'photon pressure' (Box 3.1).

In the models of Geider *et al.* (1996, 1997), the decline of ω is assumed to be accompanied by a decline in the rate of Chl*a* synthesis relative to carbon fixation, whereas in the extension of the model to include variable N:C (Geider *et al.*, 1998), the decline of ω is accompanied by a decline in Chl*a* synthesis relative to nitrogen assimilation into protein. To date, photoacclimation models based on this simple idea of a regulatory ratio (Kana *et al.*, 1997) have proven applicable to describing the effects of irradiance, temperature and nitrogen limitation (Geider *et al.*, 1996, 1997, 1998).

Physiological bases of the PE parameters

The parameters that describe the regulatory response (Box 3.1) are related to the relative rates of light absorption and carbon fixation. Under balanced growth, Chl*a*:C serves as a proxy for light absorption. Although accessory pigments also absorb light for photosynthesis, they are present in a fairly constant ratio to Chl*a* within a taxon (Goericke and Montoya, 1998; MacIntyre *et al.*, 2002). Chl*a*-specific light absorption is also modulated by the package effect, but this is secondary to the changes in cell-specific absorption due to changes in the amount of chlorophyll and associated pigments (Bricaud *et al.*, 1988; MacInyre *et al.*, 2002). The chlorophyll *a*-specific, light-limited initial slope (α^{Chl}) is the product of a^{Chl} and ϕ_m. The latter is an index of the efficiency with which energy is transferred from the pigment bed to the electron transport carriers.

Physiological acclimation of the PE curve appears to follow several simple rules under conditions of balanced growth (Geider *et al.*, 1997; Fig. 3.5). As a first approximation, α^{Chl} can be considered to be constant, with changes to the carbon-specific light saturated rate ($P_m{}^C$) inducing changes in the Chl*a*:C (Fig. 3.5). In

---

Box 3.1   Alternative ways of defining or calculating the regulatory ratio (ω).

The regulatory ratio is defined as:

$$\omega = 2\, A_{PET}/A_{hv} \tag{3.I}$$

where A_{PET} is the carbon-specific rate of photosynthetic electron transfer and A_{hv} is the carbon-specific rate of photon absorption.
    Noting that:

$$A_{hv} = a^{chl}\ \text{Chl}a{:}\text{C}\ E \tag{3.II}$$

where a^{chl} is the chlorophyll a-specific rate of light absorption, Chl$a$:C is the chlorophyll a-to-carbon ratio and E is irradiance, and defining:

$$0.5\, A_{PET} = P^{C}(E)/\phi_{m} \tag{3.III}$$

where $P^{C}(E)$ is the carbon-specific photosynthesis rate (units of inverse time) at irradiance E, and ϕ_{m} is the maximum quantum efficiency of photosynthesis, allows equation (3.I) to be rewritten as:

$$\omega = P^{C}(E)/(a^{Chl}\,\phi_{m}E\ \text{Chl}a{:}\text{C}) \tag{3.IV}$$

Noting that $\phi = P^{C}(E)/(a^{Chl}\ \text{Chl}{:}\text{C}\ E)$, where ϕ is the quantum yield of photosynthesis at irradiance E, the regulatory ratio becomes:

$$\omega = \phi/\phi_{m} \tag{3.V}$$

Alternatively, the PE curve can be treated as a Poisson model:

$$P^{C}(E) = P_{m}{}^{C}[1 - \exp(-\alpha^{Chl}\ \text{Chl}a{:}\text{C}\ E/P_{m}{}^{C})] \tag{3.VI}$$

where $\alpha^{Chl} = a^{Chl}\,\phi_{m}$.

Substituting equation (3.VI) into equation (3.IV) yields:

$$\omega = P_{m}{}^{C}[1 - \exp(-\alpha^{Chl}\ \text{Chl}\alpha{:}\text{C}\ E/P_{m}{}^{C})]/(\alpha^{Chl}\ E\ \text{Chl}{:}\text{C}) \tag{3.VII}$$

Finally, substituting $E_{K} = P_{m}{}^{C}/(\alpha^{Chl}\ \text{Chl}a{:}\text{C})$ yields:

$$\omega = E_{K}[1 - \exp(-E/E_{K})]/E \tag{3.VIII}$$

---

nutrient-replete cells at constant temperature, $P_{m}{}^{C}$ is largely independent of growth irradiance (MacIntyre *et al.*, 2002), but is highly correlated with the maximum growth rate, μ_{m} (Geider, 1993). Because temperature affects the maximum rate of enzymatic reactions $P_{m}{}^{C}$ should follow the Arrhenius equation within the optimum temperature range. Under N-limiting conditions at constant irradiance, a reduced investment of nitrogen in enzymes means that $P_{m}{}^{C}$ declines linearly with the nutrient-limited relative growth rate (μ/μ_{m}) (Geider, 1993) and the N:C ratio (Geider *et al.*, 1998). We consider these in turn, below.

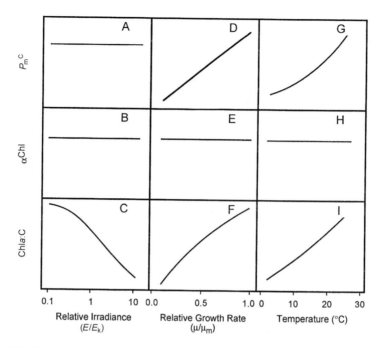

Fig. 3.5 Relationships predicted by the dynamic balance model of photoacclimation (Geider *et al.*, 1998) for light (A, B, C), nitrogen (D, E, F) or temperature (G, H, I) limitations.

Irradiance

Photoacclimation typically results in a decline of Chla:C in cells that maintain constant $P_m{}^C$ and α^{Chl} (MacIntyre *et al.*, 2002). It is less clear which catalyst limits the light-saturated rate of photosynthesis, P_m. The rate of carboxylation is proportional to the CO_2-saturated rate of RuBP consumption or to the rate of RuBP regeneration, whichever is lower (Ruuska *et al.*, 1998). At limiting internal CO_2 concentrations, photosynthesis is limited by RUBISCO and P_m is often correlated with RUBISCO concentration (Björkman, 1981). However, under high CO_2 concentrations, photosynthesis becomes sink-limited downstream of RUBISCO and P_m may not be correlated with RUBISCO concentration. Data consistent with the limiting catalyst for P_m residing in the Calvin cycle have been reported for *Dunaliella tertiolecta* (Sukenik *et al.*, 1987) and *Thalassiosira weissflogii* (Orellana and Perry, 1992) where $P_m{}^{Cell}$ was proportional to RUBISCO per cell. In contrast, a 5.3-fold increase in $P_m{}^{Chl}$ was accompanied by only a 2.3-fold increase of RUBISCO:Chla in *Tetraedron minimum* (Fisher *et al.*, 1989), and P_m was more strongly correlated with changes in the photosynthetic electron transfer chain than with RUBISCO activity in *Scenedesmus obliquus* (Fleischhacker and Senger, 1978; Senger and Fleischhacker, 1978), and *Chlorella fusca* (Wilhelm and Wild, 1984). These are characteristics of CO_2-saturated (sink-limited) growth (Anwaruzzaman *et al.*, 1995; Ruuska *et al.*, 1998).

These correlative studies do not unambiguously indicate causality, and more work is needed to determine the rate-limiting step(s) for light-saturated photosynthesis. This work is essential given the importance of the light-saturated photosynthesis rate in determining phytoplankton production, the possibility of CO_2-limitation of photosynthetic carbon fixation (section 3.5), and the increasing use of fluorescence techniques to estimate photosynthetic carbon fixation (section 3.2).

Temperature

A common simplifying assumption regarding acclimation of photosynthesis to temperature is that the activity of a single rate-limiting step varies according to an Arrhenius function (Geider *et al.*, 1997) or some other empirical function (Li, 1980). If the proportion of cell mass accounted for by the rate-limiting catalyst does not vary with temperature, then the expected response of P_m^C is that illustrated in Fig. 3.5G. If the initial slope, α^{Chl}, does not depend on temperature, then acclimation of Chl*a*:C will be as depicted in Fig. 3.5I (Geider *et al.*, 1997).

There have been few investigations that are comprehensive enough to test these theoretical predictions. Results for temperature acclimation of *Chaetoceros calcitrans* (Anning *et al.*, 2001) are largely consistent with the trends depicted in Fig. 3.5, whereas those for *Phaeodactylum tricornutum* (Li and Morris, 1982) are not. Conflicting results have been reported for acclimation of pigment contents and P_m for these two marine diatoms. Whereas a marked dependence of Chl*a*:C on temperature was found in *C. calcitrans*, there was a slight dependence in *P. tricornutum*. The slight dependence of Chl*a*:C on temperature in *P. tricornutum* contrasts with observations in other phytoplankton (Li, 1980; Geider *et al.*, 1997) and expectations based on the regulation of pigment synthesis observed in *Chlorella* (Maxwell *et al.*, 1995).

The temperature dependence of P_m in *C. calcitrans* was similar in both short-term (30 min) temperature shift experiments and long-term (four to seven day) temperature acclimation experiments (Anning *et al.*, 2001). In other words, *C. calcitrans* did not appear to compensate for a temperature-dependent decrease in catalytic rate by increasing the abundance of the rate-limiting catalyst for light-saturated photosynthesis as temperature declined. This contrasts with *P. tricornutum*, in which both P_m^C (measured at the optimum assay temperature) and carbon-specific RUBISCO activity (measured at a standard assay temperature) were higher in cells acclimated to 10°C than in cells acclimated to higher temperatures (Li and Morris, 1982). The reduced specific activity of RUBISCO due to low temperature was partially compensated by an increase in RUBISCO abundance in *P. tricornutum*. Clearly, more work is required to unravel temperature acclimation of the photosynthetic apparatus of phytoplankton. This is important, since temperature is one of the variables that provides some power in predicting chlorophyll *a*-specific light-saturated photosynthesis rates in the sea (Kyewalyanga *et al.*, 1998).

Nitrogen source

The difference in energy requirements for assimilation of nitrate and ammonium is reflected in the photosynthetic quotient, being about 1.1 mol O_2 [mol $CO_2]^{-1}$ when ammonium is assimilated and 1.4 when nitrate is assimilated (Laws, 1991). For cells with the Redfield C:N ratio of 6.6, growth on nitrate requires a 25% greater rate of photosynthetic electron transfer than does growth on ammonium. This has implications for growth rate and partitioning of resources between the thylakoid membrane components involved in light-harvesting and photosynthetic electron transfer and the Calvin cycle enzymes responsible for carbon fixation. Giordano and Bowes (1997) found that nitrogen source affected the contribution of RUBISCO to soluble protein in nutrient-replete *Dunaliella salina*. Whereas RUBISCO accounted for 30% of soluble protein in cells grown on ammonium, it only accounted for 14% in cells grown on nitrate. Despite the difference in RUBISCO content, the growth rate of *D. salina* did not depend on the nitrogen source, nor did the ratio of total chlorophyll-to-soluble protein. Thus, chlorophyll:RUBISCO ratio was greater in cells grown on nitrate than in cells grown on ammonium. The lower RUBISCO:-protein ratio in the cells that were assimilating nitrate was compensated by higher *in vivo* RUBISCO-specific CO_2 fixation rates.

Nitrogen- and phosphorus-limitation

One of the most common effects of N- or P-limitation is a linear decline of Chl$a$:C with declining growth rate (Geider, 1993). In addition to reducing pigment content, nitrogen limitation also reduces the contribution of protein to total mass, as well as the proportion of protein that is accounted for by RUBISCO. P_m^C parallels N:C in N-limited phytoplankton (Geider *et al.*, 1998), consistent with a linear covariation of P_m^C with relative growth rate (μ/μ_m) (Geider, 1993). Some of the decline of P_m^C with declining N:C can be accounted for by a reduction in the amount of RUBISCO. In addition, RUBISCO activation state is regulated. A non-linear relationship was found between the proportion of cell nitrogen accounted for by RUBISCO and growth rate in N-limited chemostat cultures of *Isochrysis galbana* (Falkowski *et al.*, 1989). RUBISCO accounted for about 23% of cell N at growth rates above 0.6 d^{-1} and about 8% of cell N at growth rates below 0.4 d^{-1} (Falkowski *et al.*, 1989). Similarly, the RUBISCO:dry weight ratio decreased abruptly by about 50% when the growth rate declined below 0.25 d^{-1} in N-limited *Chlorella emersonii* (Beardall *et al.*, 1990). Nitrogen and phosphorus limitation had similar effects on P_m^C and the contribution of RUBISCO to cell protein in *Dunaliella tertiolecta* (Geider *et al.*, 1998). Specifically, P_m^C was reduced from 2.3 d^{-1} in nutrient-replete cells ($\mu = 1.4\,d^{-1}$) to 0.4–0.6 d^{-1} in N- or P-limited *D. tertiolecta* ($\mu = 0.25\,d^{-1}$) while the contribution of RUBISCO to cell protein declined by 50%.

Iron-limitation

Iron limitation operates primarily by reducing the abundance of the Fe-containing components of the photosynthetic electron transfer chain and, as a consequence, P_m (Greene *et al.*, 1991). Fe-limitation can lead to a decrease in Chl$a$:C, but does not inevitably do so. Despite the reduction of P_m, the contribution of RUBISCO to total protein is largely unaffected by Fe-limitation (Greene *et al.*, 1991; McKay *et al.*, 1997). Most studies of the influence of iron on photosynthesis have been undertaken with Fe-starved batch cultures. Further progress in understanding the acclimation of the photosynthetic apparatus to Fe limitation requires chemostat or trace-metal buffered cultures where controlled degrees of iron limitation can be sustained (Weger and Espie, 2000).

CO$_2$ limitation

In contrast to the marked reductions of Chl$a$:C observed under N- and P-limiting conditions and to a lessor extent in Fe-limited cells, chlorosis is not observed in TCO$_2$-limited phytoplankton (Miller *et al.*, 1984; Giordano and Bowes, 1997; Clark *et al.*, 1999). For example, in *Synechococcus leopoliensis* Chl$a$:C remained constant at $0.025 \, g \, g^{-1}$ despite reductions of growth to $<0.1 \, \mu_{max}$ (Miller *et al.*, 1984). It is difficult to reconcile this invariance of Chl$a$:C with the model of pigment acclimation described previously (Geider *et al.*, 1996, 1997, 1998). This is because the regulatory ratio is expected to decline as the CO$_2$-fixation rate and thus A$_{PET}$ decline under CO$_2$-limiting conditions. Given the hypothesised regulation of pigment synthesis by the redox state of the plastoquinone pool (Huner *et al.*, 1996; Durnford and Falkowski, 1997), the invariance of Chl$a$:C under CO$_2$-limiting conditions suggests that the PQ pool remains oxidised. It is possible that there are contributions to A$_{PET}$ besides CO$_2$ fixation. These could include photorespiration, which would be stimulated by CO$_2$-limitation in the absence of a carbon-concentrating mechanism, or photosynthetic oxygen consumption (the Mehler reaction). However, given the low potential for RUBISCO oxygenation in many phytoplankton (section 3.4), another explanation must be sought. The explanation may lie in dissipation of energy to fuel a CO$_2$-concentrating mechanism. Induction of a CCM in *Chlamydomonas reinhardtii* produced alterations in the light reactions of photosynthesis that increased funnelling of excitation energy to photosystem I (Palmqvist *et al.*, 1990). These alterations were hypothesised to protect photosystem II from over-reduction and/or provide the ATP necessary for high rates of CO$_2$ transport necessary to maintain elevated intracellular CO$_2$ levels. Enhanced extracellular ferricyanide reduction in CO$_2$-limited phytoplankton (Nimer *et al.*, 1999) may indicate another mechanism for dissipating excitation energy.

Maximum cell-specific photosynthesis rates, like cell chlorophyll and β-carotene contents, were unaffected by CO$_2$-limitation of growth rate in *Dunaliella salina* cultured in air ($\mu = 0.5 \, d^{-1}$) and 5% CO$_2$ ($\mu = 1.2 \, d^{-1}$) (Giordano and Bowes, 1997).

The cell RUBISCO content increased in *Chlamydomonas reinhardtii, Chlorococcum* sp. and *Dunaliella salina* grown at high CO_2 (Yokota and Canvin, 1986; Pesheva *et al.*, 1994; Giordano and Bowes, 1997), although its contribution to soluble protein or total protein in *D. salina* and *Chlorella vulgaris* was not changed (Giordano and Bowes, 1997; Villarejo *et al.*, 1998).

3.8   Conclusion

Primary production is a deceptively simple concept that masks a complex interplay of biochemical, physiological and ecological processes. Our understanding of primary production in aquatic systems rests primarily on the observations of ^{14}C assimilation that have been collected over the past 50 years. Today, bio-optical and biophysical techniques have extended spatial and temporal coverage beyond that achieved by the ^{14}C method. For example, we now rely on remote sensing of phytoplankton pigments together with bio-optical algorithms to estimate global and regional primary production. However, bio-optical algorithms are based largely on calibration against ^{14}C measurements. Ideally, our knowledge of the biochemistry and physiology of phytoplankton photosynthesis should provide the basis for developing bio-optical algorithms from first principles. It should also provide a basis for interpreting the data generated by fluorescence techniques.

We have attempted in this chapter to emphasise the common features of the photosynthetic biochemistry and physiology amongst phytoplankton taxa whilst acknowledging the diversity in detail. In particular, we have described the role of a dynamic energy balance (Kana *et al.*, 1997) as a possible unifying concept in regulating photosynthesis on physiological time-scales from minutes to days.

Acknowledgements

This work was supported by the UK Natural Environment Research Council (RJG) and Grant OCE-9907702 from the US National Science Foundation (HLM). Horn Point Contribution No. 3412.

References

Andrews, T.J. & Ballment, B. (1984) Active-site carbamate formation and reaction-intermediate analog binding by ribulosebisphosphate carboxylase/oxygenase in the absence of its small subunits. *Proceedings of the National Academy of Science USA* **881**, 3660–64.

Anning, T., Harris, G. & Geider, R.J. (2001) Thermal acclimation in the marine diatom *Chaetoceros calcitrans* (Bacillariophycea). *European Journal of Phycology* **36**, 233–41.

Anwaruzzaman, Sawada, S., Usuda, H. & Yokota, A. (1995) Regulation of ribulose-1,5-bisphosphate carboxylase/oxygenase activation by inorganic phosphate through stimulating the binding of the activator CO_2 to the activation sites. *Plant and Cell Physiology* **36**, 425–33.

Badger, M.R., Andrews, T.J., Whitney, S.M., *et al.* (1998) The diversity and coevolution of Rubisco, plastids, pyrenoids, and chloroplast-based CO_2-concentrating mechanisms in algae. *Canadian Journal of Botany* **76**, 1052–71.

Bannister, T.T. & Laws, E.A. (1979) Modeling phytoplankton carbon metabolism. In: *Primary Productivity in the Sea* (ed. P.G. Falkowski), pp. 243–8. Plenum, New York.

Beardall, J., Mukerji, D., Glover, H.E. & Morris, I. (1976) The path of carbon in photosynthesis by marine phytoplankton. *Journal of Phycology* **12**, 409–17.

Beardall, J., Roberts, S. & Millhouse, J. (1990) Effects of nitrogen limitation on uptake of inorganic carbon and specific activity of ribulose-1,5-bisphosphate carboxylase/oxygenase in green microalgae. *Canadian Journal of Botany* **69**, 1146–50.

Belknap, W.R. & Portis, A.R. (1986) Exchange properties of the activator CO_2 of spinach ribulose 1,5-bisphosphate carboxylase/oxygenase. *Plant Physiology* **80**, 707–10.

Bidigare, R.R., Schofield, O. & Prézelin, B.B. (1989) Influence of zeaxanthin on quantum yield of photosynthesis of *Synechococcus* clone WH7803 (DC2). *Marine Ecology Progress Series* **56**, 177–88.

Björkman, O. (1981) Responses to different quantum flux densities. In: *Physiological Plant Ecology. I Responses to the Physical Environment* (eds O.L. Lange, P.S. Nobel, C.B. Osmond & H. Ziegler), pp. 57–107. Springer-Verlag, Berlin.

Bloye, S.A., Karagouni, A.D. & Carr, N.G. (1992) A continuous culture approach to the question of inorganic carbon concentration by *Synechococcus* species. *FEMS Microbiology Letters* **99**, 79–84.

Bricaud, A., Bédhomme, A.-L. & Morel, A. (1988) Optical properties of diverse phytoplankton species: experimental results and theoretical interpretation. *Journal of Plankton Research* **10**, 851–73.

Buchanan, B.B. (1992) Carbon-dioxide assimilation in oxygenic and anoxygenic photosynthesis. *Photosynthesis Research* **33**, 147–62.

Burkhardt, S., Zondervan, I. & Riebesell, U. (1999) Effect of CO_2 concentration on the C:N:P ratio in marine phytoplankton: a species comparison. *Limnology and Oceanography* **44**, 683–90.

Burns, B.D. & Beardall, J. (1987) Utilisation of inorganic carbon by marine microalgae. *Journal of Experimental Marine Biology and Ecology* **107**, 75–86.

Burris, J. (1981) Effects of oxygen and inorganic carbon concentrations on the photosynthetic quotients of marine algae. *Marine Biology* **65**, 215–19.

Clark, D.R., Merrett M.J. & Flynn K.J. (1999) Utilization of dissolved inorganic carbon (DIC) and the response of the marine flagellate *Isochrysis galbana* to carbon nitrogen stress. *New Phytologist* **144**, 463–70.

Durnford, D.G. & Falkowski, P.G. (1997) Chloroplast redox regulation of nuclear gene transcription during photoacclimation. *Photosynthesis Research* **52**, 229–41.

Edmondson, D.L., Kane, H.J. & Andrews, T.J. (1990) Substrate isomerization inhibits rubulosebisphosphate carboxylase-oxygenase during catalysis. *FEBS Letters* **260**, 62–6.

Ellis, R.J. (1979) The most abundant protein in the world. *Trends in Biochemical Science* **4** 241–4.

Falkowski, P.G. & Raven, J.A. (1997) *Aquatic Photosynthesis*. Blackwell Science, Oxford.

Falkowski, P.G., Sukenik, A. & Herzig, R. (1989) Nitrogen-limitation in *Isochrysis galbana*

(Haptophyceae) II. Relative abundances of chloroplast proteins. *Journal of Phycology* **25**, 471–8.

Finkel, Z.V. & Irwin, A.J. (2000) Modeling size-dependent photosynthesis: Light absorption and the allometric rule. *Journal of Theoretical Biology* **204**, 361–9.

Fisher, T., Shurtz-Swirski, R., Gepstein, S. & Dubinsky, Z. (1989) Changes in levels of ribulose-1,5-bisphosphate carboxylase/oxygenase (Rubisco) in *Tetraedron minimum* (Chlorophyta) during light and shade adaptation. *Plant and Cell Physiology* **30**, 221–8.

Fleischhacker, P. & Senger, H. (1978) Adaptation of the photosynthetic apparatus of *Scenedesmus obliquus* to strong and weak light conditions. II. Differences in photochemical reactions, the photosynthetic electron transport and photosynthetic units. *Physiologia Plantarum* **43**, 43–51.

Fogg, G.E. (1977) Excretion of organic matter by phytoplankton. *Limnology and Oceanography* **22**, 576–7.

Friedberg, D., Jager, K.M., Kessel, M., Silman N.J. & Bergman, B. (1993) Rubisco but not Rubisco activase is clustered in the carboxysomes of the cyanobacterium *Synechococcus* sp. PCC 7942: *Mud*-induced carboxysomeless mutants. *Molecular Microbiology* **9**, 1193–201.

Geider, R.J. (1993) Quantitative phytoplankton physiology: implications for primary production and phytoplankton growth. *ICES Marine Science Symposium* **197**, 52–62.

Geider R.J., MacIntyre H.L. & Kana, T.M. (1996) A dynamic model of photoadaptation in phytoplankton. *Limnology and Oceanography* **41**, 1–15.

Geider, R .J., MacIntyre, H.L. & Kana, T.M. (1997) A dynamic model of phytoplankton growth and acclimation: responses of the balanced growth rate and chlorophyll *a*:carbon ratio to light, nutrient-limitation and temperature. *Marine Ecology Progress Series* **148**, 187–200.

Geider, R.J., MacIntyre, H.L. & Kana, T.M. (1998) A dynamic regulatory model of phytoplankton acclimation to light, nutrients and temperature. *Limnology and Oceanography* **43**, 679–94.

Gervais, F. (1997) Light-dependent growth, dark survival, and glucose uptake by cryptophytes isolated from a freshwater chemocline. *Journal of Phycology* **33**, 18–25.

Gilbert, M., Wilhelm, C. & Richter, M. (2000) Bio-optical modelling of oxygen evolution using *in vivo* fluorescence: comparison of measured and calculated photosynthesis/ irradiance (P-I) curves in four representative phytoplankton species. *Journal of Plant Physiology* **157**, 307–14.

Giordano, M. & Bowes, G. (1997) Gas exchange and C allocation in *Dunaliella salina* cells in response to N source and CO_2 concentration used for growth. *Plant Physiology* **115**, 1049–56.

Goericke, R. & Montoya, J.P. (1998) Estimating the contribution of microalgal taxa to chlorophyll *a* in the field – variations of pigment ratios under nutrient- and light-limited growth. *Marine Ecology Progress Series* **169**, 97–112.

Goldman, J.C. (1999) Inorganic carbon availability and the growth of large marine diatoms. *Marine Ecology Progress Series* **180**, 81–91.

Grande, K.D., Marra, J., Langdon, C., Heinemann, K. & Bender, M.L. (1989) Rates of respiration in light measured in marine phytoplankton using an ^{18}O isotope-labelling technique. *Journal of Experimental Marine Biology and Ecology* **129**, 95–120.

Greene, R.M., Geider, R.J. & Falkowski, P.G. (1991) Iron limitation in a marine diatom: implications for photosynthetic energy conversion and primary productivity. *Limnology and Oceanography* **36**, 1772–82.

Hansen, P.J., Skovgaard, A., Glud, R.N. & Stoecker, D.K. (2000) Physiology of the mixotrophic dinoflagellate *Fragilidium subglobosum*. II. Effects of time scale and prey concentration on photosynthetic performance. *Marine Ecology Progress Series* **201**, 137–46.

Hein, M. & Sand-Jensen, K. (1997) CO_2 increases oceanic primary production. *Nature* **388**, 526–7.

Huner, N.P.A., Maxwell, D.P., Gray, G.R., *et al.* (1996) Sensing environmental temperature change through imbalances between energy supply and energy consumption: Redox state of photosystem II. *Physiologia Plantarum* **98**, 358–64.

Johnston, A.M., Raven, J.A., Beardall, J. & Leegood, R.C. (2001) C_4 photosynthesis in a marine diatom? *Nature* **412**, 40–41.

Jordan, D.B. & Chollet, R. (1983) Inhibition of ribulose bisphosphate carboxylase by substrate ribulose 1,5-bisphosphate. *Journal of Biological Chemistry* **258**, 13752–8.

Kana, T.M. (1992) Relationship between photosynthetic oxygen cycling and carbon assimilation in *Synechococcus* WH7803 (Cyanophyta). *Journal of Phycology* **28**, 304–8.

Kana, T.M., Geider, R.J. & Critchley, C. (1997) Dynamic balance theory of pigment regulation in microalgae by multiple environmental factors. *New Phytologist* **137**, 629–38.

Kiefer, D.A. & Mitchell, B.G. (1983) A simple, steady-state description of phytoplankton growth based on the absorption cross-section and quantum efficiency. *Limnology and Oceanography* **28**, 770–76.

Kolber, Z.S., Prasil, O. & Falkowski, P.G. (1998) Measurements of variable chlorophyll fluorescence using fast repetition rate techniques: defining methodology and experimental protocols. *Biochimica et Biophysica Acta* **1367**, 88–106.

Kozaki, A. & Takeba, G. (1996) Photorespiration protects C_3 plants from photooxidation. *Nature* **384**, 557–60.

Kyewalyanga, M., Platt, T., Sathyendranath, S., Lutz, V.A. & Stuart, V. (1998) Seasonal variations in physiological parameters of phytoplankton across the North Atlantic. *Journal of Plankton Research* **20**, 17–42.

Laws, E.A. (1991) Photosynthetic quotients, new production and net community production in the open ocean. *Deep-Sea Research* **38**, 143–67.

Lewitus, A.J. & Kana, T.M. (1994) Responses of estuarine phytoplankton to exogenous glucose: stimulation versus inhibition of photosynthesis and respiration. *Limnology and Oceanography* **39**, 182–9.

Lewitus, A.J. & Kana, T.M. (1995) Light respiration in six estuarine phytoplankton species: contrasts under photoautotrophic and mixotrophic growth conditions. *Journal of Phycology* **31**, 754–61.

Li, A.S., Stoecker, D.K. & Adolf, J.E. (1999) Feeding, pigmentation, photosynthesis and growth of the mixotrophic dinoflagellate *Gyrodinium galatheanum*. *Aquatic Microbial Ecology* **19**, 163–76.

Li, A.S., Stoecker, D.K. & Coats, D.W. (2000) Mixotrophy in *Gyrodinium galatheanum* (Dinophyceae): grazing responses to light intensity and inorganic nutrients. *Journal of Phycology* **36**, 33–45.

Li, L.-A., Gibson, J.L. & Tabita, F.R. (1993) The Rubsico activase (*rca*) gene is located downstream from *rbcS* in *Anabaena* sp. strain CA and is detected in other *Anabaena/Nostoc* strains. *Plant Molecular Biology* **21**, 753–64.

Li, W.K.W. (1980) Temperature adaptation in phytoplankton: cellular and photosynthetic characteristics. In: *Primary Productivity in the Sea* (ed. P.G. Falkowski), pp. 259–79. Plenum Press, New York.

Li, W.K.W. & Morris, I. (1982) Temperature adaptation in *Phaedactylum tricornutum* Bohlin: Photosynthetic rate compensation and capacity. *Journal of Experimental Marine Biology and Ecology* **58**, 135–50.

Li, W.K.W. & Platt, T. (1982) Distribution of carbon among photosynthetic end-products in phytoplankton of the eastern Canadian Arctic. *Journal of Phycology* **18**, 466–71.

MacIntyre, H.L., Kana, T., Anning, T. & Geider, R.J. (2002) Photoacclimation of photosynthesis irradiance response curves and photosynthetic pigment content in microalgae and cyanobacteria. *Journal of Phycology* (in press).

MacIntyre, H.L., Kana, T.M. & Geider R.J. (2000) The effect of water motion on short-term rates of photosynthesis by marine phytoplankton. *Trends in Plant Science* **5**, 12–17.

MacIntyre, H.L., Sharkey, T.D. & Geider, R.J. (1997) Activation and deactivation of ribulose-1,5-bisphosphate carboxylase/oxygenase (Rubisco) in three marine microalgae. *Photosynthesis Research* **51**, 93–106.

Marcus, Y. & Gurevitz, M. (2000) Activation of cyanobacterial RuBP-carboxylase/oxygenase is facilitated by inorganic phosphate via two independent mechanisms. *European Journal of Biochemistry* **267**, 5995–6003.

Matsunda, Y., Bozzo, G.G. & Coleman, B. (1998) Regulation of dissolved inorganic carbon transport in green algae. *Canadian Journal of Botany* **76**, 1072–83.

Maxwell, D.P., Laudenbach, D.E. & Huner, N.P.A. (1995) Redox regulation of light-harvesting complex II and *cab* mRNA abundance in *Dunaliella salina*. *Plant Physiology* **109**, 787–95.

McKay, R.M.L., Geider, R.J. & LaRoche, J. (1997) Physiological and biochemical response of the photosynthetic apparatus of two marine diatoms to Fe stress. *Plant Physiology* **114**, 615–22.

Miller, A.G., Turpin, D.H. & Canvin, D.T. (1984) Growth and photosynthesis of the cyanobacterium *Synechococcus leopoliensis* in HCO_3^--limited chemostats. *Plant Physiology* **75**, 1064–70.

Miziorko, H.M. & Lorimer, G.H. (1983) Ribulose-1,5-bisphosphate carboxylase-oxygenase. *Annual Reviews of Biochemistry* **52**, 507–35.

Moroney J.V. & Chen, Z.Y. (1998) The role of the chloroplast in inorganic carbon uptake by eukaryotic algae. *Canadian Journal of Botany* **76**, 1025–34.

Morris, I. & Glover, H. (1981) Physiology of photosynthesis by marine coccoid cyanobacteria – some ecological implications. *Limnology and Oceanography* **26**, 957–61.

Neale, P.J. & Melis, A. (1986) Algal photosynthetic membrane complexes and the photosynthesis–irradiance curve: a comparison of light-adaptation responses in *Chlamydomonas reinhardtii* (Chlorophyta). *Journal of Phycology* **22**, 531–8.

Newman, S.M., Derocher, J. & Cattolico, R.A. (1989) Analysis of chromophytic and rhodophytic ribulose-1,5-bisphosphate carboxylase indicates extensive structural and functional similarities among evolutionary diverse algae. *Plant Physiology* **91**, 939–46.

Nimer, N.A., Ling, M.X., Brownlee, C. & Merrett, M.J. (1999) Inorganic carbon limitation, exofacial carbonic anhydrase activity, and plasma membrane redox activity in marine phytoplankton species. *Journal of Phycology* **35**, 1200–205.

Noctor, G. & Foyer, C.H. (1998) A re-evaluation of the ATP:NADPH budget during C_3

photosynthesis: a contribution from nitrate assimilation and its associated respiratory activity. *Journal of Experimental Botany* **329**, 1895–1908.

Olaizola, M., La Roche, J., Kolber, Z. & Falkowski, P.G. (1994) Non-photochemical quenching and the diadinoxanthin cycle in a marine diatom. *Photosynthesis Research* **41**, 357–70.

Orellana, M.V. & Perry, M.J. (1992) An immunoprobe to measure Rubisco concentrations and maximal photosynthesis rates of individual phytoplankton cells. *Limnology and Oceanography* **37**, 978–90.

Osmond, C.B. (1981) Photorespiration and photosynthesis: some implications for the energetics of photosynthesis. *Biochmica et Biophysica Acta* **639**, 77–98.

Palmqvist, K., Sundblad, L.-G., Wingsle, G. & Samuelsson, G. (1990) Acclimation of photosynthetic light reactions during induction of inorganic carbon accumulation in the green alga *Chlamydonomas reinhardtii*. *Plant Physiology* **94**, 357–66.

Parry, M.A.J. & Gutteridge, S. (1984) The effect of SO_3^{-2} and SO_4^{-2} ions on the reaction of ribulose bis-phosphate carboxylase. *Journal of Experimental Botany* **35**, 157–68.

Pesheva, I., Kodama, M., Dionisio-Sese, M.L. & Miyachi, S. (1994) Changes in photosynthetic characteristics induced by transferring air-grown cells of *Chlorococcum littorale* to high-CO_2 conditions. *Plant Cell Physiology* **35**, 379–87.

Platt, T. & Sathyendranth, S. (1988) Oceanic primary production: estimation by remote sensing at local and regional scales. *Science* **241**, 1613–20.

Portis, A.R., Jr (1995) The regulation of rubisco by rubisco activase. *Journal of Experimental Botany* **46**, 1285–91.

Prézelin, B.B. (1981) Light reactions of photosynthesis. *Canadian Bulletin of Fisheries and Aquatic Science* **210**, 1–43.

Raven, J.A. (1990) Predictions of Mn and Fe use efficiencies of phototrophic growth as a function of light availability for growth and of C assimilation pathway. *New Phytologist* **116**, 1–18.

Raven, J.A. (1997) Inorganic carbon acquisition by marine autotrophs. *Advances in Botanical Research* **27**, 85–209.

Raven, J.A., Kübler, J.E. & Beardall, J. (2000) Put out the light, and then put out the light. *Journal of the Marine Biological Association of the United Kingdom* **80**, 1–25.

Reid, B.A. & Tabita, R. (1994) High substrate specificity factor ribulose bisphosphate carboxylase/oxygenase from eukaryotic algae and properties of recombinant cyanobacterial rubisco containing 'algal' residue modifications. *Archives of Biochemistry and Biophysics* **312**, 210–18.

Reinfelder, J.R., Kraepiel, A.M.L. & Morel, F.M.M. (2000) Unicellular C_4 photosynthesis in a marine diatom. *Nature* **407**, 996–9.

Riebesell, U., Wolf-Gladrow, D.A. & Smetacek, V. (1993) Carbon-dioxide limitation of phytoplankton growth rates. *Nature* **361**, 249–51.

Ruuska, S., Andrews, T.J., Badger, M.R., *et al.* (1998) The interplay between limiting processes in C-3 photosynthesis studied by rapid-response gas exchange using transgenic tobacco impaired in photosynthesis. *Australian Journal of Plant Physiology* **25**, 859–70.

Ruuska, S.A., Andrews, T.J., Badger, M.R., Price, G.D. & von Caemmerer, S. (2000) The role of chloroplast electron transport and metabolites in modulating rubisco activity in tobacco. Insights from transgenic plants with reduced amounts of cytochrome b/f complex or glyceraldehyde 3-phosphate dehydrogenase. *Plant Physiology* **122**, 491–504.

Sage, R.F. & Seeman, J.R. (1993) Regulation of ribulose-1,5-bisphosphate carboxylase oxygenase activity in response to reduced light-intensity in C_4 plants. *Plant Physiology* **102**, 21–8.

Sakshaug, E., Bricaud, A., Dandonneau, Y., *et al.* (1997) Parameters of photosynthesis: definitions, theory and interpretation of results. *Journal of Plankton Research* **19**, 1637–70.

Sathyendranath, S., Lazzara, L. & Prieur, L. (1987) Variations in the spectral values of specific absorption of phytoplankton. *Limnology and Oceanography* **32**, 403–15.

Seemann, J.R., Kobza, J. & Moore, B.D. (1990) Metabolism of 2-carboxyarabinitol 1-phosphate and regulation of ribulose-1,5-bisphosphate carboxylase activity. *Photosynthesis Research* **23**, 119–30.

Senger, H. & Fleischhacker, P. (1978) Adaptation of the photosynthetic apparatus of *Scenedesmus obliquus* to strong and weak light conditions. *Physiologia Plantarum* **43**, 35–42.

Sharkey, T.D. (1990) Feedback limitation of photosynthesis and the physiological role of ribulose bisphosphate carboxylase carbamylation. *Botanical Magazine of Tokyo Special Issue* **2**, 87–105.

Spreitzer, R.J. (1993) Genetic dissection of Rubisco structure and function. *Annual Review of Plant Physiology and Plant Molecular Biology* **44**, 411–34.

Stoecker, D.K. (1999) Mixotrophy among dinoflagellates. *Journal of Eukaryotic Microbiology* **46**, 397–401.

Suggett, D., Kraay, G., Holligan, P.M., Davey, M., Aiken, J. & Geider, R.J. (2001) Assessment of photosynthesis in a spring cyanobacterial bloom using a fast repetition rate fluorometer. *Limnology and Oceanography* **46**, 802–10.

Sukenik, A., Bennett, J. & Falkowski, P.G. (1987) Light-saturated photosynthesis – limitation by electron transport or carbon-fixation? *Biochimica et Biophysica Acta* **891**, 205–15.

Süss, K.H., Prokhorenko, I. & Adler, K. (1995) *In-situ* association of Calvin Cycle enzymes, ribulose-1,5-bisphosphate carboxylase/oxygenase activase, ferredoxin-NADP(+) reductase, and nitrite reductase with thylakoid and pyrenoid membranes of *Chlamydomonas-reinhardtii* chloroplasts as revealed by immunoelectron microscopy. *Plant Physiology* **107**, 1387–97.

Tabita, F.R. (1999) Microbial ribulose 1,5-bisphosphate carboxylase/oxygenase: a different perspective. *Photosynthesis Research* **60**, 1–28.

Tabita, F.R. & Coletti, C. (1979) Carbon dioxide assimilation in cyanobacteria: regulation of ribulose 1,5-bisphosphate carboxylase. *Journal of Bacteriology* **140**, 452–8.

Tchernov, D., Hassidim, M., Vardi, A., *et al.* (1998) Photosynthesizing marine microorganisms can constitute a source of CO_2 rather than a sink. *Canadian Journal of Botany* **76**, 949–53.

Tortell, P.D., Rau, G.H. & Morel, F.M.M. (2000) Inorganic carbon acquisition in coastal Pacific phytoplankton communities. *Limnology and Oceanography* **45**, 1485–1500.

Tortell, P.D., Reinfelder, J.R. & Morel, F.M.M. (1997) Active uptake of bicarbonate by diatoms. *Nature* **390**, 243–4.

Turpin, D.H. & Bruce, D. (1990) Regulation of photosynthetic light harvesting by nitrogen assimilation in the green alga *Selenastrum minutum*. *FEBS* **262**, 99–103.

Turpin, D.H., Elrifi, I.R., Birch, D.G., Weger, H.G. & Holmes, J.J. (1988) Interactions between photosynthesis, respiration, and nitrogen assimilation in microalgae. *Canadian Journal of Botany* **66**, 2083–97.

Uemura, K., Tokai, H., Higuchi, T., *et al.* (1998) Distribution of fallover in the carboxylase reaction and fallover-inducible sites among ribulose 1,5-bisphosphate carboxylase/oxyge-nases of photosynthetic organisms. *Plant Cell Physiology* **39**, 212–9.

Villarejo, A., Orus, M.I., Ramazanov, Z. & Martinez, F. (1998) A 38-kilodalton low-CO_2-inducible polypeptide is associated with the pyrenoid in *Chlorella vulgaris*. *Planta* **206**, 416–25.

Weger, H.G. & Espie, G.S. (2000) Ferric reduction by iron-limited *Chlaymdomonas* cells interacts with both photosynthesis and respiration. *Planta* **210**, 775–81.

Whitney, S.M. & Yellowlees, D. (1995) Preliminary investigations into the structure and activity of ribulose bisphosphate carboxylase from two photosynthetic dinoflagellates. *Journal of Phycology* **31**, 138–46.

Wilhelm, C. & Wild, A. (1984) The variability of the photosynthetic unit in *Chlorella*: The effect of light intensity and cell development on photosynthesis, P-700 and cytochrome f in homocontinuous and synchronous cultures of *Chlorella*. *Journal of Plant Physiology* **115**, 125–35.

Yokata, A. & Canvin, D.T. (1986) Ribulose bisphosphate carboxylase/oxygenase content detemined with [^{14}C]carboxypentitol bisphosphate in plants and algae. *Plant Physiology* **77**, 735–9.

Zhang, N. & Portis, A.R., Jr (1999) Mechanism of light regulation of Rubisco: A specific role for the larger Rubisco activase isoform involving reductive activation by thioredoxin-f. *Proceedings of the National Academy of Science USA* **96**, 9438–43.

Zhu, G., Bohnert, H.J., Jensen, R.G. & Wildner, G.F. (1998) Formation of the tight binding inhibitor, 3-ketoarabinitol-1,5-bisphosphate by ribulose-1,5-bisphosphate carboxylase/oxygenase is O_2-dependent. *Photosynthesis Research* **55**, 67–74.

Chapter 4
Approaches to the Measurement of Plankton Production

John Marra

4.1 Introduction

Measuring plant productivity on land can be as simple as mowing the lawn and weighing the clippings. In the ocean, with its added dimension of significant depth, and because the plants are microscopic and multiply very quickly, these biomass-change methods are not usually possible. The other factor is that there is no real analogue in the ocean to cows and sheep. The herbivores in the ocean are better termed omnivores, and generally ingest particles wholly and completely, instead of just consuming the smaller parts. And herbivores are similar in size compared to their food, rather than being very small (insects), or very large (cattle, etc.). Finally, the ocean constantly moves.

In observing production processes, the ocean offers two choices: sequential sampling from a ship (or other specific location), or removing a population from its environment for incubation. Neither is a good alternative, although the preponderance of measurements suggests that removing populations from their environment has been considered less of a concern than attempting to sample the same water mass. Finally, the ocean is mostly blue, reflecting the optical characteristics of the water itself rather than the populations that inhabit it. Thus, the entire structure of marine food webs, conferred by the ocean, has required approaches that seem foreign to ecologists of the terrestrial (and freshwater) biosphere. In this chapter, I briefly review the various means by which oceanic productivity can be studied and estimated, and with one illustration of how global productivity of the ocean might be calculated. In a final section, I review some of the results from the North Atlantic Bloom Experiment (NABE) which took place in 1989, a data set with which I am reasonably well acquainted.

It is typical in reviews on the productivity of the ocean to define, sometimes laboriously, what it is we measure. At best, these are operational definitions, and which therefore depend on the methods employed. I refer the reader to one of the excellent reviews by, for example, Peterson (1980), Williams (1993), Platt *et al.* (1984) and Karl (see Chapter 9). For my purpose, gross production, net production and net community production, are the appropriate variables or parameters, and these are estimated by physiological processes such as photosynthesis, nutrient uptake, and respiration, and, as I discuss below, by other attributes of the planktonic ecosystem. Also, I leave descriptions of methodological details for the reader to investigate (see, for example, Bender *et al.*, 1987).

In considering various measurement strategies for the ocean, it is useful to understand when a process comes back to an initial value. That is, at some point, the rate of production will be balanced by loss, and this can occur over time periods ranging from biochemical kinetics to at least many years. For example, we can speak of the balance of photosynthesis and respiration that occurs over a 24 h period, since irradiance varies from essentially zero to a maximal value over the day. In many instances, metabolic processes are in balance over a day, and net production over a 24 h day beginning at dawn would cycle to positive values while the sun is shining and then to negative values at night. We have observed this in the North Pacific Central Gyre (e.g. Williams *et al.*, 1983), and also often on moored bio-optical sensors (e.g. Marra, 1995).

I have argued elsewhere (Marra, 1995) that the goal should be the measurement of primary production over the period of one day – 24 h. Not only is it important in a planetary sense, but phytoplankton, at least the more numerically abundant groups, live and die at time-scales of hours to a few days. Figure 4.1 shows the time-scales of measurement and what they mean in terms of the measurement of the daily rate of primary production. Going to longer time-scales than a day necessitates measurements of the change in biomass and the dynamics of other trophic levels. Indeed, because they have longer life spans, some have argued productivity might better be observed in the biomass of zooplankton or in fish. But longer time-scales require very accurate knowledge of the efficiency of the transfer of primary production to these higher trophic levels, since to extrapolate back to a daily rate requires the biomass yield to be divided by a small number. Going to time-scales shorter than a day, on the other hand, we measure features of phytoplankton physiology, and the shorter the time-scale, the more reliant we have to be on our understanding of phytoplankton physiology among the populations. In that case, we have to assume that the physiological parameters are either constant during the day and among the populations, or that we know how to extrapolate the variability in these short-term measures (given changes in the environment) to the length of a day.

4.2   The three components to productivity

The preceding arguments suggest that there are different means for measuring productivity based on where in the food web it is being measured; and 'where', in the planktonic ecosystem, is a function of time-scale. The clues, already presented, are to observe changes in the *biomass* of the productive components, to measure a physiological *rate*, or to measure the productivity that gets passed on, or *yielded*, to

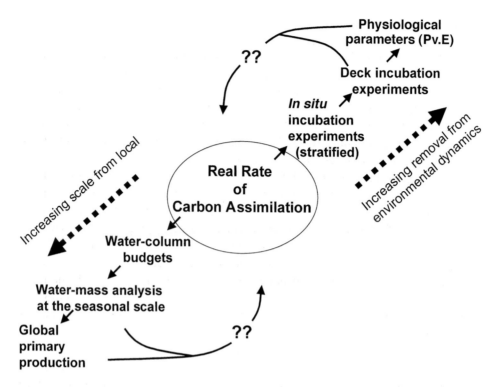

Fig. 4.1 The approaches involved in establishing the daily rate of carbon assimilation in marine planktonic ecosystems. The question marks refer to the uncertainties in referring estimates back to the daily rate of primary production. From the large scale, the uncertainties are the ecological efficiencies. From small scale the uncertainties are in phytoplankton physiology.

higher levels of the food web. Biomass, yield, and rate are the three components of productivity (Clark, 1946).

When ecologists investigate a new environment, the first thing is to figure out what is there. We then measure the community structure, or what constitutes the biomass of the ecosystem. Some of the first measurements in biological oceanography were identifications of plankton species in Kiel Bight (Mills, 1989). We are still learning about community structure as our methods become better and we see smaller and smaller components of the plankton. Another way to look at biomass is from space, which affords unprecedented geographical coverage, but with the limitation that only the surface populations can be detected. In the coming years we will have enough data to establish the statistics of variability in the biomass of the phytoplankton in the ocean from their colour signatures. Also, the three-dimensional structure of the phytoplankton community will be determined in future from sensors such as oceanographic lidars, and we will thereby obtain a more accurate measure of phytoplankton abundance from space.

Early in the history of biological oceanography, there was a need to understand the variability in fish catch (Bakun, 1996), and that meant trying to understand the mechanisms of the production of economically important species of fish. One way of defining yield is the quantity of fish, or any higher trophic level for that matter, that can be taken without the ecosystem running down. In modern terminology, yield can be equated to the 'new' production (Dugdale and Goering, 1967). New production is that which is supported by nutrients that come from outside the productive zone, for example the flux of nitrate into the upper, productive layers of the ocean. But yield is more broadly defined, and can be estimated in a variety of ways – from ecosystem community dynamics, to the sinking flux of particles out of the surface ocean, to radioisotope transformations, etc.

By themselves, biomass and yield cannot provide the complete picture of the productivity of any ecosystem. Why there exists a certain biomass in an environment, or why the yield takes on the value that it does, requires knowledge of the third property of productivity, namely the rate. All three properties provide constraints on productivity, and we must be able to reconcile the fishery yield – for example, with the rate of supply of particular kinds of food. The rate has to be consistent with the amount of productive biomass. The rate of production has assumed greater importance in marine environments compared to terrestrial because, as noted above, changes in biomass and the efficiency of plankton systems are so difficult to observe.

The three components – biomass, yield and rate – also have a time-scale associated with them. Rates are measured on time-scales of minutes to hours, and biomass changes from hours to days or weeks. Yield has a time-scale of days to many years, depending on the part of the ecosystem being examined and when that part of the system is observed to return to a zero value. For example, new production has a time-scale limit of perhaps a year in the temperate ocean where new nitrogen is supplied through wintertime mixing. Historically, the study of productivity has begun at large scales (that of oceans, and months to a year), while rate measurements are the most recent (indeed what we are celebrating in this volume). There exists perhaps no single method or series of observations that is going to provide oceanographers with an absolute measure of primary production in the ocean. All methods, all approaches, are approximations, and thus various methods have to be combined in any measurement programme.

Obviously, productivity of the ocean will be limited, and we can understand more about productivity and the approaches by examining the limits. Since the oceans are not positively productive over their entire depth, then ultimately, productivity is limited by the penetration of solar irradiance. However, solar irradiance also heats the upper layers, setting up a stratification of the water column. Under these circumstances, phytoplankton can often remove nutrients faster than the flux of nutrients upwards across a pycnocline. Thus, over longer periods (see above) nutrients will have an effect certainly on the biomass and yield of the system. There is some evidence that a lack of nutrients (e.g. iron or nitrate) can affect the rate; however, it is not entirely clear that this is not a function of the populations of phytoplankton present. Certainly other forms of nitrogen, such as ammonium, are available, and it is difficult to separate effects caused by irradiance from those of nutrients (Babin *et al.*, 1996; Marra *et al.*, 2000).

Biomass by itself has not been used to provide constraints on the productivity of the ocean separate from the associated rates; however, it has been used in this way in lakes. This is because phytoplankton biomass never itself becomes a limiting factor in the sea. Phytoplankton, although accounting for most of the variability in the optics of the upper ocean, are not any more important to light attenuation than coloured dissolved organic matter (CDOM), or the water itself. Another way of stating it is that phytoplankton compete for photons with CDOM and water. One outcome of this competition is the wavelength spectrum of absorption of phytoplankton, which shows peaks where water (and to a lesser extent CDOM) shows absorption minima (Yentsch, 1980). There is one exception to the use of biomass to constrain productivity, however, and that is the analysis by Ryther (1959), who in an estimate of the potential productivity of the sea, assumed that the light was utilised completely in the upper layers of the ocean.

Ryther (1959) also employed the rate of production in his analysis, once he established the attenuation of irradiance, but his method constrained productivity with biomass all the same. For a constraint that relies on the rate of production, we assume that there is an upper limit to the conversion of light energy. There will be an upper limit to the amount of irradiance that can be absorbed, and an upper limit to the efficiency at which that irradiance can be converted to the fixation of CO_2 into organic matter. The ability of phytoplankton to absorb irradiance varies by a factor of about 4 or 5 (Bricaud *et al.*, 1995), and the efficiency with which that irradiance is used in photosynthesis, the quantum yield, can vary by a factor of 10. Thus, the physiology of the phytoplankton needs to be better known before we can achieve useful limits of oceanic productivity based on the rate.

4.3   Biomass measurements of productivity

Biomass measurements of productivity are limited by being either non-discriminatory for phytoplankton, or else they exhibit variability to external factors that is not completely understood. For example, we can estimate changes in particulate organic carbon (POC), but POC will also include heterotrophic and detrital components. Chlorophyll *a*, on the other hand, is specific to autotrophs, yet is subject to variability from irradiance and nutrients (among other things), and its variability may bear little relationship to the daily rate of production (Marra, 1997). Yet, biomass estimators have found use in helping to understand the nature of productivity in the ocean and will be used to greater extent in the future.

Perhaps the most direct way of measuring productivity is to measure the change in particulate organic carbon over a period of time. The analytical methods are tedious, and there are no means to achieve much sampling resolution, either with depth or time. Nevertheless, Fig. 4.2 shows one set of data on the daytime increase

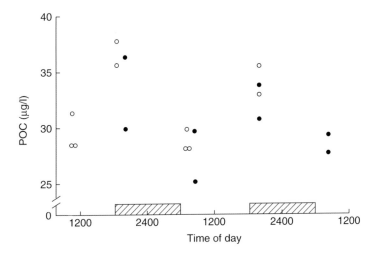

Fig. 4.2 Particulate organic carbon (POC) at 30 m depth plotted against time of day. The data are from the North Pacific central gyre, from the programme Plankton Rate Processes in the Oligo-trophic Oceans (PRPOOS). Open symbols: 20–21 August 1985; solid symbols: 26–28 August 1985. Unpublished data courtesy of J. and R.W. Eppley.

in POC (J. and R.W. Eppley, unpublished data). Since the method is non-dis-criminatory, obtaining a measure of autotrophic productivity can be difficult to achieve without a very good estimate of the initial standing crop. One assumption that can be made in this regard is that the detrital, microheterotroph, and bacterial components to the POC pool change more slowly than the phytoplankton, but it is not known whether this assumption can be accepted under most circumstances. And of course, since the initial value (B_o) appears in the denominator of the equation for exponential growth

$$B_t/B_o = e^{rt} \tag{4.1}$$

(where B is the biomass, r is the growth rate, and t is the time), true growth rates of phytoplankton, as one component of the ecosystem, will always be underestimated.

The lack of sampling resolution led in part to use of the beam attenuation of light over a defined pathlength in sea water. Total attenuation in sea water can be written as the sum of attenuation by constituents, or:

$$c_{tot} = c_w + c_p + c_y \tag{4.2}$$

where the subscripts refer to total, pure water, particle, and other attenuating components (e.g. dissolved organic matter), respectively. By carefully choosing the pathlength of the beam, the wavelength of light, and in consideration of the size spectra of particles in the ocean, these instruments measure attenuation of particles ranging in size from about 3 to 20 µm, or about the size range of the most common phytoplankton. By using longer wavelengths of light (i.e. red), the attenuation by particles is overwhelmingly dominated by scattering rather than absorption. Since absorption by dissolved organic matter occurs mostly in the blue, its effect is minimal and c_y can be neglected. The attenuation by water is fairly well known; thus it is easy to extract particle attenuation from the total. The advantage of beam attenuation is that the measurement is continuous with depth, and the data can be collected contemporaneously with hydrographic data (i.e. with a CTD). But c_p is also a function of important features of the particle population: size, shape, and refractive index. In order to interpret changes in c_p in terms of productivity, these features should not vary coincidentally (Stramski and Reynolds, 1993; Stramski *et al.*, 1995).

Beam attenuation, although solving the problem of sampling resolution, still retains the non-discriminatory nature of the biomass measurement. There is also the added problem of concocting a relationship between attenuation and POC. It is possible to obtain relationships that work for particular sites at particular times (e.g. Fig. 4.3), but a consistent relationship spanning the kinds of particles found in the ocean remains elusive. Some recent work, however, points towards greater under-standing of the nature of the particles. The particle size spectrum as well as an average refractive index can be achieved through the wavelength dependence of c_p (E. Boss and M. Twardowski, pers. comm.). These developments mean progress

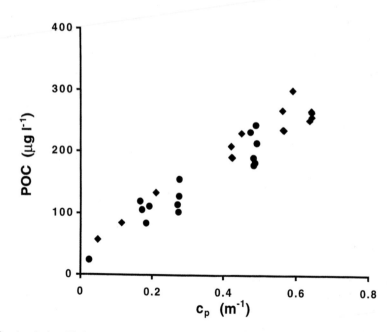

Fig. 4.3 A relationship between particulate organic carbon (POC) and beam attenuation (c_p) for samples from the North Atlantic Ocean (60° N, 20° W, see Marra *et al.*, 1995). Circles: May 1991; diamonds: August 1991.

through the use of multi-wavelength attenuation meters to understand particle dynamics in the surface ocean.

Despite the above cautions, at times there can be large diurnal variations in particle attenuation in the ocean. Siegel *et al.* (1989) were perhaps the first to exploit the use of beam attenuation for estimates of productivity in this way, from a highly resolved series of profiles in the Pacific. They noted that c_p undergoes a regular diel variability, and argued that the changes were caused by particle production during the day and loss at night. Other studies have also indicated the commonness of the diel variability in c_p, provided the water column is stable to mixing (Dickey *et al.*, 1991; Kinkade *et al.*, 1999). The diurnal signals have the proper phasing with the sun (largest amplitude near sunset) and they also attenuate with depth (Marra, 1992) as expected for a population subject to regulation by irradiance on the diel time-scale. The relationship between these changes and phytoplankton productivity would seem obvious; however, without a relationship to POC (at least), it is difficult to transpose the data into a measure of productivity. Until there is some other means to determine the composition of the particles similar to the speed and sampling resolution of the beam attenuation measurement, the method will remain a tantalising approximation, albeit one with increasing viability. Nevertheless, in Section 4.7 I show how the daily change in c_p agrees with other measures of planktonic production.

Chlorophyll *a* changes have also been measured over the diel cycle, but usually in terms of fluorescence, which, like beam attenuation, can be measured with the resolution needed to capture the diurnal variability. The interpretation of chlorophyll *a* variations has been discussed at length in Marra (1997); that publication also discusses previous work. The fluorescence properties of phytoplankton bring the added complication of the variations in fluorescence per unit chlorophyll. Also, the chlorophyll *a* concentration per cell is an adaptable quantity that can vary by itself over the diurnal period, in response to irradiance or nutrient status. Because non-photochemical quenching of fluorescence at irradiances near the surface of the ocean does not seem amenable to analysis, the *in situ* fluorescence of chlorophyll *a* has been notoriously difficult to interpret. However, there can be times, say early in the year, when fluorescence and beam attenuation have similar signals with respect to the diurnal cycle (Dickey *et al.*, 1991).

4.4   Yield measurements of productivity

The yield of the system is probably the most variable approach to productivity, and can lead to the most creativity in its estimation. I say it is the most variable, because the time-scale over which yield measurements can occur can be anything from a day to aeons. Yield is how many fish you catch, how much material ends up in a sediment trap, what zooplankton populations can be supported, how much nutrient is added to the productive layer (or conversely, removed), or the signal in isotopic transformations. The questions include 'How much zooplankton can primary productivity support?' or 'How many fish can be taken from the environment?'. For plankton studies, the most common measure of yield is new production, or the production that is supported by nutrients from depth. This is most usually nitrate, but in the coastal zone, ammonium can be considered new, from the regeneration of organic matter in sediments. It is important to recognise that new production has a time-scale associated with the cycle of mixing in the upper ocean. In places like the central gyres in summer, there may be a diffusive input of new nitrogen from below, which is compensated by a sinking flux. In the temperate ocean, the cycle would be the seasonal cycle of mixing and stratification.

The problem with yield measurements has always been trying to infer the productivity that it came from, that is, how it is scaled back to the daily rate of primary production (Fig. 4.1). As mentioned above, the yield is divided by a small number, thus slight errors or variations lead to huge changes in productivity. Even at the first trophic transition, between autotrophs and their grazers, the trophic efficiency can be as low as 10–20%, a variation that by itself leads to factor of two errors in the net primary production. Sometimes, in estimations relying on the yield, all that can be done is to set lower bounds to the net productivity. Riley (1951) based his estimate of the productivity of the Atlantic on the distribution and flux of oxygen, using the equations for the distribution of properties in the ocean, combined with an assessment of the source and sink. His estimates were later borne out in a sophisticated analysis of tritium–helium ages in the deep ocean (Jenkins, 1980). However, most of the constraints to productivity based on the yield are cautionary, or point out possible inconsistencies in other types of measurements. Furthermore, they are large-scale averages in space and time, such as the seasonal change in oxygen (Jenkins and Goldman, 1985), or the result from the distribution of variables in the ocean (Riley, 1951). Yield estimates, providing a number averaged over large regions or ocean basins, and at seasonal scales or longer, thereby say little about how planktonic systems achieve that value. Rate measurements, at shorter time-scales and observed locally, and involving the details of population growth and loss, are therefore needed to provide complementary understanding of how the yield estimate is achieved.

One goal of oceanography is to understand the productivity of the ocean as a whole, and the ocean's contribution to global biogeochemical cycles. Estimates of global production often come from the collection of data in the euphotic zone, and these will be inaccurate to the extent that they are aliassed with respect to shorter-term phenomena, because of spatial gaps, and because of changing methodologies involving the rate. Deriving estimates based on the yield provides another means. The deep ocean is largely immune from the variability at the surface, thus nutrient changes along the path of the thermohaline circulation might offer clues to the ocean's yield. Deep-water nutrient values are observed to be significantly higher in the North Pacific than anywhere else in the world ocean. And it is generally believed that deep ocean currents flow, in the mean, from the Atlantic to the Pacific

(Gordon, 1986). Assuming the overall deep-ocean current system, then, it is possible to calculate the global ocean productivity using an argument similar to that of Munk's (1966) abyssal recipes. Averaging the total inorganic carbonate species (ΣCO_2) in the deep water, from 4 km (the average depth of the ocean) to 1 km (beneath the main thermocline) depths, and looking at the increase in this value (from respiration) between the source waters in the North Atlantic to the North Pacific, it is possible to derive an equation that describes the horizontal rate of change in ΣCO_2 in deep water, while accounting for losses from vertical diffusion and upwelling.

Assuming the values of Munk (1966) for vertical diffusion and upwelling velocity, and assuming that the benthos are efficient remineralisers of organic carbon surviving to the bottom (Jahnke and Jackson, 1987), the distribution of carbon at any location along the path of the deep water circulation can be expressed as:

$$k \frac{d^2}{dz^2} C + w \frac{d_c}{dz} = v \tag{4.3}$$

The solution to this equation is given in Box 4.1. Figure 4.4 shows the distribution of ΣCO_2 in the deep ocean as a function of distance from a North Atlantic source. The linearity in the distribution is partly the result of averaging. Adopting a linear model means that it is appropriate to consider the advection of the deep ocean by a single value, or at least within the range of variability and noise represented by this tracer. Using an average horizontal velocity of 0.2 cm s^{-1} (corresponding to a residence time of 500 y), and solving for v gives a globally-averaged yield of the ocean of 27 g C m^{-2} y^{-1}. Taking into account the production of $CaCO_3$ of about 20%, the value is then reduced to about 21 g C m^{-2} y^{-1}. A global annual production is correspondingly 8 Gt C, a value that is within the range of previous work (Viecelli, 1984; Sundquist, 1985). The recently completed World Ocean Circulation Experiment will provide the data and analyses to refine the above estimate. But what is important is that the variability and circulation in the deep ocean can be applied to studies of the ocean's yield.

4.5   Rate measurements of productivity

The details of the photosynthetic process are given in Chapter 3. For my purpose here, there are two points in the overall biochemical pathway at which the rate can be measured: very early in the process in the evolution of oxygen, or late, in the fixation of carbon. In virtually all cases, the measurement of the rate of production at sea involves an incubation for the uptake of carbon or the evolution of oxygen over a period of time. Table 4.1 compares the various kinds of methods, from the partitioning of $^{18}O_2$, to the light–dark methods for fluxes of O_2 and CO_2, and the uptake of ^{14}C.

Box 4.1   The carbon flux from deep-water properties in the global ocean.

During the time it takes for a parcel of water, newly sunk in the North Atlantic, to traverse the globe in the thermohaline circulation (see Gordon, 1986), its characteristics are changed, subject to: mixing, *in situ* biological processes, the rain of particulate matter from above, and sinking out of same below. Let C be the average ΣCO_2 concentration over a specified depth range. If we assume a 'box' of unit width, no lateral gradients, and negligible loss to the benthos, the balance of the flux of C can be written as:

$$(uh)\frac{dC}{dl} = -(wC)_{\text{top}} - \left(k\frac{dC}{dz}\right)_{\text{top}} + F_s \tag{4.I}$$

The terms w and k are vertical processes, u is the flow, and h is the height of the box. The distribution of C at any locale, following Munk (1966), can be written as:

$$k\frac{d^2}{dz^2}C - w\frac{d}{dz}C = v \tag{4.II}$$

where v is the *in situ* rate of inorganic carbon formation (i.e. respiration). The solution is:

$$\frac{C - C_1}{C_2 - C_1} = \frac{e^{Q\delta} - 1}{e^Q - 1}(1 + \beta) - \beta\delta$$

where

$$q = (z_2 - z_1)\frac{w}{k}$$

$$\beta = \frac{v}{wC}$$

$$\delta = \frac{z - z_1}{z_2 - z_1}$$

$$C = \frac{C_2 - C_1}{z_2 - z_1}$$

F_s is now defined as an average utilisation over the depth of the water column:

$$F_s = (hv)$$

$$\Delta C = C_{\text{end}} - C_{\text{beginning}}$$

and equation (4.I) becomes, after combining, rearranging, and evaluating the derivative (see Munk, 1966):

$$\Delta C = -2[e^{-Q}(1 + \beta)](C_2 - C_1) + \left(\frac{lv}{u}\right) \tag{4.III}$$

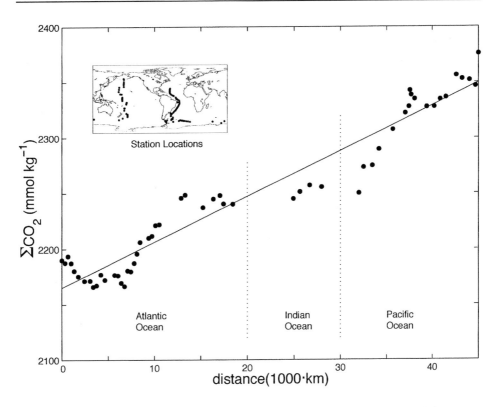

Fig. 4.4   The average value of deep-ocean ΣCO_2 (1–4 km depth) along the path of deep-water circulation (Gordon, 1986) from the North Atlantic to the North Pacific. The data are from GEOSECS (Takahashi *et al.*, 1985). The line is the linear regression.

Table 4.1   Comparison of what each rate technique measures in terms of plankton production.

Technique	Estimates
$^{18}O_2$	Gross primary production (increase in $^{18}O_2$)
CO_2, O_2 light–dark bottle method	Net community production (change in bottles kept in light) Gross primary production (change in light plus change in dark)
^{14}C assimilation	$P \gg R$ (very short time periods): gross photosynthesis $P > R$: net primary production $P = R$: net community production (at community isotopic equilibrium) and biomass increase Biomass (isotopic dilution)

For the light–dark methods, the protocol is to record the change in the dissolved oxygen or total dissolved inorganic carbon concentrations over a prescribed period in a contained water sample. To obtain the greatest amount of information, the periods of incubation are from dawn to dusk, for the light period where both photosynthetic oxygen production and respiratory consumption occur, and then to the following dawn, for the dark period of respiration. Net production is simply the change in oxygen over the period of the experiment. Gross production can be calculated by subtracting the respiratory loss of oxygen in the dark from the net production occurring in the light; however, a crucial assumption is that the respiration in the dark is equal to that in the light.

The other methods for rate measurement require the addition of a tracer, either one for the carbon fixed (^{14}C) or the oxygen produced ($^{18}O_2$). ^{14}C is added to a water sample as bicarbonate, and after a period, the sample is filtered, and the amount of ^{14}C appearing in the particulate material is assayed by standard radioisotopic techniques. For the $^{18}O_2$ method, the isotope is added as $H_2^{18}O$. After the time of incubation, the dissolved gases in the water sample are extracted by vacuum degassing. The recovered O_2, now containing $^{18}O_2$ from photosynthetic oxygen production, is combusted to CO_2, and analysed in a mass spectrometer.

The problems of incubation aside (to which I will return later), very few of these various methods give unambiguous estimates of net or gross photosynthesis. The $^{18}O_2$ method (see Bender *et al.*, 1987) probably gives a good estimate of gross photosynthesis. The production of oxygen in samples incubated in the light gives an unambiguous estimate of net production of the plankton community, over the daytime period at least. At night, respiration by heterotrophs may be such that net production in these samples is zero or negative. To estimate gross production from the O_2 light–dark method requires that respiratory processes in the dark bottle during the day are the same as those in the light, and here again we run into the problem of our poor understanding of the scale and consequences of algal light respiration – a problem that is shared with the ^{18}O and the ^{14}C techniques. Persistently, our understanding of production is held back by our lack of understanding of the consequences of respiration in its various forms. Similar concerns are visited upon the measurement of CO_2 as a light–dark incubation method.

^{14}C uptake into particulate matter shows an interesting paradox. If all the rate processes are constant (an assumption that most certainly is not true, but which we can use for illustration), then the uptake of ^{14}C will still be time-dependent, because of the pathways of fixed carbon within the cell, and the distribution of ^{14}C between the cell and the environment (see Table 4.1). If we restrict our attention to the fate of the ^{14}C label in the first few minutes of an incubation, ^{14}C should estimate gross photosynthesis, because the isotope will not have had time to enter respiratory pools. After a while longer, net photosynthesis may be approximated, because ^{14}C may also be leaving the phytoplankton through respiration (however, see below). After a longer period of time, the cells will attain isotopic equilibrium with the ambient, and any increases in ^{14}C in the particulate matter will be because of an increase in the population, or net production. Finally, after a very long period, the ^{14}C will be distributed among all the populations and components of the sample, and the technique then simply measures the amount of biomass through isotope dilution. Now the paradox is that the number we might want is net plant production, but in order to estimate that from the ^{14}C technique, you have to know the phytoplankton growth rate to get an appropriate length of incubation. The length of the incubation can be a fundamental problem with the use of ^{14}C for planktonic production.

However, the knowledge of the paradox, or the appreciation of it at any rate, has not meant that ^{14}C measurements have historically been done with care. There are three reasons for this. First, the technique is relatively easy: ^{14}C, as a radioactive species, is safe to handle, and easy to obtain. All you need to do is add an aliquot of known activity to a sample, after a period retrieve the particulate material in the sample, and assay the radioactivity. Second, it is impossible to get a negative result. There will always be uptake even though in actuality it may have nothing to do with photosynthetic fixation. Despite sometimes dreadful techniques, you always get a positive answer. Finally, the method is far and away the most sensitive method available, and will probably not be replaced in the near future. Thus, historically, since the method is simple, it has been easy to obtain lots of data. Since there was always a positive result, a researcher could console himself that the method was working and all the data of value. And finally, since the method was so sensitive, there was no means to check the results independently. At worst then, we have an abundance of meaningless data of a rate that occurs in nature at the level of change of parts per billion.

The other problems with rate measurements have to do with the necessity for incubation, and for ^{14}C, filtration of the sample. Incubation means several things. First, it means removal from the environment. Even if the samples are returned to the depth from which they were sampled, they have been effectively removed from the inherent variability that exists in euphotic zone in terms of the quantity and quality of irradiance. Incubations on deck have further problems. Heterotrophic populations may suffer from confinement; and temperature control, especially in the tropics, can be difficult to achieve. We experienced both of these artefacts during the PRPOOS programme in 1982 (see Marra *et al.*, 1988). Destruction of heterotrophic populations was thought to be the cause of the phenomenal population increases in autotrophs during one incubation. Lack of proper temperature control was implicated in the exponential decline of all populations during another incubation. Second, filtration of sea water is never perfect. Populations may be missed by the filter and, on the other hand, dissolved organic matter released during the incubation may be adsorbed (Maske and Garcia-Marquez, 1994; Karl *et al.*, 1998), hindering the interpretation of the measurements.

In terms of the characteristics of an ideal isotope, ^{14}C should qualify. It is added in small concentrations; the form of its addition is specific to one metabolic pathway, and in the phytoplankton, being single cells, one might assume that the isotope is quickly mixed through cellular organelles. However, the latter two reasons are precisely where questions arise in the interpretation of the data. Through what metabolic pathways within the phytoplankton does ^{14}C proceed? And given that phytoplankton are only one component of the community in an incubated sample, what happens to the phytoplankton cells themselves?

Treating ^{14}C uptake in accordance with isotopic uptake theory began with Conover and Francis (1973), and was adapted and reformulated by Hobson et al. (1976). Dring and Jewson (1982) modified the Hobson et al. (1976) model, and Williams and Lefevre (1996) considered further modifications. Marra et al. (1981) and Marra et al. (1988) developed numerical models based on the equations for use with the microbial food web. My analysis here follows most closely that of Dring and Jewson (1982).

The basic equations (Hobson et al., 1976) are:

$$d^T C/dt = P^T C - R^T C \qquad (4.4)$$

$$d^{12} C/dt = P^T C/(1 + K) - R^{12} C \qquad (4.5)$$

$$d^{14} C/dt = K P^T C/(1 + K) - R^{14} C \qquad (4.6)$$

where ^nC refers to the isotopes of C as well as the total (T) amount (units of mass, M), P is the photosynthetic rate (time^{-1}), R the respiration rate (time^{-1}), and K the specific activity of the isotope. K is a very small number (<0.001, typically), thus for all practical purposes, equations (4.4) and (4.5) are identical.

Dring and Jewson (1982) considered modifications to the basic equations above ((4.4)–(4.6)) in light of time courses of ^{14}C uptake in incubations in natural populations as well as in culture. It turns out that photosynthetic carbon uptake in their experiments, and in a variety of other environments as well (Fig. 4.5, Li and Goldman, 1981; Marra, unpublished data), is strikingly linear. The first modification considered by Dring and Jewson (1982) is that the photosynthetic rate does not depend on the carbon content of the cells. In other words, the metabolic pathways are such that photosynthesis did not produce new photosynthetic machinery during the day. The second modification to the basic model is either that recent photosynthate is not available for respiration, or else respired CO_2 is entirely reassimilated. With the two modifications, the equations for the assimilation of C become:

$$d^{12} C/dt = P^T Co/(1 + K) - [R^T Co + qR(^{12}C - {}^T Co)] \qquad (4.7)$$

$$d^{14} C/dt = K P^T C/(1 + K) - qR^{14} C \qquad (4.8)$$

where Co is the initial carbon content of the cells, q is the factor representing the reassimilation of respired CO_2, and $q = 0$ means that all C is reassimilated. Dring and Jewson (1982) concluded that these two modifications were required to produce

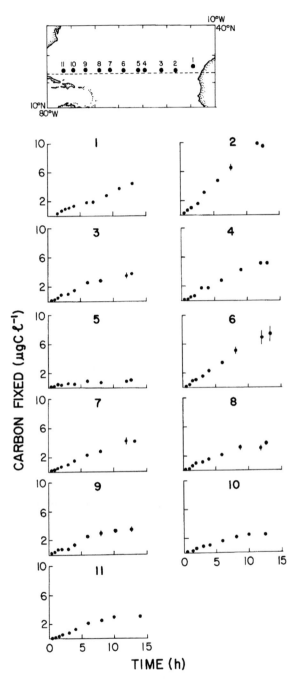

Fig. 4.5 Time courses of ^{14}C assimilation (cumulative) for several stations across the tropical North Atlantic (see inset) in August 1981. All samples were collected by bucket from the windward bow of the ship, except for experiment No. 5, which was collected using a Niskin sampler tripped at 50 m depth. The time origin is dawn of the day of each experiment. Error bars are the SE of the mean of three replicates. (Reprinted from *Journal of Experimental Marine Biology and Ecology* **115** 263–80, by Marra *et al.* (1988) with permission from Elsevier Science.)

a linear time course for the assimilation of ^{14}C. Without the first modification (uptake proportional to the initial value of carbon) the time course would be exponential. Without the second modification, uptake would saturate within a short time. Therefore, agreement with the experimental data required both modifications, and further that respired carbon is entirely refixed in photosynthesis (i.e. $q = 0$). From these observations, Dring and Jewson (1982) concluded that ^{14}C uptake approximated gross photosynthesis.

In Fig. 4.6, I have compiled the published Joint Global Ocean Flux Study (JGOFS) measurements comparing gross photosynthesis based on ^{18}O, with the uptake of ^{14}C. These measurements are data from oceanic populations and are thus thought to be more representative of open ocean conditions. I have chosen to use *in situ* data only in this comparison to lessen the likelihood of artefacts arising from a different light environment on the deck of the ship, and possible differential effects on carbon and oxygen metabolism (e.g. Grande *et al.*, 1989; Barber *et al.*, 1997).

The data are revealing in several respects. First, almost universally, ^{14}C underestimates gross production as measured by the ^{18}O method (Fig. 4.6(a)). Agreement with gross ^{18}O production would require photosynthetic quotients that almost certainly could not be maintained in growing populations of phytoplankton. Second, there is a smaller data set for which comparisons can be made with ^{14}C uptake from dawn to dusk (Fig. 4.6(b)). (^{18}O should not decline overnight.) Here as well there remains the underestimate in ^{14}C uptake. Only when the comparison is made between net production over 24 h and ^{14}C uptake are the ^{14}C uptake values within prescribed limits of the photosynthetic quotient (Fig. 4.6(c)) and the agreement acceptable. According to these data, then, ^{14}C uptake looks to approximate net community production. (That conclusion depends on how much heterotrophic respiration is occurring, and this problem is reconsidered below.)

Nevertheless, we are left with two observations to resolve. First, the uptake of carbon, in general, agrees more closely with net oxygen production on a daily basis. Second, we can expect that carbon assimilation is linear, over periods of hours. As indicated above, a linear time course implies (Dring and Jewson, 1982) both that uptake is not dependent on the instantaneous carbon concentration in the cell, and that most, if not all, carbon respired is refixed in photosynthesis. At first site, these facts seem to be at odds with one another since refixation of respired carbon suggests that C assimilation should estimate gross photosynthesis (Dring and Jewson, 1982). However, the metabolism of carbon and oxygen in the cell occurs over different pathways, and it is probably wrong to assume that gross carbon fixation should be equated to gross oxygen production, at least over physiological timescales.

Ryther (1956) encountered the same dilemma in explaining culture experiments. He concluded that respired CO_2 is reassimilated into photosynthetic pathways, whereas O_2 released is not similarly reassimilated in respiration. Thus, there is an imbalance between the CO_2 and O_2 dynamics, at least as it is measured by ^{18}O. In effect, from the external medium, the cells are using proportionately more H_2O

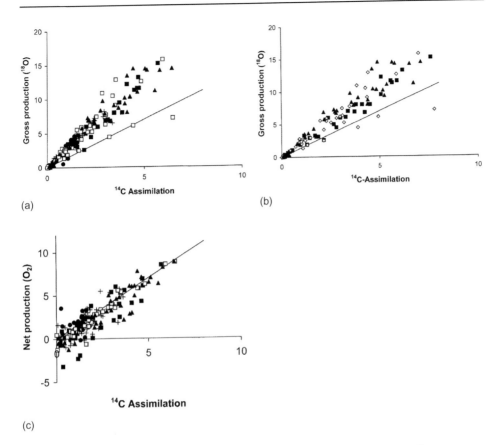

Fig. 4.6   Graphs showing: (a) the daily rate of gross production as measured by the $^{18}O_2$ technique compared to daily (24 h) ^{14}C assimilation; (b) the daily rate of gross production as measured by the ^{18}O technique compared to daytime ^{14}C assimilation ($^{18}O_2$ should not decrease overnight); and (c) the daily net production as measured by the light–dark O_2 method compared to daily ^{14}C assimilation. Solid squares: NABE; solid circles: EqPacTT08; plus sign: EqPacTT12; open squares and triangles: AS043; solid triangle: AS049; line: PQ = 1.4.

(and $H_2^{18}O$) than CO_2, since CO_2 also has an internal source. This would elevate ^{18}O gross photosynthesis relative to ^{14}C uptake. In all, CO_2 is taken up within the cell, while O_2 continues to be produced and released. Thus, O_2 measures gross oxygen photosynthesis, and ^{14}C will be closer to net, relative to that measured by oxygen.

Figure 4.7 shows the result of a simple spreadsheet-style accounting which illustrates these points. The graph shows the end-point of linear uptake over several hours, assuming that respired CO_2 is immediately refixed while respired O_2 comes from the ambient medium. I have set the PQ = 1 merely to show how CO_2 and O_2 scale with one another. Over a varying range of respiration, represented here as net/gross photosynthesis, carbon assimilation will always approximate net oxygen production, while carbon assimilation may approach gross oxygen production if

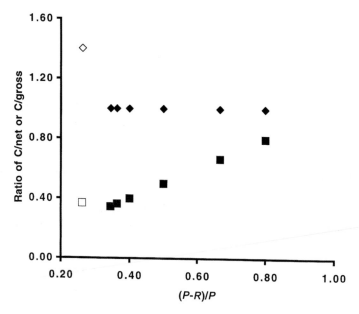

Fig. 4.7 A simple spreadsheet style model of the end-point result of carbon assimilation where all respired CO_2 is refixed by photosynthesis. C assimilation is normalised to net oxygen production, and plotted on an axis of decreasing respiration (increasing net/gross oxygen production, or $(P-R)/P$). The open symbols at the far left of the plot refer to a low value of $(P-R)/P$ and where additional respiratory loss has been added such that respiration *exceeds* photosynthetic production. Squares: C/gross production; diamonds: C/net production.

respiration is a small loss. The interpretation in Fig. 4.7, as the differences between the metabolic pathways of carbon and oxygen for photosynthesis, does not consider what occurs among the populations within the sample. There are grazing effects, as well as the possibility for differential growth between producers and consumers that might not exist outside the bottle incubation. Also, I have not explicitly considered the release of dissolved organic matter by the cells. Both of these are losses that can also be thought of in terms of respiration. Overall, simply requiring respired carbon to be refixed resolves much of the uncertainty about what the ^{14}C technique actually measures. If there is little respiratory loss, the ^{14}C technique can be considered to estimate gross production, but only because net and gross are approximately equal. Otherwise, carbon assimilation will always be closer to net production. Williams *et al.* (1996) came to the same conclusion, by means of observational data being consistent with the complete refixation of respired CO_2.

Before, I take a retrospective look at some older data, there are two final points to be made. First, if respiration is greater than photosynthesis, because of microbial community processes, such that net production is negative after 24 h, the refixation model predicts that ^{14}C will be greater than net, but something less than gross production, depending on the respiratory losses. To illustrate this, the point to the farthest left of the graph in Fig. 4.7, where net/gross is lowest, is an outcome for which I have added in additional respiration. Only here does the carbon assimilation exceed net production.

The second point is that I have suppressed the differences between net primary and net community production in terms of what the ^{14}C technique estimates. The analysis here and the model in Dring and Jewson (1982) consider only phytoplankton processes, so to a first approximation, ^{14}C equals net primary production. However, net primary production is extremely difficult to observe in nature. The problem in interpreting Fig. 4.6(c), showing ^{14}C assimilation to be equal to net community production, is that the PQ cannot be determined with precision. Also, the variability in the data may be such as to obscure the magnitude of heterotrophic respiration.

Some of the above issues are indicated in the data collected as part of the PRPOOS project in 1985, in the North Pacific central gyre. Figure 4.8 shows the

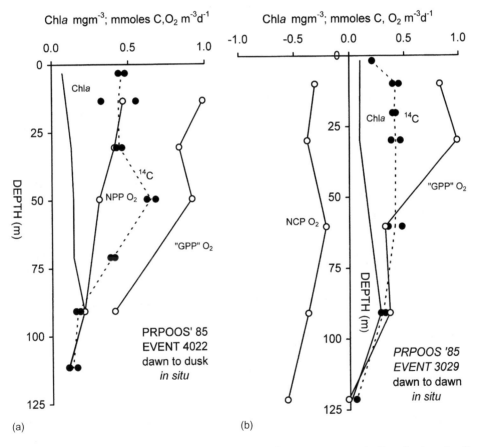

Fig. 4.8 Carbon assimilation compared to gross and net oxygen production for two *in situ* experiments during the PRPOOS programme in 1985 (see, for example, Williams and Purdie, 1991). Chlorophyll *a* is also plotted as a function of depth. Event 4022: (a) dawn-to-dusk experiment; and Event 3029: (b) 24 h experiment.

results of two *in situ* experiments, comparing net and gross O_2 production (the light–dark bottle technique) with ^{14}C assimilation. Chlorophyll *a* is plotted to give some sense of the planktonic structure of the euphotic zone. For the dawn-to-dusk experiment (Fig. 4.8(a)), ^{14}C uptake agrees best with net production (except for the sample near 50 m). Near the base of the euphotic zone where turn-over times are expected to be long, there is a convergence of ^{14}C assimilation with net and gross production. For the 24 h experiment (Fig. 4.8(b)), ^{14}C uptake is less than or equal to gross O_2 production. The two experiments can be interpreted to be similar, with the 24 h experiment a continuation overnight of the one in Fig. 4.8(a). ^{14}C assimilation retains its relationship to net and gross production as predicted by the refixation of respired CO_2 and as shown in Fig. 4.7.

The same group of investigators also compared production methodologies at the Marine Ecosystem Research Laboratory (MERL) at the University of Rhode Island in 1983 (see Bender *et al.*, 1987). The experiments reported by Bender *et al.* (1987) are all on-deck and the samples are from tower tanks filled from the Narragansett Bay. They may not be representative of pelagic ecosystems. Nevertheless, Fig. 4.9 shows the results of two experiments, and these are interesting because respiration for each was a different fraction of gross production. In the 'Tank 9' experiment (Fig. 4.9(a)), respiration was much larger, and ^{14}C assimilation compared well with net oxygen production. For the 'Tank 5' experiment (Fig. 4.9(b)), ^{14}C assimilation looks closest to gross production. However, since respiration is such a small loss, net and gross production are barely distinguishable, and it is possible to conclude that ^{14}C is consistent with net production as well. These results agree with the prediction of the refixation model of carbon assimilation (Fig. 4.7) and, according to that analysis, ^{14}C assimilation can also be considered to estimate net production.

Neither of these examples is conclusive regarding refixation. There are anomalous values in both that suggest other processes such as release of dissolved organic matter or differing metabolic pathways for recently fixed and older carbon. Accepting the refixation of respired CO_2 solves most of the problems in the interpretation of the ^{14}C method and does not require ad hoc explanations of respiratory losses (e.g. Laws *et al.*, 2000). Nevertheless, the scheme described here does not explain the results of shipboard incubations of Williams *et al.* (1983) nor that of Grande *et al.* (1989). Grande *et al.* (1989) explained their anomalous results in terms of differences in the spectral quality of light. We have since learned (Barber *et al.*, 1997) that deck incubator design can also lead to artefacts in the total irradiance received on the deck of a ship compared to *in situ*.

4.6   Measurements of fluorescence properties

Before proceeding to an application of the concepts presented above, I discuss a newer type of measurement, based on the fluorescent properties of photosynthesis

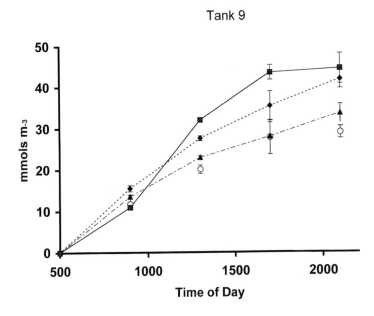

(a)

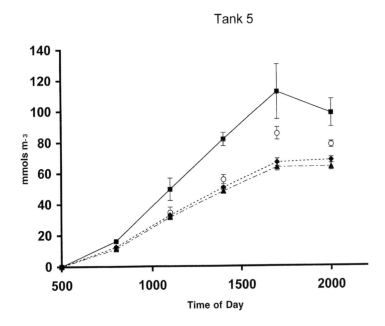

(b)

Fig. 4.9 Cumulative carbon assimilation over the course of a day in two separate experiments, designated as MERL tank No. 9 (a) and tank No. 5 (b) in 1983, compared to gross and net oxygen production. (CO_2 fluxes were very close to the oxygen fluxes and are omitted for clarity.) Open circles: ^{14}C assimilation; triangles: net O_2; diamonds: gross O_2; squares: $^{18}O_2$ gross.

in phytoplankton. Fluorescence properties do not fit neatly into a measurement of biomass, yield, or rate as discussed earlier. However, it is likely that measurements of the fluorescence yield will support, and perhaps supplant, the rate measurements, and afford a means to make rapid assessments of photosynthetic potential. Space does not permit a full treatment here, so the reader should consult the primary references for completeness.

Fluorescence is an optical phenomenon where light is absorbed by certain molecules, and is re-emitted at a longer wavelength. When photons strike the chlorophylls of plants, they have three fates: the bulk of the energy is lost as heat, and the rest used in photochemistry (photosynthesis), or else re-emitted as fluorescence. Chlorophyll *a* absorbs light in the blue, and depending on its particular chemical form, fluoresces at various wavelengths in the red. It would seem that fluorescence would be inversely proportional to photosynthesis; however, that is only partly true. Much depends on the wavelength and intensity of the light, for example. Weak light, continuously applied, such as that used in standard laboratory fluorometers, will show a rough correlation with the amount of chlorophyll *in vivo*. Stronger light can induce a quenching of the fluorescent response but which is non-photochemical in nature (see, for example, Marra, 1997). Fluorescence also shows a complex time-dependent behaviour in the initial milliseconds of light exposure. The kinetics of fluorescence over these time periods has been exploited to learn much about the series of chemical events which constitute the photosynthetic process, as well as provide information needed to ground fluorescence kinetics in terms of an overall photosynthetic rate (see, for example, Krause and Weis, 1991).

According to models, photosynthetic pigments can be thought of as an array of 'traps' for light. When light intense enough to saturate photosynthesis strikes the photosynthetic units, these traps close, and much of the light is re-emitted as fluorescence. If the light is weak, the traps remain open for photosynthesis to proceed, and fluorescence occurs at a baseline value. The fluorescence with the traps closed is termed the 'maximum' fluorescence or F_m; the baseline value fluorescence yield is termed F_o; and the difference between the two is called the 'variable' fluorescence or F_v. The first method, developed to understand the environmental influences on fluorescence (and therefore photosynthesis), is the 'pump-and-probe' fluorometer (Mauzerall, 1972; Falkowski *et al.*, 1988). Here, an intense ('pump') flash of light is used to close the traps, producing F_m, and this is followed by weak, 'probe' light, after a delay of 80–100 µs. The low intensity light reopens the traps, and yields F_o. $(F_m - F_o)/F_m = F_v/F_m$, a dimensionless number that can be equated to the quantum yield of photochemistry, and therefore related to photosynthetic parameters and productivity. A variant of the pump-and-probe technique is pulse-amplitude-modulated (PAM) fluorometry (Schreiber and Shliwa, 1986) which uses repeating intense flashes of light against a continuous background of low light to achieve F_m and F_o, respectively, and from which F_v/F_m can be calculated.

PAM and pump-and-probe fluorometry, however, have some disadvantages in being able to describe the other important features of the photosynthetic units that are important to productivity studies, such as the cross-section for absorption of irradiance. 'Fast repetition rate' (FRR) fluorometry, a succeeding development, overcomes these limitations. The basis of FRR is a series of flashes, or 'flashlets' at sub-saturating intensities. The rapid flashlet sequence produces an overall rise in fluorescence as photosynthesis proceeds and traps become closed, and at F_m, the frequency of the flashlets is decreased, and fluorescence then declines. The rate at which fluorescence increases to F_m can be related to the cross-section of the photosynthetic unit, while the rate at which F_m decays after saturation is related to the turn-over time of photosynthesis. It is from these properties of photosynthesis that quantum yields can be calculated and related to productivity. Probably because of the differences in the time-scales, measurements of the quantum yield of photosynthesis (F_v/F_m) have not been directly interpreted in terms of rates obtained from incubations. However, the FRR fluorometry has been used notably in the Ironex-II experiment in 1994 (Kolber *et al.*, 1994), and it indicated an immediate and expected physiological response to iron enrichment.

Fluorescence from phytoplankton can also be detected passively, that is, detected as an upwelling light signal at 683 nm, as a result of solar stimulation. Solar-stimulated fluorescence has the same caveats associated with other weak-light methods for interpreting the fluorescent signal. However, newer satellite ocean colour sensors such as MODIS and the MERIS sensor on Envisat have wavelength bands designed to detect phytoplankton fluorescence, and thus provide a new tool for investigating phytoplankton properties from space.

4.7   Comparing rate measurements and biomass: NABE

I now illustrate some of the concepts presented on the approaches to the measurement of productivity at sea with a discussion of the results of the first Joint Global Ocean Flux Study (JGOFS) field expedition, the North Atlantic Bloom Experiment (NABE) (see Ducklow and Harris, 1993). NABE included many of the measurements I have discussed, and the sampling regimen consisted of repeated sampling in a localised area. (This is not a true time series at a single station because the ship relocated from time to time to accommodate different scientific programs, e.g. floating sediment trap and drifter deployments.) NABE was a fairly comprehensive experiment, however limited by the fact that it is not representative of many ocean regions. Still, the initiation of the spring bloom was observed, accomplishing a major objective of the experiment. The increase in phytoplankton biomass is a useful backdrop to the several kinds of approaches to the problem of productivity.

NABE began in April 1989 and continued throughout the summer and autumn of that year. I will focus on the second cruise of the programme, since that is the one I am most familiar with (having been the Chief Scientist). It is where the initial increase in phytoplankton biomass was observed, and that is where the most techniques were utilised, notably, the first estimate of daily water column fluxes of CO_2 (Chipman *et al.*, 1993).

First we can look at the biomass signals at the initiation of the spring bloom (Fig. 4.10). Chlorophyll *a* and POC in the mixed layer both increase exponentially over the course of the observations, albeit with some variability caused by the differential movements of the ship and ocean. There is a hint that POC increases more rapidly than chlorophyll *a*, but the ratio of POC/Chl *a* is relatively constant, indicating that heterotrophic activity is minor, or else that micro-zooplankton populations are increasing in tandem with the autotrophs.

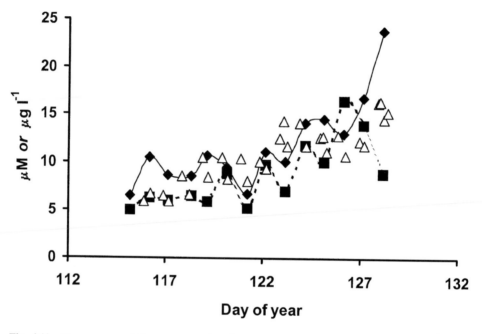

Fig. 4.10 Time course of the average value (in the mixed layer) of chlorophyll *a* ($\mu g\ l^{-1}$) (diamonds) particulate organic carbon (POC) ($\mu mol\ l^{-1}$) (squares); and POC derived from c_p ($\mu mol\ l^{-1}$) (triangles) for the North Atlantic Bloom Experiment, Leg 2, April 1989. The last variable comes from the regression of POC on c_p, and thus is not independent of the POC values. It is shown here for an indication of more highly resolved time variability.

Figure 4.11 shows a comparison of the daily rate of production in the mixed layer from net oxygen production (Kiddon *et al.*, 1995), the *in situ* drawdown of ΣCO_2, ^{14}C assimilation (in bottles suspended *in situ*), and the diurnal change in beam attenuation. The agreement between the methods is very good, and although it is difficult to assign errors, it is doubtful that there are significant differences between the various methods. As suggested above (and in Chipman *et al.*, 1993), ^{14}C assimilation is estimating net production. As shown in Fig. 4.6, the comparison with the estimates of gross production from ^{18}O evolution shows that ^{14}C was an

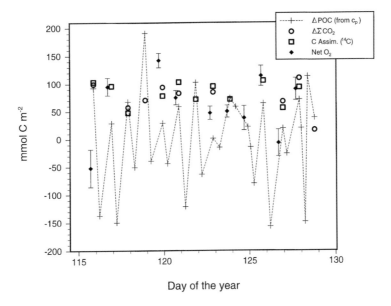

Fig. 4.11   The rate of primary production in the surface mixed layer by four different methods, during Leg 2 of the North Atlantic Bloom Experiment. The various methods are given in the inset. ^{14}C assimilation and the change in ΣCO_2 are from Chipman *et al.* (1993). The net O_2 values are from Kiddon *et al.* (1995). The daily change in particulate organic carbon (ΔPOC) is found from the daily change in beam attenuation, multiplied by the factor 31.5 mmol C m^{-2} found by Gardner *et al.* (1993). Plus signs: ΔPOC (from c_p); open circles: $\Delta\Sigma CO_2$; open squares: C assimilated (^{14}C); diamonds: net O_2.

estimator for net production as well. If heterotrophic respiration is a minor component to total respiration during this time, then it is reasonable to conclude that ^{14}C assimilation is estimating net primary production.

The difference in the cumulative rate of production and the biomass increase observed is the yield. The rate measurements average about 75 mmol C m^{-2} d^{-1}, or about 970 mmol C m^{-2} over the 13-day period. The increase in POC calculated from beam attenuation over the same period is only about 520 mmol C (see Marra, 1995), thus the yield is slightly less than one half the rate of production. Another measure of the yield, the sediment trap flux, has been estimated to be 39 mmol C m^{-2} d^{-1} (at 35 m depth) (Martin *et al.*, 1993), consistent with these results. The difficulty in estimating the yield in pelagic systems is what to do about productivity below the mixed layer. Production occurs, but there is no physical boundary for either the flux of nutrients upwards or particles downwards. At least for the mixed layer, however, we can see a consistency between the three components of productivity – biomass, rate and yield.

4.8   Summary and a thought for the future

^{14}C uptake

Early in the study of the productivity of the ocean's plankton, the only technique available to measure the rate of production was the Winkler method for oxygen analysis. The light–dark bottle oxygen technique was combined with censuses of phytoplankton throughout the seasons and with the seasonal depletion of nutrients in the water column to assess overall productivity. However, as early researchers used it, the Winkler method was not sensitive enough and the rate had poor accuracy and precision. Population censuses suffered from oceanic variability. It was not possible to connect data points in time, and investigators then could not hope to sample the same population repeatedly or with resolution. Nutrient utilisation was an indication of the yield of the system, but after the spring burst, nutrients did not change very much, while at the same time there were other indications of growth and succession. The solution to these early problems was to enlarge the scale of the study. The ocean's motion could then be averaged out in the study of oxygen and nutrient utilisation over entire ocean basins.

The introduction of the ^{14}C technique changed the study of productivity, eventually pointing the way to reconcile the yield of the ocean with the biomass. And today we are fortunate not only to witness the development of technique refinements, as well as newer methodologies (e.g. based on fluorescence), but also to have the means to understand the connections between rate, biomass and yield.

The argument here is that the ^{14}C technique which measures the rate of net production independently of the rate of autotrophic respiration will help us to understand oceanic productivity. At least now we understand what the technique will measure under most conditions and without resorting to locale-specific explanations of respiratory losses. Certainly, the effect of heterotrophic respiration remains an unknown, and if net production is negative after 24 h, according to the argument here, the method becomes difficult to interpret. Nevertheless, autotrophic respiration is a difficult line of study by itself. It is population-dependent (Langdon, 1993), and the environmental effects on photorespiration and the Mehler reaction are little understood. That ^{14}C uptake seems not to depend on these processes will mean that greater emphasis can be placed on the productive ecology of the plankton and the relationship between the ocean's biota and the ocean's carbon cycle.

The results that I have analysed also reinforce the need for *in situ* measurements of primary production. These have long been criticised because they are an expensive use of ship time. However, if primary production values are considered important enough to an overall programme, ship costs should not be a factor, and this argument rings hollow. Also, there are savings in ship time in employing *in situ* incubations instead of deck incubations. There is no need to spend the extra wire-time it takes to determine light depths (to correspond with those in a deck incubator). For incubations of 24 h, there is more flexibility in scheduling ship time than is the case for deck incubations or physiological assays, such as the Pvs.E response. As a final point, the use of *in situ* incubations does not introduce a new variable, 'incubator design', into the interpretation of the data.

Population-based estimates of productivity

For too long we have considered phytoplankton populations in the ocean to be nothing more than chlorophyll *a*. While this has been a powerful simplification in all sorts of studies in biological oceanography, further understanding can only come with an identification of the kinds of populations that are responsible for production at any one locale and time. Phytoplankton variability, in both community structure and total biomass, is strongly linked to sharp gradients in either time or space, such as fronts, eddies, and other mesoscale phenomena. To understand the regulation of net production and yield requires study at these scales. Episodic or mesoscale variability often produces characteristic populations, 'bloom-formers', that are typically large cells that grow fast and that also contribute disproportionately to losses through grazing and sinking.

Models are not yet able to capture mesoscale or episodic variability at basin and global scales, at least not for the ecology. Certainly, we have not incorporated such variability into algorithms to estimate productivity from space-borne sensors. One question we can ask is whether the inclusion of mesoscale variability will rectify, and therefore enhance productivity seen at the surface, or if mesoscale variability can be averaged out. And it may turn out that even if small-scale variability is just 'noise', the rectification of productivity will come in terms of the yield, the production exported to depth.

Acknowledgements

I thank Peter for inviting me to prepare this chapter and for his comments on the manuscript. I would also like to thank Dick Eppley for comments, and his many insights to the problems of plankton growth and productivity in the ocean. I am grateful to C. Ho for help with the global estimate of yield, and to J. Aiken for input into the section on fluorescence kinetics.

References

Babin, M., Morel, A., Claustre, H., Bricaud, A., Kolber, Z. & Falkowski, P.G. (1996) Nitrogen- and irradiance-dependent variations of the maximum quantum yield of carbon fixation in eutrophic, mesotrophic and oligotrophic marine systems. *Deep-Sea Research* **43**, 1241–72.

Bakun, A. (1996) *Patterns in the Ocean*, 323 pp. California Sea Grant College System of NOAA.

Barber, R.T., Borden, L., Johnson, Z., Marra, J. & Trees, C. (1997) Ground-truthing modeled

k_{PAR} and on deck primary productivity incubations with *in situ* observations. In: Ocean Optics XIII (eds S.G. Ackleson & R. Frouin). *Proceedings of SPIE 2963*, pp. 834–9.

Bender, M., Grande, K., Johnson, K. *et al.* (1987) A comparison of four methods for determining planktonic community production. *Limnology and Oceanography* **32**, 1085–98.

Bricaud, A., Babin, M., Morel, A. & Claustre, H. (1995) Variability in the chlorophyll-specific absorption coefficients of natural phytoplankton: analysis and parameterisation. *Journal of Geophysical Research* **100**, 13321–32.

Chipman, D., Marra, J. & Takahashi, T. (1993) Primary production at 47° N, 20° W: A comparison between the ^{14}C incubation method and mixed layer carbon budget observations. *Deep-Sea Research* **40**, 151–69.

Clarke, G.L. (1946) The dynamics of production in a marine area. *Ecological Monographs* **16**, 321–35.

Conover, R.J. & Francis, V. (1973) The use of radioactive isotopes to measure the transfer of materials in aquatic foodchains. *Marine Biology* **18**, 272–83.

Dickey, T., Marra, J., Granata, T., *et al.* (1991) Concurrent high resolution bio-optical and physical time series observation in the Sargasso Sea during spring of 1987. *Journal Geophysical Research* **95**, 8643–63.

Dring, M.J. & Jewson, D.H. (1982) What does ^{14}C uptake by phytoplankton really measure? A theoretical modelling approach. *Proceedings of the Royal Society*, London **B 214**, 351–68.

Ducklow, H.W. & Harris, R.P. (1993) Introduction to the JGOFS North Atlantic Bloom Experiment. *Deep-Sea Research II*, **40**, 1–8.

Dugdale, R. & Goering, J.J. (1967) Uptake of new and regenerated forms of nitrogen in primary productivity. *Limnology and Oceanography* **12**, 196–206.

Falkowski, P.G., Kolber, Z.S. & Fujita, Y. (1988) Effect of redox state on the dynamics of photosystem II during steady-state photosynthesis in eucaryotic algae. *Biochimica et Biophysica Acta* **933**, 432–43.

Gardner, W.S., Walsh, I.D. & Richardson, M.J. (1993) Biophysical forcing of particle production and distribution during a spring bloom in the North Atlantic. *Deep-Sea Research* **40**, 171–95.

Gordon, A.L. (1986) Interocean exchange of thermocline water. *Journal of Geophysical Research* **91**, 5037–46.

Grande, K., Williams, P.J.le B., Marra, J. *et al.* (1989) Primary production in the North Pacific Central Gyre: a comparison of rates determined by the ^{14}C, O_2 concentration and ^{18}O methods. *Deep-Sea Research* **36**, 1621–34.

Hobson, L.A., Morris, W.J. & Pirquet, K.T. (1976) Theoretical and experimental analysis of the ^{14}C technique and its use in studies of primary production. *Journal of the Fisheries Research Board Canada* **33**, 1715–21.

Jahnke, R. & Jackson, G.A. (1987) The role of seafloor organisms in oxygen consumption in the deep North Pacific Ocean. *Nature* **329**, 621–3.

Jenkins, W. (1980) Tritium and ^{3}He in the Sargasso Sea. *Journal of Marine Research* **38**, 533–69.

Jenkins, W. & Goldman, J.C. (1985) Seasonal oxygen cycling and primary production in the Sargasso sea. *Journal of Marine Research* **43**, 465–92.

Karl, D.M., Hebel, D.V., Björkman, K. & Letelier, R.M. (1998) The role of dissolved organic matter release in the productivity of the oligotrophic North Pacific Ocean. *Limnology and Oceanography* **43**, 1270–86.

Kiddon, J.M., Bender, M. & Marra, J. (1995) Production and respiration in the 1989 North Atlantic spring bloom: an analysis of irradiance-dependent changes. *Deep-Sea Research* **42**, 553–76.

Kinkade, C.S., Marra, J., Dickey, T., Langdon, C., Sigurdson, D.E. & Weller, R. (1999) Diel bio-optical variability in the Arabian Sea as observed from moored sensors. *Deep-Sea Research II* **46**, 1813–31.

Kolber, Z.S., Barber, R.T., Coale, K.H. *et al.* (1994) Iron limitation of phytoplankton photosynthesis in the equatorial Pacific Ocean. *Nature* **371**, 145–9.

Krause, G.H. & Weis, E. (1991) Chlorophyll fluorescence and photosynthesis – the basics. *Annual Review of Plant Physiology and Plant Molecular Biology* **42**, 313–49.

Langdon, C. (1993) The significance of respiration in production measurements based on oxygen. In: *Measurement of Primary Production from the Molecular to the Global Scale*, pp. 69–78. International Council for the Exploration of the Sea, Copenhagen.

Laws, E.A., Landry, M.R., Barber, R.T., Campbell, L., Dickson, M.-L. & Marra, J. (2000) Carbon cycling in primary production bottle incubations: inferences from grazing experiments and photosynthetic studies using ^{14}C and ^{18}O in the Arabian Sea. *Deep-Sea Research II* **47**, 1339–52.

Li, W.K.W. & Goldman, J.C. (1981) Problems in estimating growth rates of marine phytoplankton from short-term ^{14}C assays. *Microbial Ecology* **7**, 113–21.

Marra, J. (1992) Diurnal variability in chlorophyll fluorescence: observations and modelling. *Proceedings of SPIE, Ocean Optics XI*, Vol. 1750, pp. 233–44.

Marra, J. (1995) Primary production in the North Atlantic: measurements, scaling, and optical determinants. *Philosophical Transactions of the Royal Society, London* **B 348**, 153–60.

Marra, J. (1997) Analysis of diel variability in chlorophyll *a* and particulate attenuation. *Journal of Marine Research* **55**, No. 4, 767–84.

Marra, J., Haas, L.W. & Heinemann, K.R. (1988) Time course of C assimilation and microbial food-web. *Journal of Experimental Marine Biology and Ecology* **115**, 263–80.

Marra, J., Landriau, G. Jr. & Ducklow, H.W. (1981) Tracer kinetics and phytoplankton rate processes in oligotrophic oceans. *Marine Biology Letters* **3**, 215–23.

Marra, J., Langdon, C. & Knudson, C. (1995) Primary production, water column changes and the demise of a *Phaeocystis* bloom at the ML-ML site in the northeast Atlantic Ocean. *Journal of Geophysical Research* **100**, 6633–43.

Marra, J., Trees, C.C., Bidigare, R.R. & Barber, R.T. (2000) Pigment absorption and quantum yields in the Arabian Sea. *Deep-Sea Research II* **47**, 1279–99.

Martin, J., Fitzwater, S.E., Gordon, R.M., Hunter, C.N. & Tanner, S.J. (1993) Iron, primary production and carbon-nitrogen flux studies during the JGOFS North Atlantic Bloom Experiment. *Deep-Sea Research* **40**, 115–33.

Maske, H. & Garcia-Marquez, E. (1994) Adsorption of dissolved organic matter to the inorganic filter substrate and its implications for ^{14}C uptake measurements. *Applied Environmental Microbiology* **60**, 3887–9.

Mauzerall, D. (1972) Light-induced changes in *Chlorella*, and the primary photoreaction for the production of oxygen. *Proceedings of the National Academy of Science, USA* **69**, 119–40.

Mills, E.L. (1989) *Biological Oceanography: An Early History, 1870–1960*. Cornell University Press, Ithaca, NY.

Munk, W.H. (1966) Abyssal recipes. *Deep-Sea Research* **13**, 707–730.

Peterson, B.J. (1980) Aquatic primary productivity and the ^{14}C-CO$_2$ method: A history of the productivity problem. *Annual Review of Ecology and Systematics* **11**, 359–85.

Platt, T., Harrison, W.G., Lewis, M.R., *et al.* (1984) Biological production of the oceans: a case for consensus. *Marine Ecology Progress Series* **52**, 37–88.

Riley, G.A. (1951) Oxygen, phosphate and nitrate in the Atlantic Ocean. *Bulletin of the Bingham Oceanographic Collection* **13**, 1–126.

Ryther, J.H. (1956) The measurement of primary production. *Limnology and Oceanography* **1**, 72–84.

Ryther, J.H. (1959) Potential productivity of the sea. *Science* **130**, 602–608.

Schreiber, U. & Shliwa, U. (1986) Continuous recording of photochemical and non-photochemical chlorophyll fluorescence quenching with a new type of modulation fluorometer. *Photosynthesis Research* **48**, 395–410.

Siegel, D.A., Dickey, T.D., Washburn, L., Hamilton, M.K. & Mitchell, B.G. (1989) Optical determination of particulate abundance and production variations in the oligotrophic ocean. *Deep-Sea Research* **36**, 211–22.

Stramski, D. & Reynolds, R. (1993) Diel variations in the optical properties of a marine diatom. *Limnology and Oceanography* **38**, 1347–64.

Stramski, D., Shalapyonok, A. & Reynolds, R.A. (1995) Optical characterization of the oceanic unicellular cyanobacterium *Synechococcus* grown under a day–night cycle in natural irradiance *Journal of Geophysical Research* **100**, No. C7, 13 295–308.

Sundquist, E.T. (1985) Geological perspectives on carbon dioxide and the carbon cycle. In: *The Carbon Cycle and Atmosphere CO$_2$: Natural Variations Archaen to Present* (eds E.T. Sundquist & W.S. Broecker), pp. 5–59. American Geophysical Union, Washington, DC.

Takahashi, T., Broecker, W.S. & Bainbridge, A.E. (1985) The alkalinity and total carbon dioxide concentration in the world ocean. In: *Carbon Cycle Modeling* (ed. B. Bolin), pp. 271–86), John Wiley & Sons, New York.

Viecelli, J.A. (1984) Analysis of a relationship between the vertical distribution of inorganic carbon and biological productivity in the oceans. *Journal of Geophysical Research* **89**, 8194–6.

Williams, P.J.le B. (1993) Chemical and tracer methods of measuring plankton production. *ICES Marine Science Symposium* **197**, 20–36.

Williams, P.J.le B. & LeFevre, D. (1996) Algal ^{14}C uptake and total carbon matabolisms. 1. Models to account for the physiological processes of respiration and recycling. *Journal of Plankton Research* **18**, 1941–59.

Williams, P.J.le B., Heinemann, K.R., Marra, J. & Purdie, D.A. (1983) Comparison of ^{14}C and O$_2$ measurements of phytoplankton production in oligotrophic waters. *Nature* **305**, 49–50.

Williams, P.J.le B. & Purdie, D.A. (1991) *In vitro* and *in situ* derived rates of gross production, net community production and respiration of oxygen in the oligotrophic subtropical gyre of the North Pacific Ocean. *Deep-Sea Research* **38**, 891–910.

Williams, P.J.le B., Robinson, C., Sondergaard, M., *et al.* (1996) Algal ^{14}C and total carbon metabolisms. 2. Experimental observations with the diatom *Skeletonema costatum*. *Journal of Plankton Research* **18**, 1961–74.

Yentsch, C.S. (1980) Light attenuation and phytoplankton photosynthesis. In: *The Physiological Ecology of Phytoplankton* (ed. I. Morris) Vol. I, pp. 95–127. University of California Press, 1980 (Blackwell Scientific Publications).

Chapter 5
Supply and Uptake of Inorganic Nutrients

Ulf Riebesell and Dieter A. Wolf-Gladrow

5.1 Introduction

From the perspective of this chapter, the 50th jubilee of Steemann Nielsen's seminal work on carbon assimilation in aquatic systems comes at least five years too late. Well before Steemann Nielsen published the radioisotope method for measuring carbon fixation in aquatic systems in 1952, he was among the first to shed light on two other important processes preceding carbon fixation, namely the supply of inorganic carbon to and it's acquisition by photoautotrophs. In a chapter entitled 'The layer across which the diffusion of CO_2 and HCO_3^- has to take place and its importance for the photosynthesis', Steemann Nielsen (1947) recognised that under certain conditions 'it must be the rate of diffusion of the CO_2 molecules into the plant that limits the photosynthetic intensity'. He further concludes that: 'Submerged aquatic plants with a thin layer across which the diffusion of CO_2 to the chloroplasts has to take place, must also be expected to manage at low CO_2 tensions. This must especially apply to very small plankton algae.' More than 50 years later these conclusions still hold true. Moreover, in view of the present increase in atmospheric CO_2 levels, many of the questions originally raised by Steemann Nielsen have gained a new urgency.

While Steemann Nielsen based his conclusions largely on experimental evidence, Munk and Riley (1952) were the first to do a systematic analysis of the potential limitation of phytoplankton by the rate of nutrient supply to the cell surface. On the size scale of planktonic microalgae, nutrient supply occurs primarily by way of molecular diffusion. The rate of diffusive transport thereby depends on the diffusion coefficient of the nutrient, the concentration gradient from the bulk medium to the algal surface, and the thickness of the diffusive boundary layer (DBL) surrounding the alga. The fact that diffusion-limited transport can set bounds to the growth of large phytoplankton in the sea was first shown by Pasciak and Gavis (1974, 1975). This was extended to CO_2 limitation by Gavis and Ferguson (1975).

Although caught in an almost stagnant microenvironment akin to a fly in a jar of honey, phytoplankton are able to affect the flow of inorganic carbon and nutrients through the DBL in at least two ways. By controlling nutrient concentrations at the cell surface, phytoplankton determine the concentration gradient and thus, the rate of diffusive transport across the DBL. Furthermore, the thickness of the DBL is to a certain extent under the influence of the organism. Cell shape, motility and sinking, or the formation of cell-chains and colonies which can increase phytoplankton size to the scales of turbulent motion, are suitable mechanisms to change DBL thickness. Today, a solid theoretical understanding of the relevant processes allows a quantitative description of inorganic carbon and nutrient supply to planktonic micro-algae (Karp-Boss *et al.*, 1996; Wolf-Gladrow and Riebesell, 1997).

Early on in his work, Steemann Nielsen addressed the question of which form of inorganic carbon (CO_2, HCO_3^-, or CO_3^{2-}) is primarily utilised by aquatic photo-autotrophs (Steemann-Nielsen, 1947, 1960, 1963, 1975), a question that is still under investigation today. Based on experimental work on the freshwater phanerogamic *Microphyllum spigatum*, Steemann Nielsen (1947) concluded that both CO_2 and HCO_3^-, but not CO_3^{2-}, are used as inorganic carbon sources. With the exception of the coccolithophorid *Coccolithus huxleyi* (now *Emiliania huxleyi*), which was extensively investigated by Paasche (1964), experimental evidence for HCO_3^- utilisation in marine phytoplankton was still lacking well into the 1980s.

Although he was well aware of the possible use of HCO_3^- by aquatic photo-autotrophs, in developing the radioactive carbon isotope method Steemann Nielsen (1952) implicitly assumed that only CO_2 was taken up during photosynthetic carbon fixation. While he corrected for the difference in the rate of photosynthetic assimilation between $^{14}CO_2$ and $^{12}CO_2$, he did not account for the isotopic offset between CO_2 and HCO_3^-. In fact, it was not until 1974 that Mook and co-workers measured the difference in the ^{13}C isotopic composition between CO_2 and HCO_3^- to be of the order of 1% (Mook *et al.*, 1974). With the isotope effects between ^{12}C and ^{14}C being twice as high as between ^{12}C and ^{13}C (Craig, 1954), this infers a difference of about 2% between $^{14}CO_2$ and $^{14}HCO_3^-$ (with HCO_3^- being enriched in ^{14}C relative to CO_2). HCO_3^- uptake therefore yields a 2% higher rate of ^{14}C incorporation relative to CO_2 uptake. In comparison, to account for the slower assimilation of $^{14}CO_2$ relative to $^{12}CO_2$, Steemann Nielsen (1952) suggested a 6% correction of ^{14}C-based carbon assimilation estimates – a correction still commonly applied in ^{14}C primary production measurements today. Although this value is probably too high (see 'Carbon isotope fractionation' in section 5.4 below), the possible error introduced by either of these two isotope effects is small compared to the overall uncertainty in the ^{14}C technique: 'We should not consider measurements of primary production to be correct by more than about ±30%.' (Steemann Nielsen, 1975)

The objective of this chapter is to describe some of the processes determining the supply and utilisation of inorganic nutrients by phytoplankton. This topic has been dealt with extensively in numerous review papers over the past few decades, most of them focusing on inorganic nitrogen, phosphorus, and silicon, and more recently on trace elements. This chapter offers a slightly different perspective. Therefore, and as a tribute to Steemann Nielsen, who throughout his career focused on inorganic carbon as a main resource in primary production, this chapter will be biased in favour of the C-nutrient. We wish to emphasise, however, that this is not a comprehensive review on this subject. Instead, we have chosen to explore some aspects which we feel may be of relevance for future investigations on this subject.

5.2 Elemental composition and nutrient availability

The main building blocks for the production of organic matter are the elements C, H, O, N, S, P, K, Na, Ca, Mg and Cl. Eight of these are among the ten most abundant elements in sea water. In addition to these bulk elements, photoautotrophs have an absolute requirement for various trace elements. These include Fe, Mn, Cu, Zn, Co, and Mo as well as a few essential vitamins which some phytoplankton species are unable to synthesise themselves. Among the bulk elements, only N and P can be depleted through biological utilisation to levels which can limit biomass production. Research on phytoplankton–nutrient dynamics has therefore traditionally focused on these two elements. However, some of the trace elements are also present in critically small concentrations and may become limiting in certain areas of the ocean.

Aside from N and P, only one other bulk element – inorganic carbon – has been suggested to potentially limit phytoplankton growth. Although its concentration in sea water (and for the most part also in freshwater) exceeds the carbon requirement by photoautotrophic organisms by far, its biological availability can become rate-limiting under certain conditions due to the rather slow conversion between HCO_3^- and CO_2 (see below). In this context it is important to note the difference between *biomass-* and *rate-limiting* nutrients. The former only applies to nutrients which can undergo complete exhaustion in the growth environment, thereby preventing further biomass production. This is the case for nitrate and phosphate, and in the case of diatoms for silicate. The latter refers to nutrients for which the rate of supply can limit the rate of primary production. This can be the case for various elements, such as N, P, Si, C, as well as some of the trace elements.

Over large scales in space and time, phytoplankton stoichiometry is generally considered to be rather uniform. For the major elements C, N and P, it is defined by the Redfield ratio as C:N:P = 106:16:1 (Redfield *et al.*, 1963). This ratio is sometimes referred to as one of the few 'magic numbers' in biological oceanography. The attribute 'magic' relates to the fact that, despite the large variability in phytoplankton elemental composition, there appears to be a remarkable uniformity in this ratio across time, space, and species boundaries in the ocean. This is astonishing considering that unlike the N:P ratio, which due to the biomass-limiting nature of both these elements is largely under geochemical control (see Falkowski, 1997), the C:N and C:P ratios are under complete biological control. It appears that the considerable small-scale variability in phytoplankton stoichiometry, both due to species- and group-specific differences and in response to variable environmental conditions, averages out over larger scales. However, large-scale changes in environmental conditions and/or long-term shifts in the phytoplankton species distribution and succession have the potential to modulate the Redfield ratio. The widely used assumption of a spatially and temporarily constant Redfield ratio in the world ocean may therefore not hold the test of time.

Distinct differences in elemental composition between major taxonomic groups of phytoplankton originate to a large extent from the production of different cell coverings such as silica frustules in diatoms, calcium carbonate platelets in coccolithophorids, and organic scales in dinoflagellates. While the formation of these protective structures represents an energetic cost for all of these groups, only diatoms can be deprived of the raw material needed to build their frustules. In contrast, sea water contains essentially non-limiting supplies of raw material for both calcium carbonate and organic carbon production. A question immediately arising from this is why silicon, rather than calcium or carbon, is the element preferred by marine phytoplankton to build protective hard structures. The answer to this question may eventually arise from an improved understanding of the biomineralisation processes involved in the formation of silica and calcite coverings.

Due to the difference in cell composition, the major taxonomic groups of phytoplankton strongly differ with regard to their effects on large-scale elemental cycling. This is represented in the so-called 'functional group' concept. According to this concept, the major functional groups in the marine environment are the diatoms, the coccolithophorids, nitrogen-fixing cyanobacteria, and flagellates. Each of these groups has its own distinct effects on the nutrient cycling and vertical flux of particulate material in the ocean. Our understanding of the processes controlling the distribution and succession of the phytoplankton functional groups is still in its infancy. Reconstructing past variability in phytoplankton distribution and succession, and in particular predicting future alterations in response to global change, will be a major challenge in the decades to come.

5.3   Nutrient supply

Most inorganic nutrients in sea water occur as more than one chemical species (Table 5.1). However, not all nutrient species are readily available for phytoplankton utilisation. Selective uptake of one species causes a chemical disequilibrium in the DBL followed by conversion of the non-utilised form(s) into the form that is taken up by the alga. This applies for the different species of inorganic carbon, silicon, and phosphorus. It also applies for NH_4^+ and NH_3. Other forms of inorganic nitrogen, such as NO_3^-, NH_4^+, and N_2 can be converted into each other only by complex biochemical (catalysed) reactions.

Nutrient replenishment in the microenvironment of a microalgal cell can occur in three ways:

Table 5.1 Elements which can potentially limit phytoplankton growth and the chemical forms in which they occur in sea water.

Bio-elements	Nutrient species	Remarks
C	CO_2, HCO_3^-, CO_3^{2-}	Slow conversion between CO_2 and HCO_3^-
N	NO_3^-, NO_2^-, NH_4^+, NH_3, urea, N_2	
P	HPO_4^{2-}, PO_4^{3-}, $H_2PO_4^-$	—
Si	H_4SiO_4, $H_3SiO_4^-$, $H_2SiO_4^{2-}$	$H_2SiO_4^{2-}$ only at pH > 11
Fe	$Fe' = Fe(OH)_3 + Fe(OH)_2^+ + Fe(OH)^{2+}$ $+ Fe^{3+}$	$> 99.9\%$ organically complexed
Zn	Zn^{2+}, $ZnCl^+$	$> 99.9\%$ organically complexed

(1) by diffusive transport from the bulk medium to the cell surface,
(2) by advective transport due to water motion in the microenvironment of the cell, and
(3) by chemical reactions in the DBL when the system is out of equilibrium for a given nutrient species.

The relative importance of diffusive versus advective transport depends on the size, form and velocity of the algae. Ocean surface waters are in a turbulent state where eddies are fed by energy from wind or buoyancy forcing. The kinetic energy is transported down to smaller and smaller scales by a cascade of eddies ('big whirls have little whirls that feed on their velocity, and little whirls have lesser whirls and so on to viscosity.' Richardson, 1922). It is finally dissipated at the size L_d of the smallest eddies, which is around one order of magnitude larger than the Kolmogorov scale $L_K = (v^3/\varepsilon)^{1/4}$, where $v = 10^{-6}$ m^2 s^{-1} is the kinematic viscosity of sea water and ε is the energy dissipation rate. Typical values of ε between 10^{-6} to 10^{-10} m^2 s^{-3} lead to $L_k = 1$ to 10 mm and $L_d = 1$ to 10 cm. On spatial scales immediately below L_d, the flow is laminar with a certain shear motion induced by the turbulence (think of the slow stirring of honey with a spoon. NB: the flow of honey is not a property of honey and if we would impose a faster stirring motion the flow can become turbulent.) On even smaller scales, the mixing of fluid properties is purely diffusive.

Diffusive boundary layer

Like the spoon in the honey, any object in the fluid (single algal cells, algal chains or aggregates) is surrounded by a diffusive boundary layer in which advective transport is negligible compared to diffusive transport. The DBLs of small spherical particles with radius a at rest with respect to a much larger surrounding water parcel possess a spherical symmetric DBL with an effective thickness a (this is the only length scale available). The concentration c of nutrient species c in the microenvironment of the sphere (e.g. an idealised single-celled alga) obeys the diffusion equation:

$$\frac{\partial c}{\partial t} = D_c \nabla^2 c \tag{5.1}$$

where t is time, D_c is the diffusion coefficient of c in sea water, $\nabla = (\frac{\partial}{\partial x}, \frac{\partial}{\partial y}, \frac{\partial}{\partial z})$ is the nabla operator, and x,y,z are Cartesian co-ordinates. Assuming spherical symmetry and steady state, the diffusion equation written in spherical co-ordinates reduces to the following ordinary differential equation:

$$\frac{d}{dr}\left(r^2 \frac{dc}{dr}\right) = 0 \tag{5.2}$$

(where r is the radial distance from the centre of the sphere) which can be easily solved given the boundary conditions $c(r = a) = c_{\text{surface}}$ and $c(r \rightarrow \infty) = c_{\text{bulk}}$

$$c(r) = c_{\text{bulk}} - (c_{\text{bulk}} - c_{\text{surface}})\frac{a}{r} \tag{5.3}$$

From this solution one can calculate the flux of species c towards the cell:

$$F = 4\pi a^2 D_c \frac{dc}{dr}|_{r=a} = 4\pi a D_c (c_{\text{bulk}} - c_{\text{surface}}) \tag{5.4}$$

If the cell takes up all the inwardly diffusing molecules, the surface concentration c_{surface} decreases to zero and (for a given bulk concentration) the flux becomes maximal:

$$F_{\text{max}} = 4\pi a D_c c_{\text{bulk}} \tag{5.5}$$

For non-spherical algae the diffusion equation needs to be solved in the appropriate geometry. For prolate and oblate spheroidal geometry, the stationary diffusion equation can be solved analytically (Pasciak and Gavis, 1975). Wolf-Gladrow and Riebesell (1997) have calculated the solution for a general spheroid. All fluxes are of the same form as equation (5.4) except that a is replaced by the shape radius r_{shape}:

$$F = 4\pi r_{\text{shape}} D_c (c_{\text{bulk}} - c_{\text{surface}}) \tag{5.6}$$

(For explicit expressions of r_{shape} see Wolf-Gladrow and Riebesell, 1997.) For more realistic algal forms, the solution of the diffusion equation is tedious. Wolf-Gladrow and Riebesell (1997) showed that for prolate and oblate spheroids the surface equivalent radius (the radius of a sphere with the same surface as the non-spherical alga) is very close to the exact shape radius. It can be assumed that the shape radius for other algal forms can also be estimated from the surface area. (Actually, this is a hidden assumption when scaling nutrient fluxes by surface-to-volume ratios.)

The effect of motion

A particle in motion experiences a decrease in the DBL thickness and increase in solute fluxes relative to a non-motile particle. The effect of motion on nutrient supply to aquatic plants, such as for sinking, rising or swimming phytoplankton, was first discussed by Munk and Riley (1952). Their estimates were based on a small number of observations and on rather crude approximations of the solutions to the flow equations available at that time. The following discussion is based on the work of Karp-Boss *et al.* (1996), which is highly recommended for further details. Only spherical particles will be considered.

The distribution of a nutrient species in the vicinity of a cell is ruled by the advection-diffusion equation:

$$\frac{\partial c}{\partial t} + \boldsymbol{u}\nabla c = D_{\mathrm{c}}\nabla^2 c \qquad (5.7)$$

in which the flow velocity of water, $\boldsymbol{u}$, is a solution of the hydrodynamic (Navier–Stokes) equations. The advection–diffusion equation is difficult to solve even for the stationary state. Much insight can be gained, however, by rewriting the problem in non-dimensional variables and by discussing the characteristic non-dimensional parameters. The non-dimensional Navier–Stokes equations contain only one characteristic quantity, namely the Reynolds number:

$$Re = \frac{VL}{\nu} \qquad (5.8)$$

where V is a characteristic velocity (for example, the sinking velocity of the cell), L is a characteristic length scale (the radius of the cell), and $\nu = 10^{-6}$ m^2 s^{-1} is the kinematic viscosity of sea water (at 20°C). Re is a measure of the (non-linear) advection term compared to the viscous term; flows with $Re < 1$ are laminar, flows with $Re > 1000$ are turbulent. Phytoplankton cells with $L < 100$ μm and $V < 10$ m d^{-1} are in a laminar flow regime ($Re \leq 0.01$). For a given flow field $\boldsymbol{u}$ (x, y, z, Re), the solutions of the stationary advection–diffusion equation can be characterised by the Péclet number:

$$Pe = \frac{VL}{D_{\mathrm{c}}} \qquad (5.9)$$

which is the ratio between the characteristic scales of the advective transport of tracers ($|\boldsymbol{u}\nabla C|$) and the diffusive transport ($|D_{\mathrm{c}}\nabla^2 C|$). The Sherwood number, Sh, is the ratio between the total flux of a nutrient species arriving at the cell surface in the presence of motion and the purely diffusional flux.

For particles in motion at small Reynolds numbers, the dependence of *Sh* on *Pe* (represented by the solid line in Fig. 5.1) is given by:

$$Sh = 1 + \frac{1}{2}Pe + \frac{1}{2}Pe^2 \ln Pe \qquad Pe < 0.01 \qquad (5.10)$$

$$Sh = \frac{1}{2} + \frac{1}{2}(1 + 2Pe)^{1/3} \qquad 0.01 \leq Pe \leq 100 \qquad (5.11)$$

$$Sh = 0.6245\, Pe^{1/3} + 0.461 \qquad Pe > 100 \qquad (5.12)$$

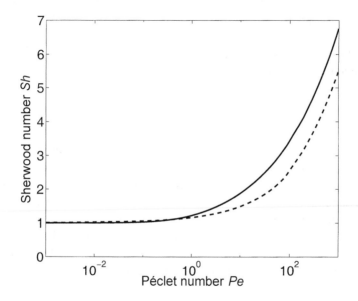

Fig. 5.1 The Sherwood number as a function of the Péclet number (for small Reynolds numbers) for: (a) steady, uniform flow past the particle, as experienced by sinking, rising, and swimming phytoplankton (solid line); and (b) turbulent shear (dashed line; after Karp-Boss *et al.*, 1996).

The influence of turbulence on the nutrient supply of spherical algae has been studied by Lazier and Mann (1989). Kiørboe (1993) has used the Sherwood number to assess the effect of cell size on nutrient uptake in phytoplankton communities. As in the case of particles in motion, the influence of turbulence on nutrient supply can be estimated by a Sherwood scaling. According to Karp-Boss *et al.* (1996):

$$Sh = 1 + 0.29Pe^{1/2} \qquad \text{for} \quad Pe < 0.01 \qquad (5.13)$$

$$Sh = 1.014 + 0.15Pe^{1/2} \qquad \text{for} \quad 0.01 \leq Pe \leq 100 \qquad (5.14)$$

$$Sh = 0.55Pe^{1/3} \qquad \text{for} \quad Pe > 100 \qquad (5.15)$$

with

$$Pe = \frac{a^2 E}{D_c} \qquad (5.16)$$

where $E = \sqrt{\varepsilon/v}$ is the turbulent shear rate, ε is the energy dissipation rate, v is the kinematic viscosity of sea water, and a is the particle radius. The relationship (represented by the dashed line in Fig. 5.1) is valid for small Reynolds numbers ($Re < 1$).

Non-spherical algae and algal chains undergo non-stationary rotation (so-called Jeffrey' orbits) that may further increase nutrient supply. The discussion of this effect is beyond the scope of this chapter, and for further information see Pahlow *et al.* (1997) and Karp-Boss and Jumars (1998). In essence, phytoplankton size and form, chain- and colony-formation, elasticity of chains and colonies and the spacing of the cells in them, all affect the flow of nutrients experienced by phytoplankton cells in turbulent environment.

Chemical reaction in the diffusive boundary layer

When a nutrient is taken up by an alga, its concentration $c(r)$ in the DBL is reduced relative to the bulk concentration. Selective uptake of a certain nutrient species also creates a chemical disequilibrium in the DBL. The extent to which equilibrium is re-established depends on the relationship between the characteristic time constants of diffusion and chemical reactions in the DBL (Table 5.2). The best-studied example is the uptake of CO_2 by diatoms (Gavis and Ferguson, 1975; Wolf-Gladrow and Riebesell, 1997). A reduction of the CO_2 concentration in the DBL initiates net conversion of HCO_3^- to CO_2 according to the reactions:

$$HCO_3^- + H^+ \rightarrow H_2CO_3 \rightarrow CO_2 + H_2O \qquad (5.17)$$

and

$$HCO_3^- \rightarrow CO_2 + OH^- \qquad (5.18)$$

The concentration $[CO_2](r)$ in the DBL obeys the following diffusion–reaction equation:

$$\frac{D_{CO_2}}{r^2} \frac{d}{dr} \left(r^2 \frac{d[CO_2]}{dr} \right) + k_c([CO_2]_{bulk} - [CO_2]) = 0 \qquad (5.19)$$

where k_c is a rate constant. With the same boundary conditions as above, i.e. for a given CO_2 concentrations at the cell surface ($[CO_2]_{surface}$) and in the bulk medium ($[CO_2]_{bulk}$), the solution reads:

Table 5.2 Characteristic length and time-scales relevant for the supply of inorganic nutrients to phytoplankton cells.

Length scale, time-scale	Mathematical expression/notation	Characteristic value	Meaning/remarks
Algal size	a	$10\,\mu m$	—
Thickness of DBL	$d = a$	$10\,\mu m$	Effective thickness
Reacto-diffusive length scale	a_k	$200\,\mu m$	CO_2 supply is doubled through conversion of HCO_3^- to CO_2 if $d = a_k$
Kolmogorov scale	$L_K = (\nu^3/\varepsilon)^{1/4}$	$1–10\,mm$	—
Diffusion time scale	$\tau_d = \dfrac{a^2}{D_c}$	$0.1\,s$	$a = 10\,\mu m$ $D_c = 10^{-9}\,m^2\,s^{-1}$
Reacto-diffusive time-scale	$\tau_{rd} = \dfrac{1}{k_c}$	$30\,s$	$k_c = 0.03\,s^{-1}$ see Zeebe *et al.* (1999) for further details on kinetic time-scales
Kolmogorov time-scale	$\tau_K = \dfrac{L_K^2}{\nu}$	$1–100\,s$	$L_K = 1–10\,mm$ $\nu = 10^{-6}\,m^2\,s^{-1}$

$$[CO_2](r) = [CO_2]_{bulk} - ([CO_2]_{bulk} - [CO_2]_{surface})\frac{a}{r}\exp\left(\frac{a-r}{a_k}\right) \qquad (5.20)$$

where a_k is the reacto-diffusive length scale (Wolf-Gladrow and Riebesell, 1997), which is given as:

$$a_k = \sqrt{D_{CO_2}/k_c} \approx 200\,\mu m$$

The flux towards the cell is defined as:

$$F = 4\pi D_{CO_2} a\left(1 + \frac{a}{a_k}\right)([CO_2]_{bulk} - [CO_2]_{surface}) \qquad (5.21)$$

Comparing equations (5.4) and (5.21) shows that due to the reactions in the DBL the total flux to the cell surface increases relative to the diffusive flux by the factor $(1 + a/a_k)$. For an algal cell with radius $a = 20\,\mu m$, for example, the CO_2 flux is increased by 10%.

Correction factors

Summing up over all factors influencing nutrient supply to a phytoplankton cell, the flux of nutrient species c can be written as follows:

$$F = \gamma_{turb}\, \gamma_{shape}\, \gamma_{conversion}\; 4\pi a D_c (c_{bulk} - c_{surface}) \tag{5.22}$$

where a is the surface equivalent spherical cell radius, D_c is the diffusion coefficient of species c in sea water, and c_{bulk} and $c_{surface}$ are the concentrations of c in the bulk medium and at the surface of the cell, respectively. The three correction factors are:

$$\gamma_{turb} = Sh \tag{5.23}$$

$$\gamma_{conversion} = 1 + a/a_k \tag{5.24}$$

$$\gamma_{shape} = \frac{\arctan(\frac{\sigma}{\sqrt{1-\sigma^2}})}{\sigma (r_1/r_2)^{1/3}} \tag{5.25}$$

for oblate spheroids, where r_1 and r_2 are the major and minor radii, respectively, and

$$\sigma = \sqrt{1 - r_2^2/r_1^2} \tag{5.26}$$

The correction factors γ_{turb} and γ_{shape} principally apply to all nutrients; $\gamma_{conversion}$, which accounts for the chemical conversion between different nutrient species in the DBL, mainly applies to inorganic carbon, but may also be of relevance for the supply of P, Si, and Fe (Völker and Wolf-Gladrow, 1999). As no spontaneous conversion occurs between the N nutrient species NO_3^-, NH_4^+, and N_2, no such correction needs to be applied in calculating the fluxes of these nutrients.

It is important to note that the correction factor γ_{shape} only accounts for the purely static aspect of a non-spherical shape. In a shear field, the dynamical effect on nutrient uptake due to time-dependent movements of non-spherical cells or cell chains in Jeffrey orbits can not be described by a simple function of the cell geometry because the increase in nutrient supply also depends on the uptake kinetics of the phytoplankton (Pahlow *et al.*, 1997). The latter is known to vary depending on the nutrient.

Diffusion limitation and the effect of cell size

The size range of phytoplankton, from single-celled picoplankton to chain or colony-forming microplankton, spans several orders of magnitude. A terrestrial analogue covering a similar size range would extend from mosses to trees. Over this size range, potential constraints on nutrient and inorganic carbon supply exerted by diffusive transport differ vastly. As indicated in equations (5.4) and (5.21), both the diffusive flux across the DBL and the relative contribution of chemical conversion within the DBL to the total nutrient supply increase linearly with the cell radius a. The nutrient content of algal cells, which can be taken as a measure for nutrient demand, increases with a^b, with $2 < b < 3$ (Montagnes *et al.*, 1994). Thus, for a given bulk concentration c_{bulk}, growth rate and nutrient affinity, the ratio of nutrient demand to nutrient supply rapidly increases with increasing cell size. Assuming that phytoplankton are perfect nutrient absorbers (i.e. zero concentration at the cell surface), we have calculated the ratio of dissolved inorganic nitrogen (DIN) demand versus supply for phytoplankton cells as a function of cell sizes, growth rates, and DIN concentrations (Fig. 5.2).

A typical microalga with a cell radius $a = 10\,\mu m$ growing at a rate of $1\,d^{-1}$ is diffusion-limited at DIN concentrations $\leq 1\,\mu mol\ kg^{-1}$. At DIN concentrations prevailing in the surface waters of oligotrophic oceans of $\leq 0.1\,\mu mol\ kg^{-1}$ (Eppley and Koeve, 1990), the maximum growth rate of this alga supported by diffusive DIN supply would be $\leq 0.1\ d^{-1}$. Thus, even with the most efficient uptake kinetics for inorganic nitrogen sources, phytoplankton in this size range are poor competitors in low nutrient environments. A nanoplankton cell with $a = 1\,\mu m$, on the other hand, could maintain a growth rate of $1\,d^{-1}$ at DIN concentrations as low as $10\,nmol\ kg^{-1}$ without suffering DIN diffusion limitation. The same conclusions can be drawn regarding the potential for diffusion limitation with inorganic phosphorus. In a similar analysis on iron and zinc availability in oligotrophic oceans, Hudson and Morel (1991) and Morel *et al.* (1991) concluded that diffusion limitation also imposes constraints on cell size and/or on cell quotas of these two trace elements.

This simple analysis can be used to explain the predominance of small phytoplankton in oligotrophic waters. But how do large phytoplankton cells, which commonly occur in small but significant numbers in low nutrient environments, manage to grow at appreciable rates? There are a number of strategies, which may partly alleviate diffusion limitation of large phytoplankton.

As described above, cell shapes that deviate from a sphere can increase the nutrient flux per unit of cell volume by increasing the surface-to-volume ratio. Likewise, sinking, swimming, and fluid motion can reduce the effect of diffusion limitation for larger cells. Assuming phytoplankton sinking or swimming rates and turbulent water motion typical of the marine environment, the flux of nutrients to the cell surface can increase by as much as a factor of 2 (see below). In a turbulent environment, elongated algal cells and cell chains can experience increased nutrient supply due to non-stationary rotation (Pahlow *et al.*, 1997; Karp-Boss and Jumars, 1998). For large spherical colonies of the plankton alga *Phaeocystis* sp., on the other hand, Ploug *et al.* (1999) showed nutrient uptake and cellular growth of colonial cells to be lower relative to free-living cells in a turbulent environment over a wide range of shear rates.

The above estimates of nutrient supply assume that nutrient uptake is restricted to the time of photosynthetic activity. Large cell vacuoles present in most phytoplankton provide the capacity for luxury consumption and storage of inorganic nutrients during periods of low demand, while supplementing nutrient supply during times of high rates of biosynthesis (Raven, 1984). In terms of increasing nutrient flux, an interesting phenomenon is also given by algae incorporated in porous marine snow aggregates. The nutrient supply within large, sinking aggregates can be

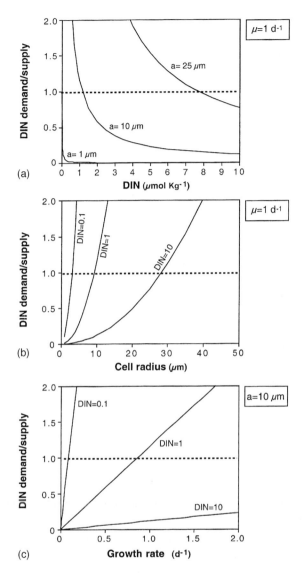

Fig. 5.2 Dissolved inorganic nitrogen (DIN) demand versus supply for phytoplankton at: (a) varied DIN concentration; (b) cell radius; and (c) growth rate. Nitrogen demand is calculated as the product of growth rate and the cellular N-quota, assuming the cell volume to nitrogen content relationship of Montagnes *et al.* (1994). Nutrient uptake only occurs during the light period of a 12 h/12 h light/dark cycle. Assuming spherical cell geometry and diffusive transport only (i.e. no advection), the maximum DIN supply is calculated from equation (5.5). Values of DIN demand/supply above 1 (dashed line) indicate diffusive transport limitation.

significantly higher than experienced by non-aggregated phytoplankton (Logan and Alldredge, 1989). An effective mechanism for overcoming nutrient limitation in oligotrophic areas of the ocean has been developed by some large oceanic diatom species, which regulate their buoyancy to commute between the deep nutricline and the sunlit surface layer (Villareal *et al.*, 1999). Above all, the primary advantage of large phytoplankton occurring at low densities in nano- and picoplankton-dominated, nutrient-poor environments is in the reduced loss term due to the comparatively low grazing pressure in this size class. A similar conclusion was drawn by Ploug *et al.* (1999) regarding possible advantages of large colonies relative to free-living cells in the colony-forming alga *Phaeocystis*.

CO$_2$ limitation

As discussed in section 5.2, the concentration of dissolved inorganic carbon (DIC) in sea water and in most freshwater environments exceeds the phytoplankton carbon requirement by far. However, less than 1% of DIC occurs as aqueous CO_2, which due to its electric neutrality is the only form which passively crosses cell membranes. Due to the slow conversion of HCO_3^- to CO_2, the one reaction in the carbonate system which phytoplankton would prefer to be fast, replenishment of CO_2 from the large pool of inorganic carbon is rather slow. For large algal cells relying on CO_2 diffusive transport and non-catalysed conversion of HCO_3^- to CO_2 in the diffusive boundary layer, carbon requirement can therefore surpass the flux of CO_2 to the cell surface (Gavis and Ferguson, 1975; Riebesell *et al.*, 1993).

By comparing the maximum flux of CO_2 calculated from equation (5.5) (i.e. assuming zero concentration at the cell surface) with the phytopankton carbon requirement, we have estimated (following Chisholm, 1992) the CO_2 concentration above which the cellular carbon demand exceeds the non-catalysed flux of CO_2 to the cell (Fig. 5.3). Whereas phytoplankton smaller than 10 µm in radius are unlikely to experience any shortage in CO_2 supply at typical sea water CO_2 concentrations (ranging between 8 and 25 µmol kg^{-1}), a large phytoplankton of radius $a = 25$ µm growing at a rate of 1 d^{-1} requires an ambient CO_2 concentration of 12 µmol kg^{-1} or higher to satisfy its carbon demand via CO_2 diffusive transport. This value reduces to about 11 to 11.5 µmol kg^{-1} if the effects on nutrient supply by turbulent shear, chemical conversion within the boundary layer, and non-spherical cell shapes, respectively, are considered. Combining all three effects yields a critical concentration for CO_2 limitation of about 9 µmol CO_2 kg^{-1}. Thus, phytoplankton with a radius greater than 25 µm and/or growing at a rate faster than 1 d^{-1} can experience CO_2 diffusion limitation in the sea. To avoid inorganic carbon shortage, these phytoplankton must draw from the abundant pool of HCO_3^-.

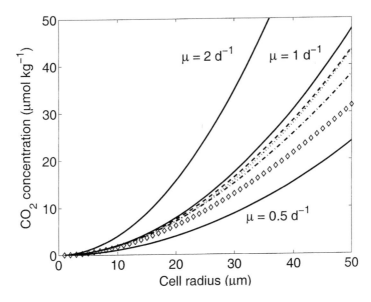

Fig. 5.3 Ambient CO_2 concentration above which the CO_2 flux to the cell surface exceeds the carbon requirement of a phytoplankton cell of given radius and growth rate of μ. Broken lines indicate the effect of turbulent shear (dashed line; assuming $E = 0.2\,s^{-1}$), chemical conversion in the DBL (dash-dotted line), and cell shape (dotted line; assuming $r_1/r_2 = 3$) on the critical CO_2 concentration at $\mu = 1\,d^{-1}$. Diamonds indicate the combination of these three effects. Carbon requirement is calculated as the product of growth rate and the cellular C-quota, assuming the cell volume to carbon content relationship of Montagnes *et al.* (1994). Assuming spherical cell geometry and non-catalysed diffusive transport only (i.e. no advection), maximum CO_2 flux is calculated from equation (5.5).

5.4 Nutrient utilisation

The nutrient requirement of a phytoplankton cell is ultimately determined by its chemical composition and specific growth rate. Under steady-state conditions, the nutrient supply ratio must be equal to the elemental ratio of the cell. In the natural environment, nutrient concentrations are highly variable in space and time. In addition, the contribution of advective flux to total nutrient supply can vary on short time-scales. Phytoplankton therefore experience considerable variability both in the absolute nutrient supply and in the supply ratio for different nutrients. Large variations also occur in nutrient assimilation. Fluctuation in the light environment due to vertical mixing, superimposed on the diurnal cycle, causes considerable variability in photosynthetic carbon fixation and, albeit to a lesser degree, in overall biosynthesis.

In order to cope with a nutrient supply intermittently out of line with nutrient assimilation, phytoplankton have developed various strategies. Internal pools for nutrients or the assimilation in storage compounds can buffer the intrinsic environmental variability in carbon and nutrient fluxes. On short time-scales of the order of the doubling time of a cell or less, temporary nutrient storage can serve to synchronise nutrient supply and assimilation. On longer time-scales, however, the supply of inorganic carbon and nutrients to the cell surface must match the assimilation in biosynthetic processes. To achieve this delicate equilibrium requires a considerable degree of flexibility by the phytoplankton to fine-tune their uptake kinetics for the various nutrients.

Due to their ability to dissolve in the lipid portion of the membrane, many biologically relevant solutes can cross lipid bilayers by molecular diffusion. In principle, this holds true for all neutral solute species such as CO_2, NH_3, H_4SiO_4, O_2, N_2, H_2O, urea, and $B(OH)_3$ (Raven, 1980). The transmembrane flux of these solutes is a function of the permeability coefficient of the solute species and its concentration gradient across the membrane. Permeability coefficients are determined by the effective diffusivity of the solute species in the membrane material and the length of the diffusive path, i.e. the thickness of the membrane. For all electrically charged nutrient solutes, such as HCO_3^-, CO_3^{2-}, NO_3^-, NO_2^-, NH_4^+, $H_2PO_3^-$, HPO_4^{2-}, $H_3SiO_4^-$, the lipid membrane constitutes an effective barrier. Utilisation of these nutrients requires facilitated transport by way of specific protein porters and channels located within the lipid membrane or by means of an energy-consuming active transport.

Concentrating mechanisms

An efficient assimilation of inorganic nutrients often requires that their concentrations be actively raised at the sites of nutrient utilisation inside the algal cell compared to the surrounding water. This applies to potentially biomass-limiting nutrients such as phosphate, nitrate, and silicate, particularly when their bulk water concentrations are becoming depleted. However, this also holds true for nutrients whose concentrations are well in excess of algal requirements. In the coccolith-producing vesicles of coccolithophorids, for instance, calcium and carbonate ion concentrations must be raised well above bulk water levels in order to permit calcite precipitation. To operate efficiently, photosynthetic carbon fixation also requires inorganic carbon concentrations above those present in the surrounding water at the site of carboxylation (see Chapter 3). The dominant carboxylating enzyme in eukaryotic mircoalgae, Ribulose-1,5 bisphosphate carboxylase/oxygenase (RUBISCO), uses CO_2 as the only carbon substrate and has a half-saturation constant of 20 to 70 μmol kg^{-1} of CO_2, depending on the species (Badger *et al.*, 1998). Typical sea water CO_2 concentrations, ranging between 10 and 25 μmol kg^{-1}, are insufficient to ensure effective operation of RUBISCO carboxylation.

To overcome the low CO_2 affinity of their main carboxylating enzyme, algae have developed mechanisms to actively increase the CO_2 concentration at the site of carboxylation. These include active uptake of CO_2 and HCO_3^- into the algal cell and/or into the chloroplasts. In concert with active carbon transport, algae catalyse the conversion between HCO_3^- and CO_2 intra- and extracellularly with the enzyme carbonic anhydrase. Whichever the pathways for this so-called carbon concentrating mechanism (CCM) may be, operation of the CCM requires additional energy (Thoms *et al.*, 2001). Since marine phytoplankton live in an environment in which light (energy) is one of the primary limiting resources, the energy available for active transport of inorganic carbon and other nutrients is often limited. Energy limitation may be one of the reasons why phytoplankton do not rely entirely on the active uptake of inorganic carbon from the abundant pool of HCO_3^-, but are drawing part of their carbon from the small and variable pool of CO_2.

At the same time that evidence for a CCM in marine phytoplankton is accumulating (Kaplan and Reinhold, 1999), recent work by Reinfelder *et al.* (2000) suggests the operation of a C_4 photosynthetic pathway in the marine diatom *Thalassiosira weissflogii* when grown under CO_2 or Zn-limiting conditions. According to Reinfelder *et al.* (2000), carbon acquisition in *T. weissflogii* is facilitated by the enzyme phosphoenolpyruvate carboxylase (PEPCase), which catalyses the carboxylation of HCO_3^- to generate the C_4-compound malate in the cytoplasm. This is followed by transport of malate into the chloroplasts and release of CO_2 at the site of RUBISCO carboxylation. For the same diatom species, *T. weissflogii*, Burkhardt *et al.* (2001) found evidence for a highly efficient CCM involving simultaneous uptake of CO_2 and HCO_3^-. But why would this diatom have two means (CCM and C_4 pathway) to the same end? A possible advantage in operating a C_4 pathway may be in the temporal storage of carbon in malate, providing a means to intermittently supplement the flow of carbon to the site of RUBISCO carboxylation (Riebesell, 2000). The relevance of C_4 photosynthesis in marine phytoplankton under natural conditions and its contribution to marine carbon fixation remains to be seen.

CO_2 versus HCO_3^- utilisation

The fundamental question of which inorganic carbon species is predominantly used by algae, an issue which has preoccupied Steemann Nielsen throughout his career (Steemann Nielsen, 1947, 1960, 1963, 1975), is still under investigation today. HCO_3^- utilisation occurs through active HCO_3^- uptake and/or the extracellular conversion of HCO_3^- to CO_2 by the enzyme carbonic anhydrase (CA). To maintain charge balance in the cell during HCO_3^- uptake requires that equimolar amounts of protons are taken up or hydroxyl ions are extruded. In either case the carbonate chemistry in the DBL is shifted towards higher pH values, thereby diminishing the concentration of CO_2 and increasing that of CO_3^{2-} in the microenvironment of the cell (Wolf-Gladrow and Riebesell, 1997). Similar changes in the carbonate chemistry occur when CO_2 is taken up by the cell. To re-establish chemical equilibrium, CO_2 uptake is followed by (spontaneous or CA-catalysed) conversion of HCO_3^- to CO_2 within the DBL, which also releases hydroxyl ions (or consumes protons) and shifts the carbonate system to higher pH values, higher CO_3^{2-} and lower CO_2 concentrations. In both cases, sea water alkalinity remains constant. Thus, measurements of bulk sea water parameters can not be used to decipher which inorganic carbon species is taken up. This principle also applies to other nutrients such as phosphate and silicate. On the other hand, the uptake of nitrate, ammonium, or urea *can* be discerned from changes in bulk water alkalinity (Goldman and Brewer, 1980).

In principle, maintaining charge balance during HCO_3^- uptake could be achieved through co-transport of other ions, e.g. other nutrient species with a positive charge. This is ruled out, however, because most major nutrient species are anions (see Table 5.1; here NH_4^+ is an exception). Another possibility would be a co-transport with salt ions such as Na^+ or Cl^-. A co-transport of HCO_3^- and Cl^- (antiport) is well-known for erythrocytes. However, in the case of erythrocytes HCO_3^- is not a nutrient, and co-transport of HCO_3^- and Cl^- goes both ways. Over time the net transport is zero and Cl^- is neither accumulated nor depleted in the cells. The situation is different in algae. The net accumulation of carbon in the cell would lead to a total depletion of intercellular Cl^-. This mechanism is therefore only feasible over short time-scales.

Various methods have been employed to distinguish between CO_2 and HCO_3^- uptake. The isotope disequilibrium method takes advantage of the slow conversion between HCO_3^- and CO_2. After the addition of isotopically labelled carbon to the growth medium, either in the form of $^{14}CO_2$ or $H^{14}CO_3^-$, it takes about 60 s before all carbon species are in isotopic equilibrium (Espie and Colman, 1986). Short-term (10–30 s) experiments therefore allow the distinction between $^{14}CO_2$ and $H^{14}CO_3^-$ uptake. An essential condition for this method is that conversion between HCO_3^- and CO_2 external to the algal cell is purely spontaneous, i.e. not catalysed by CA. Thus, for species with extracellular CA activity, application of membrane impermeable CA inhibitors is indispensable. Application of this method by Sikes *et al.* (1980) on two strains of coccolithophorids and by Korb *et al.* (1997) on three species of marine diatoms demonstrated uptake of both CO_2 and HCO_3^- in these species. Using a slight modification of this approach, Elzenga *et al.* (2000) measured the exclusive utilisation of CO_2 in four species of marine phytoplankton and HCO_3^- use in three others. A surprising discovery of Elzenga and co-workers was the use of different carbon sources by two species of the same genus (*Thalassiosira punctigera* and *T. pseudonana*).

The proportion in which CO_2 and HCO_3^- are taken up and the extent to which carbon uptake is affected by changes in CO_2 supply, however, are still poorly quantified in marine microalgae. Evidence for HCO_3^- utilisation is often provided by methods which do not differentiate between CA-catalysed HCO_3^- use (with subsequent CO_2 uptake) and direct HCO_3^- uptake across the plasmalemma (Burns and Beardall 1987; Korb *et al.*, 1997). Other studies infer direct HCO_3^- transport from pH drift experiments (Tortell *et al.*, 1997) and the inhibition of light-dependent carbon utilisation by 4'4'-diisothiocyanatostilbene-2,2-disulfonic acid (DIDS), which is suggested to inhibit anion exchange processes (Nimer *et al.*, 1997). However, no quantitative estimates of carbon fluxes can be derived from these approaches. Separate quantification of CO_2 and HCO_3^- uptake, on the other hand, is usually based on measurements under conditions at which photosynthetic carbon fluxes were not at steady state (Colman and Rotatore, 1995; Rotatore *et al.*, 1995).

A method proposed by Badger *et al.* (1994) introduces a technique which overcomes the above-mentioned shortcomings. Using the chemical disequilibrium between CO_2 and HCO_3^- during light-dependent carbon uptake, this mass spectrometric procedure allows to differentiate between CO_2 and HCO_3^- fluxes across the plasmalemma and to quantify these fluxes during steady-state photosynthesis. While this method has been successfully applied to different strains of the cyanobacterium *Synechococcus* and several freshwater microalgae (Badger, *et al.* 1994; Tchernov *et al.*, 1997; Sültemeyer *et al.*, 1998), it has only recently been used to examine representative species of dominant eukaryotic marine phytoplankton taxa (Burkhardt *et al.*, 2001). These studies indicate large differences in the efficiency of inorganic carbon acquisition mechanisms and in the response to changes in CO_2 supply between different taxonomic groups (diatoms, coccolithophorids, and flagelates) as well as between closely related taxa.

The use of extracellular carbonic anhydrase

Carbonic anhydrase, a zinc-metallo enzyme, is ubiquitously used in plant and animal cells to catalyse the conversion between HCO_3^- and CO_2. Carbonic anhydrase activity is also measured on the outer surface of most phytoplankton species (Sültemeyer, 1998), where it is thought to speed up the rate of conversion of HCO_3^- to CO_2. An increase in extracellular CA activity in response to decreasing bulk water CO_2 concentrations is observed in many microalgae (Nimer *et al.*, 1997; Elzenga *et al.*, 2000; Burkhardt *et al.*, 2001) and suggests a vital role of extracellular CA in inorganic carbon acquisition. It should be kept in mind, however, that CA activity merely enhances the re-establishment of equilibrium in the carbonate system. Since deviations from equilibrium in the DBL of small phytoplankton cells are minor (see Wolf-Gladrow and Riebesell, 1997), it is not at all clear how small cells benefit from extracellular CA. For example, for a cell with radius $a = 2.5\,\mu m$ an enhancement of HCO_3^- to CO_2 conversion in the DBL of 25 times the non-catalysed reaction (mean enhancement obtained by Elzenga *et al.* (2000)) increases the cell surface CO_2 concentration by $0.04\,\mu mol\ kg^{-1}$ over that of a cell with no external CA activity (assuming a growth rate of $1\,d^{-1}$, the cell volume to carbon content relationship of Montagnes *et al.* (1994), and a net carbon uptake given by the product of growth rate and cell carbon quota). This number increases to $0.25\,\mu mol\ kg^{-1}$ for a cell with radius $a = 5\,\mu m$.

However, even for larger phytoplankton, the benefit of expressing high extracellular CA activity under low CO_2 concentrations is small. At a CO_2 concentration of $1\,\mu mol\ kg^{-1}$, the level at or below which external CA activity reaches its maximum (Elzenga *et al.*, 2000), the non-catalysed CO_2 flux to a cell with radius $a = 25\,\mu m$ amounts to 7% of its value at $15\,\mu mol\ kg^{-1}$ bulk CO_2 concentration. With a 25 times CA-enhanced HCO_3^- conversion, this value increases to 21% of the flux in air-equilibrated sea water. Here we assume that the cell is able to lower the cell surface CO_2 concentration to zero, i.e. maintaining maximum disequilibrium in the DBL. The beneficial effect of extracellular CA is even lower when deviations from equilibrium are smaller. This calculation indicates that CA activities of this magnitude are insufficient to compensate for CO_2 diffusion limitation under highly reduced CO_2 concentrations (note that a cell of this size growing at a rate of $1\,d^{-1}$ is diffusion-limited at concentrations below around $12\,\mu mol\,kg^{-1}$ CO_2; see Fig. 5.3). Why extracellular CA activity is so widespread in microalgae and closely attuned to CO_2 availability remains an enigma.

Carbon isotope fractionation

During carbon acquisition, and particularly during carbon fixation, the lighter ^{12}C is favoured over the heavier ^{13}C and ^{14}C isotopes. This was accounted for by Steemann Nielsen in the development of the ^{14}C technique. Based on work by Urey (1948) and Anderson and Libby (1951), Steemann Nielsen (1952) suggested that 'it is reasonable to expect that the assimilation of $^{14}CO_2$ during photosynthesis is about 6 per cent slower than that of $^{12}CO_2$'. In a later publication he gives a value of 5% (Steemann Nielsen, 1975). Extensive work on ^{13}C isotope fractionation, particularly during the past ten years, has shown that:

(1) isotope fractionation strongly varies with algal growth conditions such as CO_2 and nutrient availability, light intensity, and diurnal light cycles, and
(2) large species-specific differences in isotope fractionation are related to phytoplankton cell size, cellular carbon contents, carbon assimilation rate, and carbon acquisition mechanisms.

Calculating isotope fractionation (ε_p) from the isotopic composition of the primary produced organic matter $(\delta^{13}c_{POC})$ and the dissolved CO_2 $(\delta^{13}c_{CO_2})$ according to Freeman and Hayes (1992):

$$\varepsilon_p = \frac{\delta^{13}c_{CO_2} - \delta^{13}c_{POC}}{1 + \dfrac{\delta^{13}c_{POC}}{1000}} \tag{5.27}$$

yields ε_p values in the ocean ranging between 8 and 20‰ (Goericke and Fry, 1994). Considering that the isotope discrimination between ^{12}C and ^{14}C is twice as high as between ^{12}C and ^{13}C (Craig, 1954), discrimination against ^{14}C would be in the range 1.5–4%. Part of the reason that the values are lower than expected by Steemann Nielsen (1952, 1975) may be due to the effect of HCO_3^- utilisation on ε_p. Since HCO_3^- is isotopically enriched in ^{14}C by about 2% relative to CO_2, uptake of HCO_3^- yields a 2% higher rate of ^{14}C incorporation relative to CO_2 uptake. Note that due to isotope fractionation during CA-catalysed conversion between HCO_3^-

and CO_2 (Riebesell and Wolf-Gladrow, 1995), this does not apply if HCO_3^- is utilised through extracellular conversion of HCO_3^- to CO_2.

Based on a model by Farquhar *et al.* (1982), Sharkey and Berry (1985) argued that photosynthetic isotope fractionation is ultimately determined by the processes of carbon acquisition and the source(s) of inorganic carbon taken up by the phytoplankton. Knowing the fluxes of CO_2 and HCO_3^- in and out of the cell and the isotope fractionation associated both with these fluxes and due to carboxylation, it is possible to predict the isotopic signal incorporated during photosynthetic carbon fixation. The reverse, i.e. the deduction of cellular carbon fluxes from the carbon isotope fractionation of phytopankton, is feasible only if at least one of the parameters – (a) the ratio of carbon influx to efflux, or (b) the contribution of HCO_3^- uptake to total carbon acquisition – can be specified.

Accounting for the relative proportion of CO_2 and HCO_3^- fluxes, Burkhardt *et al.* (1999) described isotope fractionation (ε_p) by:

$$\varepsilon_p = a(\varepsilon_3 + \varepsilon_4) + (1 - a)\varepsilon_1 + (\varepsilon_2 - \varepsilon_{-1})\frac{F_{-1}}{F_1 + F_4} \tag{5.28}$$

where F_1 and F_{-1} are CO_2 fluxes in and out of the cell, respectively; F_4 is the flux of HCO_3^- into the cell; and a is the fractional contribution of HCO_3^- flux to gross total carbon uptake ($F_t = F_1 + F_4$) into the cell. Enzymatic fractionation during C fixation is denoted by ε_2; equilibrium fractionation between CO_2 and HCO_3^- is denoted by ε_3. Fractionation associated with the influx and outflux of CO_2 ($\varepsilon_1, \varepsilon_{-1}$) and with the uptake of HCO_3^- (ε_4) is often assumed to be small (i.e. $\sim 1‰$; Raven, 1997; Keller and Morel, 1999) compared to the isotope fractionation of around 30‰ of the main carboxylating enzyme RUBISCO (Guy *et al.*, 1993), and of the equilibrium fractionation between CO_2 and HCO_3^- of ca. 10‰ (Mook *et al.*, 1974). If the small fractionation during inorganic carbon fluxes is ignored, equation (5.28) reduces to:

$$\varepsilon_p = a\varepsilon_3 + \varepsilon_2\frac{F_{-1}}{F_t} \tag{5.29}$$

The term F_{-1}/F_t denotes the diffusive loss of CO_2 from the cell back to the medium relative to gross total carbon uptake and is often termed 'CO$_2$ leakage'. In this model HCO_3^- efflux is assumed to be negligible.

Based on the consideration that carbon flux into the cell (F_t) may be related to external CO_2 concentration, the carbon isotope composition of sedimentary organic matter ($\delta^{13}C_{org}$) has been proposed as a proxy for ancient CO_2 concentrations (Freeman and Hayes, 1992). Support for this approach has come from theoretical considerations and experimental evidence (Laws *et al.*, 1995; Rau *et al.*, 1996). Since ε_p is determined by the CO_2 leakage, which is equal to the difference between total carbon flux into the cell and net fixation, for a given carbon influx ε_p should be proportional to net carbon fixation. As the latter is equivalent to the carbon specific growth rate, ε_p may also serve as a proxy for the phytoplankton growth rate (Laws *et al.*, 1997).

Thus far, attempts to determine the relationship of ε_p with CO_2 concentration and growth rate have yielded equivocal results. Species-specific differences in isotope fractionation (Popp *et al.*, 1998; Burkhardt *et al.*, 1999) preclude the use of bulk organic matter $\delta^{13}C$ in ε_p applications. In addition, factors other than CO_2 and growth rate, such as light intensity, the light–dark cycle, and the growth-rate limiting resource, affect ε_p and severely complicate the interpretation of carbon isotope data (Thompson and Calvert, 1995; Riebesell *et al.*, 2000; Rost *et al.*, 2002). Moreover, the validity of present models of carbon isotope fractionation may be compromised by the fact that cell compartmentation is not accounted for. Measurements of unidirectional carbon fluxes into and out of the cell by means of membrane inlet mass spectrometry, combined with measurements of ε_p over a wide range of environmental conditions, would provide the ultimate test for our models and may help to clarify some of the uncertainties in phytoplankton isotope fractionation.

5.5   Modelling nutrient uptake

Various theories on the uptake of nutrients by phytoplankton have illustrated the important role of nutrients in determining phytoplankton species distribution and succession and maintaining phytoplankton species diversity. In 1840, Liebig hypothesised that plants are limited by the one nutrient with the lowest concentration relative to the plants' requirement for growth (Liebig, 1840). Models of nutrient uptake established to understand species interaction were henceforth built on the assumption of a single limiting nutrient. The simplest description of nutrient uptake is given by the Michaelis–Menten function:

$$U = \frac{U_{max}R}{K_R + R} \qquad (5.30)$$

proposed by Michaelis and Menten (1913) as an empirical description for the representation of enzyme kinetics. Later it was shown that this function can be derived from a certain reaction scheme, with U_{max} being the maximum uptake rate, R the extracellular concentration of the rate-limiting nutrient ('resource'), and K_R the half-saturation constant for uptake (also called the Michaelis constant). Aksnes and Egge (1991) have given an interpretation of the Michaelis–Menten uptake kinetics in terms of number and area of uptake sites and handling time of nutrient molecules. Although this interpretation may provide some insights into the regulation of cellular nutrient uptake, currently there are no methods available for measuring these parameters. Hence, based on this approach, no estimates of U_{max} and K_R can be derived.

In 1942, Monod proposed a function for the growth of micro-organisms which is of the same form as the Michaelis–Menten function (Monod, 1942):

$$\mu = \frac{\mu_{max}R}{K_\mu + R} \tag{5.31}$$

However, whereas nutrient uptake can probably be adequately simulated by a simple reaction scheme, this is not the case for the growth of whole cells. Substitution of growth for nutrient uptake is valid only in the steady state of continuous cultures, as was the case in Monod's organic carbon-limited bacterial cultures. Application to natural populations shows that growth and nutrient uptake are generally uncoupled. Nevertheless, the rectangular hyperbolic function used in the Michaelis–Menten and the Monod model, with:

$$f(x) = \frac{f_{max}x}{k + x} \tag{5.32}$$

captures typical features of nutrient uptake and growth kinetics:

(1) $f(x)$ vanishes at $x = 0$,
(2) $f(x)$ increases monotonously with increasing x,
(3) increase of $f(x)$ is maximal at small values of x, and
(4) at very high values of x the function $f(x)$ goes into saturation (because it is limited by factors other than x).

The hyperbolic function is sometimes modified in two ways in order to obtain better fits to empirical data. The function:

$$g(x) = \left(\frac{g_{max}x}{k_g + x}\right)^n \qquad (n > 1) \tag{5.33}$$

has the same properties listed above for the hyperbolic function, whereas the Hill function:

$$h(x) = h_{max}\frac{x^m}{k_h^m + x^m} \qquad (m > 1) \tag{5.34}$$

shows maximum increase at $x_c > 0$ (Fig. 5.4).

Cell quota models

A major step forward in modelling nutrient uptake and growth came with the proposition of cell quota models by Droop (1968) and Caperon (1968). These authors hypothesised that growth rate is most likely related to the cellular nutrient

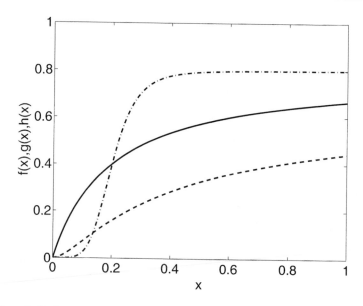

Fig. 5.4 Hyperbolic function $f(x)$ (solid line, $f_{max} = 0.8$, $k = 0.2$) and modifications $g(x)$ (dashed line, $g_{max} = 0.8$, $k_g = 0.2$, $n = 2$) and $h(x)$ (dash-dotted line, $h_{max} = 0.8$, $k_h = 0.2$, $m = 5$).

content, termed cell quota Q. In this case, the nutrient history of a phytoplankton becomes important. According to this model, the specific growth rate can be described as:

$$\mu = \mu_{max} \frac{Q - Q_0}{k_Q + Q - Q_0} \tag{5.35}$$

in which μ_{max} (d^{-1}) is the maximum growth rate, k_Q (mol cell^{-1}) is a half saturation constant with respect to $Q - Q_0$, and Q_0 (mol cell^{-1}) is the subsistence quota which can be interpreted as an non-reactive or structural component. The cell quota Q is not a simple function of the concentration of the limiting nutrient in sea water, R (mol kg^{-1}; resource); however, for a batch culture it can be calculated from the following system of differential equations (Davidson and Gurney, 1999):

$$\frac{dR}{dt} = -UA \qquad \text{(depletion of nutrient)} \tag{5.36}$$

$$\frac{dQ}{dt} = U - \mu Q \qquad \text{(growth of single cell)} \tag{5.37}$$

$$\frac{dA}{dt} = -\mu A \qquad \text{(growth of population)} \tag{5.38}$$

in which A is the population density (cells kg^{-1}). The nutrient uptake rate, U (mol cell^{-1} s^{-1}), depends on the limiting resource in a Michaelis–Menten fashion (equation (5.30)) or on the limiting resource and the cell quota according to:

$$U = U_{max} \frac{R}{k_R + R} \frac{Q_{max} - Q}{Q_{max} - Q_0} \tag{5.39}$$

in which Q_{max} is the maximum cell quota (Lehman *et al.*, 1975). Davidson and Gurney (1999) found this representation of uptake rate U (equation (5.39)) to provide an improved fit to data obtained from batch cultures of N-limited *Thalassiosira pseudonana*. Fig. 5.5 shows results of an integration of the above system of differential equations using model parameters from Davidson and Gurney (1999). The cell quota varies due to uptake of nutrients combined with cell growth and division. Cell division continues even after the limiting nutrient in the medium has been depleted; this is in contrast to growth according to the Monod model. The parameters U_{max}, k_R, μ_{max}, k_Q, Q_0 and Q_{max} of the cell quota model must be derived from data as obtained; for example, in time series experiments monitoring the temporal development of R and A. Although cell quota models provide a more realistic representation of nutrient-dependent growth, their greater success in fitting data of population development compared with simple Monod functions may be partly due to the higher number of model parameters.

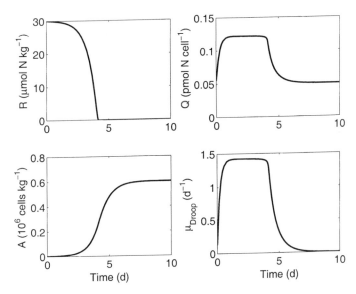

Fig. 5.5 Simulation of algal growth in a batch culture with the cell quota model (numerical integration of the system of differential equations (equations (5.36) to (5.38)); model parameters for the diatom *Thalassiosira pseudononana* under high light conditions as reported by Davidson and Gurney (1999). Cell division continues, accompanied by a decrease in the cell quota, even after the resource (nitrate) in seawater is depleted (day 4). This is in contrast to growth according to the Monod function in which cell division ends after the limiting resource is depleted.

Co-limitation

The availability of nutrients may influence the species composition and succession of phytoplankton in an obvious way (e.g. no growth of diatoms when Si is depleted) but also in more subtle ways (e.g. the dominance of different diatom species depending on the N:Si ratio in the growth medium; Sommer, 1986). While cases of one limiting nutrient can often be described reasonably well by cell quota models, the situation is more complicated if two or more nutrients are simultaneously in short supply. No simple mathematical description has been derived which would be applicable for all nutrients. In fact, because of different interdependencies of various nutrients in phytoplankton, no such description can be expected. Droop (1983) distinguishes between *interactive nutrients*, where both nutrients contribute to the reduction of the growth rate, and *non-interactive nutrients*, where one or the other nutrient leads to a reduction of maximum growth rate. In the case of non-interactive nutrients and Monod–type growth dependencies for each individual resource, this can be described by the minimum of the two growth rates according to:

$$\mu(R_1, R_2) = \min\left(\mu_{max,1}\frac{R_1}{k_1 + R_1}, \mu_{max,2}\frac{R_2}{k_2 + R_2}\right) \tag{5.40}$$

This description closely adheres to Liebig's law of the minimum (Liebig, 1840). The simplest approach for describing the growth dependence on two interactive nutrients is to combine models derived for the limitation by a single nutrient, such as by the product of two Monod functions:

$$\mu(R_1, R_2) = \mu_{max}\frac{R_1}{k_1 + R_1}\frac{R_2}{k_2 + R_2} \tag{5.41}$$

This approach is frequently used in global models of biogeochemical cycles to simulate co-limitation by two nutrients and, for the time being, may provide an adequate approximation for this particular purpose. However, it gives no satisfactory description of the temporal development of species populations observed in the laboratory or in the field.

A somewhat more elaborate description of co-limitation by different resources would be the multiplication of two cell quota functions:

$$\mu = \mu_{max}\frac{Q_1 - Q_{0,1}}{k_{Q,1} + Q_1 - Q_{0,1}}\frac{Q_2 - Q_{0,2}}{k_{Q,2} + Q_2 - Q_{0,2}} \tag{5.42}$$

However, Davidson and Gurney (1999) found that multiplicative cell quota models are unable to simulate batch culture data of *Thalassiosira pseudonana* grown under N- and Si-limitation. They proposed a modified model that takes into account the specific interaction between nitrogen and silicon as essential nutrients in diatoms. The discussion of this and other more complex models on nutrient uptake in phytoplankton (Lancelot *et al.*, 2000) is beyond the scope of this chapter.

Despite the potential importance of nutrient co-limitation in natural phyto-plankton populations, its quantitative description based on models beyond Monod formulations is still in its infancy. Among the various algal nutrients (C, N, P, Si, trace elements and vitamins), only a small number of nutrient pairs have been classified as interactive or non-interactive. But even in cases where interactions between different nutrients have been suggested, as for example in the case of C and Zn (Morel *et al.*, 1994), the underlying mechanisms are hardly understood to date.

5.6   Future perspectives

The large variety of life forms of phytoplankton was viewed by Margalef (1978) as adaptive strategies to varied nutrient and flow environments, with phytoplankton distribution and succession being under the control of the physical environment. Obviously, the ability of a species to compensate for changes in the nutrient environment while allowing minimal disruption of its own biosynthetic processes can determine its success in a fluctuating environment. Over the past decades it has become increasingly evident that the processes determining phytoplankton species distribution and succession are multifactorial, extending from short-term acclima-tion to a highly fluctuating light environment, to the reduction of grazing pressure and the development of specific life cycle strategies. This highlights the need to understand nutrient utilisation by phytoplankton in a much larger context.

Although nutrient supply and utilisation and their effect on phytoplankton growth have been studied for over a hundred years (Baar, 1994), our knowledge in this field is still rather limited. Significant progress in this field can be expected within the next decade for at least two reasons:

(1)   In an effort to understand and predict possible responses of the marine biota to global environmental change, new questions are being posed and new approaches are being taken. This includes the need to understand the role of phytoplankton functional groups in biogeochemical cycles of C, N, P and Si. Any progress in this area crucially depends on the ability to integrate the processes determining phytoplankton resource utilisation into higher-order trophic interactions.

(2)   Improved technology is becoming available on all levels of scientific endea-vour, extending from molecular techniques determining the cell's genetic and biochemical capacities to the incorporation of a more realistic representation of biological processes into ocean-wide general circulation biogeochemical models.

Our predictive capabilities of 'Biogeochemical controls and feedbacks on ocean primary production' (Falkowski *et al.*, 1998) will partly depend on a realistic representation of phytoplankton resource utilisation. Ultimately, improvements in our understanding of the role of phytoplankton in the marine environment vitally hinge upon our ability to take an integrated-systems approach.

Acknowledgements

We wish to thank U. Passow, S. Thoms, and C. Völker for critically reading the manuscript. Special thanks go to M. Botros for helpful comments and editorial revision of the manuscript.

References

Aksnes, D.L. & Egge, J.K. (1991) A theoretical model for nutrient uptake in phytoplankton. *Marine Ecology Progress Series* **70**, 65–72.

Anderson, E.C. & Libby, W.F. (1951) World-wide distribution of natural radiocarbon. *The Physical Review* **81**, 64.

Baar, H.J.W., de (1994) von Liebig's law of the minimum and plankton ecology (1899–1991). *Progress in Oceanography* **33**, 347–86.

Badger, M.R., Palmqvist, K. & Yu, J.-W. (1994) Measurement of CO_2 and HCO_3^- fluxes in cyanobacteria and microalgae during steady-state photosynthesis. *Physiologia Plantarum* **90**, 529–36.

Badger, M.R., Andrews, T.J., Whitney, S.M., *et al.* (1998) The diversity and coevolution of Rubisco, plastids, pyrenoids and chloroplast-based CO_2 concentrating mechanisms in algae. *Canadian Journal of Botany* **76**, 1052–71.

Burkhardt, S., Amoroso, G., Riebesell, U. & Sültemeyer, D. (2001) CO_2 and HCO_3^- uptake in marine diatoms acclimated to different CO_2 concentrations. *Limnology and Oceanography* **46**, 1378–91.

Burkhardt, S., Riebesell, U. & Zondervan, I. (1999) Effects of growth rate, CO_2 concentration, and cell size on the stable carbon isotope fractionation in marine phytoplankton. *Geochimica et Cosmochimica Acta* **63**, 3729–41.

Burns, B.D. & Beardall, J. (1987) Utilization of inorganic carbon by marine microalgae. *Journal of Experimental Marine Biology and Ecology* **107**, 75–86.

Caperon, J. (1968) Population growth response of *Isochrysis galbana* to nitrate variation at limiting concentrations. *Ecology* **49**, 866–72.

Chisholm, S. (1992) Phytoplankton size. In: *Primary Productivity and Biogeochemical Cycles in the Sea* (eds P.G. Falkowski & A.D. Woodhead), pp. 213–327. Plenum Press, New York.

Colman, B. & Rotatore, C. (1995) Photosynthetic inorganic carbon uptake and accumulation in two marine diatoms. *Plant and Cell Environment* **18**, 919–24.

Craig, H. (1954) Carbon 13 in plants and the relationship between carbon 13 and carbon 14 variations in nature. *Journal of Geology* **62**, 115–52.

Davidson, K. & Gurney, W.S.C. (1999) An investigation of non-steady-state algal growth. II. Mathematical modelling of co-nutrient-limited algal growth. *Journal of Plankton Research* **21**, 839–58.

Droop, M.R. (1968) Vitamin B_{12} and marine ecology. IV. The kinetics of uptake, growth and inhibition in *Monochrysis lutheri*. *Journal of the Marine Biological Association of the United Kingdom* **48**, 689–733.

Droop, M.R. (1983) 25 years of algal growth kinetics. *Botanica Marina* **26**, 99–112.

Elzenga, J.T., Prins, H.B.A. & Stefels, J. (2000) The role of extracellular carbonic anhydrase activity in inorganic carbon utilization of *Phaeocystis globosa* (Prymnesiophyceae): A comparison with other marine algae using the isotopic disequilibrium technique. *Limnology and Oceanography* **45**, 372–80.

Eppley, R.W. & Koeve, W. (1990) Nitrate use by plankton in the eastern subtropical North Atlantic, March–April 1989. *Limnology and Oceanography* **35**, 1781–8.

Espie, G.S. & Colman, B. (1986) Inorganic carbon uptake during photosynthesis. I. A theoretical analysis using the isotope disequilibrium technique. *Plant Physiology* **80**, 863–9.

Falkowski, P.G. (1997) Evolution of the nitrogen cycle and its influence on the biological sequestration of CO_2 in the ocean. *Nature* **387**, 272–5.

Falkowski, P.G., Barber, R.T. & Smetacek, V. (1998) Biogeochemical controls and feedbacks on ocean primary production. *Science* **281**, 200–206.

Farquhar, M.R., O'Leary, M.H. & Berry, J.A. (1982) On the relationship between carbon isotope discrimination and the intercellular carbon dioxide concentration in leaves. *Australian Journal of Plant Physiology* **9**, 121–37.

Freeman, K.H. & Hayes, J.M. (1992) Fractionation of carbon isotopes by phytoplankton and estimates of ancient CO_2 levels. *Global Biogeochemical Cycles* **6**, 185–98.

Gavis, J. & Ferguson, J.F. (1975) Kinetics of carbon dioxide uptake by phytoplankton at high pH. *Limnology and Oceanography* **20**, 211–21.

Goericke, R. & Fry, B. (1994) Variations of marine plankton $\delta^{13}C$ with latitude, temperature, and dissolved CO_2 in the world ocean. *Global Biogeochemical Cycles* **8**, 85–90.

Goldman, J.C. & Brewer, P.G. (1980) Effect of nitrogen source and growth rate on phytoplankton mediated changes in alkalinity. *Limnology and Oceanography* **25**, 352–7.

Guy, R.D., Fogel, M.L. & Berry, J.A. (1993) Photosynthetic fractionation of the stable isotopes of oxygen and carbon. *Plant Physiology* **101**, 37–47.

Hudson, R.J. & Morel, F.M.M. (1991) Trace metal transport by marine microorganisms: Implications of metal coordination kinetics. *Deep-Sea Research I* **40**, 129–50.

Kaplan, A. & Reinhold, L. (1999) CO_2 concentrating mechanisms in photosynthetic microorganisms. *Annual Review of Plant Physiology and Plant Molecular Biology* **50**, 539–70.

Karp-Boss, L., Boss, E. & Jumars, P.A. (1996) Nutrient fluxes to planktonic osmotrophs in the presence of fluid motion. *Oceangraphy and Marine Biology* **34**, 71–107.

Karp-Boss, L. & Jumars, P.A. (1998) Motion of diatom chains in steady shear flow. *Limnology and Oceanography* **43**, 1767–73.

Keller, K. & Morel, F.F.M. (1999) A model of carbon isotopic fractionation and active carbon uptake in phytoplankton. *Marine Ecology Progress Series* **182**, 295–8.

Kiørboe, T. (1993) Turbulence, phytoplankton cell size, and the structure of pelagic food webs. *Advances in Marine Biology* **29**, 1–72.

Korb, R.E., Saville, P.J., Johnston, A.M. & Raven, J.A. (1997) Sources of inorganic carbon for photosynthesis by three species of marine diatoms. *Journal of Phycology* **33**, 433–40.

Lancelot, C., Hannon, E., Becquevort, C., Veth, C. & de Baar, H.J.W. (2000) Modeling phytoplankton blooms and carbon export in the Southern Ocean: dominant controls by light and iron in the Atlantic sector in Austral spring 1992. *Deep-Sea Research I* **47**, 1621–62.

Laws, E.A., Popp, B.N. & Bidigare, R.R. (1997) Effect of growth rate and CO_2 concentration on carbon isotopic fractionation by the marine diatom *Phaeodactylum tricornutum*. *Limnology and Oceanography* **42**, 1552–60.

Laws, E.A., Popp, B.N., Bidigare, R.R., Kennicutt, M.C. & Macko, S.A. (1995) Dependence of phytoplankton carbon isotopic composition on growth rate and $[CO_2]$aq: Theoretical considerations and experimental results. *Geochimica et Cosmochimica Acta* **59**, 1131–8.

Lazier, J.R.N. & Mann, K.H. (1989) Turbulence and the diffusive layers around small organisms. *Deep-Sea Research I* **36**, 1721–33.

Lehman, J.T., Botkin, D.B. & Likens, G.E. (1975) The assumptions and rationales of a computer model of plankton population dynamics. *Limnology and Oceanography* **20**, 343–64.

Liebig, J., von (1840) *Organic Chemistry and its Application to Agriculture and Physiology*, 1st edn, 387 pp. Taylor and Walton, London.

Logan, B.E. & Alldredge, A.L. (1989) Potential for increased nutrient uptake by flocculating diatoms. *Marine Biology* **101**, 443–50.

Margalef, R. (1978) Life-forms of phytoplankton as survival alternatives in an unstable environment. *Oceanologica Acta* **1**, 439–509.

Michaelis, L. & Menten, M.M.L. (1913) Die Kinetik der Invertinwirkung. *Biochemische Zeitschrift* **49**, 333–69.

Monod, J. (1942) *Réchèrche sur la Croissance des Cultures Bactériennes*. Hermann, Paris.

Montagnes, D.J.S., Berges, J.A., Harrison, P.J. & Taylor, F.J.R. (1994) Estimating carbon, nitrogen, protein, and chlorophyll *a* from volume in marine phytoplankton. *Limnology and Oceanography* **39**, 1044–60.

Mook, W.G., Bommerson, J.C. & Staverman, W.H. (1974) Carbon isotope fractionation between dissolved bicarbonate and gaseous carbon dioxide. *Earth and Planetary Science Letters* **22**, 169–76.

Morel, F.M.M., Hudson, R.J. & Price, N.M. (1991) Trace metal limitation in the sea. In: What controls phytoplankton production in nutrient rich areas of the open sea? (eds S.W. Chisholm & F.M.M. Morel). *Limnology and Oceanography* **36**, 1742–55.

Morel, F.M.M., Reinfelder, J.R., Roberts, S.B., Chamberlain, C.P., Lee, J.G. & Yee, D. (1994) Zinc and carbon co-limitation of marine phytoplankton. *Nature* **309**, 740–42.

Munk, W.H. & Riley, G.A. (1952) Absorption of nutrients by aquatic plants. *Journal of Marine Research* **11**, 215–40.

Nimer, N.A., Iglesias-Rodriguez, M.D. & Merrett, M.J. (1997) Bicarbonate utilization by marine phytoplankton species. *Journal of Phycology* **33**, 625–31.

Paasche, E. (1964) A tracer study of the inorganic carbon uptake during coccolith formation and photosynthesis in the coccolithophorid *Coccolithius huxleyi*. *Physiologia Plantarum*, Suppl. III, 1–82.

Pahlow, M., Riebesell, U. & Wolf-Gladrow, D.A. (1997) Impact of cell shape and chain formation on nutrient acquisition by marine diatoms. *Limnology and Oceanography* **42**, 1660–72.

Pasciak, W.J. & Gavis, J. (1974) Transport limitation of nutrient uptake in phytoplankton. *Limnology and Oceanography* **19**, 881–9.

Pasciak, W.J. & Gavis, J. (1975) Transport limited nutrient uptake rates in *Ditylum brightwellii*. *Limnology and Oceanography* **20**, 604–17.

Ploug, H., Stolte, W. & Jørgensen, B.B. (1999) Diffusive boundary layers of the colony-forming plankton alga *Phaeocystis* sp. – implications for nutrient uptake and cellular growth. *Limnology and Oceanography* **44**, 1959–67.

Popp, B.N., Laws, E.A., Bidigare, R.R., Dore, J.E., Hanson, K.L. & Wakeham, S.G. (1998) Effect of phytoplankton cell geometry on carbon isotopic fractionation. *Geochimica et Cosmochimica Acta* **62**, 69–77.

Rau, G.H., Riebesell, U. & Wolf-Gladrow, D. (1996) A model of photosynthetic ^{13}C fractionation by marine phytoplankton based on diffusive molecular CO_2 uptake. *Marine Ecology Progress Series* **133**, 275–85.

Raven, J.A. (1980) Nutrient transport in microalgae. *Advances in Microbial Physiology* **21**, 47–226.

Raven, J.A. (1984) *Energetics and Transport in Aquatic Plants*. Alan R. Liss, Inc., New York.

Raven, J.A. (1997) Inorganic carbon acquisition by marine autotrophs. *Advances in Botanical Research* **27**, 85–209.

Redfield, A.C., Ketchum, B.H. & Richards, F.A. (1963) The influence of organisms on the composition of sea water. In: *The Sea*, Vol. 2 (ed. M.N. Hill), pp. 26–77. Interscience, New York.

Reinfelder, J.R., Kraepiel, A.M.L. & Morel, F.M.M. (2000) Unicellular C_4 photosynthesis in a marine diatom. *Nature* **407**, 996–9.

Richardson, L.F. (1922) *Weather Prediction by Numerical Process*, 236 pp. Cambridge University Press, London.

Riebesell, U. (2000). Carbon fix for a diatom. *Nature* **407**, 959–60.

Riebesell, U. & Wolf-Gladrow, D. (1995). Growth limits on phytoplankton. *Nature* **373**, 28.

Riebesell, U., Wolf-Gladrow, D. & Smetacek, V. (1993). Carbon dioxide limitation of marine phytoplankton growth rates. *Nature* **361**, 249–51.

Riebesell U., Dauelsberg, A., Burkhardt, S. & Kroon, B. (2000) Carbon isotope fractionation by a marine diatom: Dependence on the growth rate limiting resource. *Marine Ecology Progress Series* **193**, 295–303.

Rost, B., Zondervan, I. & Riebesell, U. (2002) Light-dependent carbon isotope fractionation in the coccolithophorid *Emiliania huxleyi*. *Limnology and Oceanography* **47** (in press).

Rotatore, C., Colman, B. & Kuzma, M. (1995) The active uptake of carbon dioxide by the marine diatoms *Phaeodactylum tricornutum* and *Cyclotella* sp. *Plant and Cell Environment* **18**, 913–18.

Sharkey, T.D. & Berry, J.A. (1985) Carbon isotope fractionation of algae as influenced by an inducible CO_2 concentrating mechanism. In: *Inorganic Carbon Uptake by Aquatic Photosynthetic Organisms* (eds W.J. Lucas & J.A. Berry), pp. 389–401. American Society of Plant Physiology, Rockville, MD.

Sikes, C.S., Roer, R.D. & Wilbur, K.M. (1980) Photosynthesis and coccolith formation: inorganic carbon sources and net inorganic reaction of deposition. *Limnology and Oceanography* **25**, 248–61.

Sommer, U. (1986) Nitrate- and silicate-competition among Antarctic phytoplankton. *Marine Biology* **91**, 345–51.

Steemann Nielsen, E. (1947) Photosynthesis of aquatic plants with special reference to the carbon-sources. *Dansk Botanisk Ark*iv **12**, 1–71.

Steemann Nielsen, E. (1952) The use of radio-active carbon (C^{14}) for measuring organic production in the sea. *Journal du Conseil Permanent International pour l'Exploration de la Mer* **18**, 117–40.

Steemann Nielsen, E. (1960) Uptake of CO_2 by the plant. *Encyclopedia of Plant Physiology* (new series), **5**, 70–84.

Steemann Nielsen, E. (1963) On bicarbonate utilization by marine phytoplankton in photosynthesis, with a note on carbamino carboxylic acids as a carbon source. *Physiologia Plantarum* **16**, 466–9.

Steemann Nielsen, E. (1975) *Marine Photosynthesis: With Special Emphasis on the Ecological Aspects*, 141 pp. Elsevier, Amsterdam.

Sültemeyer, D. (1998) Carbonic anhydrase in eukaryotic algae: characterization, regulation, and possible function during photosynthesis. *Canadian Journal of Botany* **76**, 962–72.

Sültemeyer, D., Klughammer, B., Badger, M.R. & Price, G.D. (1998) Fast induction of high-affinity HCO_3^- transport in cyanobacteria. *Plant Physiology* **116**, 183–92.

Tchernov, D., Hassidim, M., Luz, B., Sukenik, A., Reinhold, L. & Kaplan, A. (1997) Sustained net CO_2 evolution during photosynthesis by marine microorganisms. *Current Biology* **7**, 723–8.

Thompson, P.A. & Calvert, S.E. (1995) Carbon isotope fractionation by *Emiliania huxleyi*. *Limnology and Oceanography* **40**, 673–9.

Thoms, S., Pahlow, M. & Wolf-Gladrow, D.A. (2001) Model of the carbon concentrating mechanism in chloroplasts of eukaryotic algae. *Journal of Theoretical Biology* **208**, 295–313.

Tortell, P.D., Reinfelder, J.R. & Morel, F.M.M. (1997) Active bicarbonate uptake by diatoms. *Nature* **390**, 243–4.

Urey, H.C. (1948) Oxygen isotope in nature and in the laboratory. *Science* **108**, 489.

Villareal, T.A., Pilskaln, C., Brzezinski, M., Lipschultz, F. Dennett, M. & Gardener, G.B. (1999) Upward transport of oceanic nitrate by migrating diatom mats. *Nature* **397**, 423–5.

Völker, C. & Wolf-Gladrow, D.A. (1999) Physical limits on iron uptake mediated by siderophores or surface reduction. *Marine Chemistry* **65**, 227–44.

Wolf-Gladrow, D.A. & Riebesell, U. (1997) Diffusion and reactions in the vicinity of plankton: a refined model for inorganic carbon transport. *Marine Chemistry* **59**, 17–34.

Zeebe, R.E., Wolf-Gladrow, D.A. & Jansen, H. (1999). On the time required to establish chemical and isotopic equilibrium in the carbon dioxide system in seawater. *Marine Chemistry* **65**, 135–53.

Chapter 6
Variability of Plankton and Plankton Processes on the Mesoscale

Marlon R. Lewis

6.1 Introduction

'There are considerable variations in the productivity of the different areas of the oceans. The tropical parts are often – but far from always – relatively poor.'-
Steemann Nielsen, 1952a

The pioneering cruise of the *Galathea*, on which Steemann Nielsen first employed radiotracers to measure the productivity of the sea (Steemann Nielsen, 1951, 1952b), set both a standard for the field and initialised 50 years of arguments and discussion concerning the scale of primary production in the oligotrophic ocean gyres. The argument continues to the present, albeit with a focus on new and exported production (Jenkins, 1988; Platt *et al.*, 1989; McGillicuddy *et al.*, 1998). The pressing need to evaluate and predict the role of the ocean in the global carbon cycle has provided strong impetus for continued attempts to resolve the apparent and persistent discrepancies concerning the rate of the primary production of organic matter in the open ocean, and its export to the deep sea.

An interesting feature of the initial arguments was the astonishingly low number of data points on which the proponents based their respective cases. The statement attributed to Steemann Nielsen, 'An impeccable experiment is more worth than a lousy experiment with lots of replicates' (see Chapter 1), is elegant in its simplicity, but misguided when used to extrapolate impeccable experiments to the regional and global scale. Consequently, the initial arguments had as much to do with severe under-sampling as specific methodologies. The situation has improved over the years with new geochemical methods that average over long space- and time-scales, but it has only been the routine availability of synoptic-scale satellite observations and advanced numerical models of the physical and biological dynamics of the open ocean that have crystallised the problem.

As a result, a view has emerged over the past decade of an oligotrophic ocean region, rich in a wide spectrum of variability in both time and space (Platt and Harrison, 1985), particularly at the so-called 'mesoscale'. This variability must be either explicitly resolved or sensibly parameterised to even begin to address the larger-scale evolution of the oceanic carbon cycle. New modelling approaches which resolve or permit mesoscale eddy scales (McGillicuddy and Robinson, 1997), the routine availability of satellite observations of physical (e.g. altimetry) and biological (e.g. ocean colour) processes, long-term continuous moored and drifting buoy observations (Dickey *et al.*, 1998; Abbott *et al.*, 2001), and advanced data assimilation schemes which tie the models and data together (Doney *et al.*, 2001), have all contributed significantly towards bringing this new view of the open ocean into widespread acceptance.

Here, I present some of the early discussions concerning the productivity of the open ocean, and trace the arguments through to the present. A resolution to ongoing differences is then presented which derives from explicit consideration of the role of mesoscale variations. The focus is on the North Atlantic, the battleground for many past conflicts, although conclusions likely apply on a global scale.

6.2   Large-scale estimation of primary productivity in the open ocean

The initial arguments concerning the level of primary production in the sea were most intense between Steemann Nielsen, who favoured the measured lower rates based on the then new method of measurement of the uptake of inorganic ^{14}C, and scientists at the Woods Hole Oceanographic Institution, notably Gordon Riley and John Ryther, who favoured higher rates estimated on the basis of oxygen changes in incubation bottles (Petersen, 1980). From the outset, the scales were mismatched: Steemann Nielsen's arguments were initially based on only three stations in the Sargasso Sea occupied on a single transect during the Northern hemisphere summer (Steemann Nielsen and Jensen, 1957–1959), and most of the Woods Hole Oceanographic Institute (WHOI) work carried out over several years (which may, in fact, have been during an anomalous climatic state), but in a limited geographical range along the north-western boundary of the Sargasso Sea (Plate 6.1).

As the years unfolded, new questions arose based on results of geochemical techniques which indicated rates of organic input to water masses removed from the sea surface that were apparently inconsistent with estimates of total primary production made in the upper ocean (Schulenberger and Reid, 1981; Jenkins,1982). The downward organic flux requires a source of nutrients to the surface to support these rates of export production; rates of turbulent vertical transport based on measurements of small-scale mixing are at least an order of magnitude too small (Lewis *et al.*, 1986). Again, scales are severely mismatched. The geochemical tracer techniques average over time-scales of years, and over unknown, but large, spatial scales. The incubation methods in the upper ocean sample a fixed point in space, and on time-scales of hours. The estimates of vertical turbulent transport of nutrients represent scales of weeks to months, and again, are limited in the geographical range encompassed by the measurements.

Clearly, different oceans are being sampled, but several key questions remain, including:

(1) Why are these estimates so different?
(2) What is the large spatial-, and long time-scale-averaged new production in the open ocean?
(3) How might the small-scale variability in our large-scale climate models be sensibly parameterised?

6.3 Physical mechanisms and scales of mesoscale variability

The mesoscale fluctuation spectrum in the open ocean spans length scales of the order of 100 km, and time-scales ranging between 10 and 100 days. These scales dominate the frequency–wavenumber spectra of eddy kinetic energy as seen in both moored instrumentation (Wyrtki *et al.*, 1976; Dickey *et al.*, 1998), and satellite observations (Le Traon, 1991; Stammer, 1997).

The mesoscale eddy kinetic energy distributions are aligned in time and space with the low-frequency (mean) baroclinic flow and are highly correlated with the primary mode internal Rossby radius of deformation (Stammer, 1997). The dynamical conclusion is that instabilities in the mean flow are a dominant source of eddy energy in the oceans. For example, high levels of mesoscale energy are found along boundary currents at mid-latitudes, and along frontal structures and current systems (Stammer and Wunsch, 1999). Minima are found in the interior of the ocean gyres. Decorrelation time-scales are proportional to $1/f$, where f is the Coriolis parameter, and length scales vary as the Rossby radius, c_n/f, where c_n is the nth mode internal gravity wave phase speed (Stammer, 1997). In general, eddies are smaller, and have faster time-scales as the latitude increases.

The eddy energy cascades down the frequency–wavenumber spectra in a fashion that can be described on dimensional grounds based on assumptions of geostrophic turbulence. Eddy variance is related to wavenumber to a power of approximately -3 (Holladay and O'Brien, 1975; Gower *et al.*, 1980), and a similar estimation yields a frequency dependence of the order of -1 in the neighbourhood of the local mesoscale frequency. At higher frequencies and wavenumbers, where dissipation plays a more important role, the slopes, again on dimensional grounds, reach -2 and $-5/3$ respectively. These results are largely consistent with observation on the global scale based on satellite altimetric observations (Stammer, 1997).

The surface kinetic energy distributions are dominated by the first baroclinic mode, and hence the variance in the sea-surface height observed by altimetry should directly reflect the vertical displacements of isopycnals in the ocean interior (Gill, 1982). Siegel *et al.* (1999) analysed a combined TOPEX/POSEIDON-ERS altimeter data set on sea level height anomalies, and related these to subsurface displacement of isopycnals in cyclonic eddies found in the north-western Sargasso Sea. The transfer relationship yielded a value of 400 m of subsurface displacement for every 1 m of change in sea-surface height (at 200 m), with the gain function diminishing linearly towards the surface. Turk *et al.* (2001) have examined inter-annual variability in sea surface height anomalies in the equatorial region from TOPEX/POSEIDON altimetry, and also related this to anomalies in the displacement of the subsurface isopycnal surface characterised by the 20° isotherm. In this case and at these scales, the gain function is ~ 200 m of subsurface displacement for every metre of sea surface height variation, a result consistent with that of Siegel *et al.* (1999). The altimetric data sets therefore provide a useful remote means for evaluating variation in the uplift of isopycnal surfaces at the mesoscale. The importance of this lies in the fact that below depths of significant photosynthetic activity, the main gradients in nutrients are approximately vertical, and subsurface isopycnals typically contain elevated nutrients. By examining the isopycnal displacements associated with mesoscale activity, one can infer an effective nutrient flux into the well-lit surface waters.

Consequences for transport of nutrients

Mesoscale variations are also clearly seen in the distribution of the scalar chlorophyll *a* fields, as derived from satellite observations of the relative amplitude of the spectral radiance leaving the ocean interior (Gower *et al.*, 1980). An outstanding question is the significance of, and the mechanism by which, variations in the physical dynamics at the mesoscale translate into variability in biological processes. Given that the primary source of variability in primary production on the large scale is due to variations in nutrient input, then the focus becomes the role of mesoscale eddies in moderating the nutrient flux into the well-lit euphotic zone. As this represents a 'new' source of nutrients (*sensu* Eppley and Peterson, 1979), it is relevant to estimates of the export of organic material that is thought to be responsible for geochemical changes in deeper waters.

Mesoscale processes potentially contribute to nutrient enrichment in a number of ways. These include the uplift of isopycnals with high nutrient concentrations into the euphotic zone ('eddy pumping'), enhanced shear and vertical mixing associated with eddy dynamics, and frontal processes that input new nutrients into potentially productive surface waters, both vertically and in the horizontal.

Eddy pumping

Cyclonic eddies exhibit marked doming of isopycnals. This displacement has the potential of bringing nutrient-rich water into the well-lit surface waters. The nutrients are then available for uptake by the resident phytoplankton population. As the isopycnals subside with the passing (or decay) of the eddy, the nutrient concentrations are diminished (although must be restored on rather fast time-scales, see below). This process has been referred to as eddy pumping (see Falkowski *et al.*,

1991); it is analogous to turbulent transport as the magnitude is the covariance of isopycnal vertical velocities and nutrient concentrations, $(\overline{w'N'}$, where w' is the vertical fluctuation in isopycnal surfaces about the reference level, N' is fluctuation in nutrient (nitrate) concentration, and the overbar indicates time average).

Jenkins (1988) proposed this mechanism to explain the source of new nutrient input required to balance the high rates of organic matter flux to deeper horizons in the north-western Sargasso Sea. Falkowski *et al.* (1991) evaluated topographically generated eddy pumping in the North Pacific, and concluded that the upward doming of isopycnals had a strong and positive influence on photosynthetic performance, presumably because of the enhanced nutrient levels in this otherwise nutrient-impoverished region. They furthermore attempted to extend the analysis by reference to the statistical distribution of such eddies and concluded that approximately 20% of new nutrient injection in this region could be accounted for by reference to the eddy pumping mechanism. Further analysis in the north-western Sargasso Sea, including modelling studies (McGillicuddy and Robinson, 1997), direct field observations (McNeil *et al.*, 1999; McGillicuddy *et al.*, 1998, 1999) and satellite-based inferences (Siegel *et al.*, 1999) have indicated that the eddy pumping process associated with mesoscale features plays a large, and perhaps dominant (50% in Siegel *et al.*, 1999), role in supplying new nitrate to the euphotic zone.

An issue in this approach is that for the mechanism to provide a sustained source, then there must also be a restoring term which would bring nutrient concentration on an isopycnal surface back to equilibrium on return to the reference depth on an appropriate time-scale. Garçon *et al.* (2001) have called attention to this and point out that the effectiveness of the eddy pumping mechanism depends on the relative time and vertical length scales of regeneration, and the period of eddy recurrence. If there is insufficient supply of organic material to the isopycnal, and if it is not resident for a sufficient time to remineralise it between eddy events, then the eddy pumping mechanism only 'works' once, and serves to push the nitracline to a deeper isopycnal surface that would not be brought into the euphotic zone. McGillicuddy and Robinson (1997) have argued for a three-month restoring time-scale for nitrate, and Siegel *et al.* (1999) have assumed a restoring time-scale of ten days consistent with other estimates (Christian *et al.*, 1997), much shorter than an Eulerian measure of eddy recurrence time-scale of the order of 60 days. The difficulty lies in an apparent inconsistency inherent in a purely one-dimensional approach. If nitrate introduced by eddy uplift is a strong contributor to new production (and by extension, exported production), but nitrate is required to be remineralised to restore the concentration on the uplifted isopycnal over short time-scales, then there is a balance between uptake and remineralisation over shallow depths. Consequently, there is little remaining organic matter available to be exported to the deeper depth horizons to satisfy the requirements inherent in the results from the geochemical estimation of the rate of apparent oxygen utilisation.

Another difficulty is the appropriate perspective to take with respect to this process. A wave-like view, such as taken for the larger-scale coherence between

Rossby waves and pigment distribution (Uz *et al.*, 2001), requires a different remineralisation time-scale (and interpretation) than necessary if the eddy vortice retains its water mass as it propagates across the basin, where it subsequently decays. In the latter case, the nutrient input to the eddy is a single event, and the remineralisation necessary is spatially uncoupled from the injection.

Clearly, if the eddy-pumping mechanism is to solve the apparent discrepancies between estimations of new production based on measurements made in the upper ocean and those made over longer time-scales in the deep sea, some sort of enhanced diapycnal mixing, lateral stirring on isopycnals, or a full decoupling of carbon and nitrogen remineralisation, is required. Adopting the Siegel *et al.* (1999) assumption of complete removal on an isopycnal surface after introduction into the euphotic zone, a horizontal eddy length scale of the order of 100 km, and an iso-pycnal eddy diffusivity of $100 \, \text{m}^2 \, \text{s}^{-1}$, then a mixing time-scale along the isopycnal is approximately 1000 days, far too long to provide the required restoring of nitrate. Typical values of turbulent diapycnal transport of nitrate (Lewis *et al.*, 1986; Ledwell *et al.*, 1993) are clearly too small as well. Decoupling of carbon and nitrogen remineralisation exists and results in different vertical length scales (370 m for C, 250 m for N in Christian *et al.*, 1997), but the differences are not sufficient to provide a full explanation. For the eddy-pumping mechanism to be the driving force for export production, then there must exist a means to restore the nutrient con-centrations on relatively short time-scales. This is presently uncertain.

Shear-induced mixing, fronts, jets and horizontal transport

There are other mechanisms that provide for enhanced primary production (and new production) in association with mesoscale features. Enhanced shear associated with the baroclinic currents could conceivably lead to enhanced vertical stirring of water masses. Variations in Richardson numbers $(N^2/|\partial u/\partial z|^2$; where N^2 is the Brunt Väisälä frequency, and u is the horizontal velocity) in the world's oceans (Stammer, 1998) show reduced values along all ocean frontal systems. These lower values imply increased vertical stirring of water masses along mesoscale features, particularly at the periphery where the combination of isopycnal uplift and enhanced shear can give rise to higher turbulent mixing rates (Woods, 1988). These can translate directly into higher productivity as seen in the north-western Atlantic and Mediterranean Sea (Lohrenz *et al.*, 1988, 1993). Using a novel bio-optical drifting buoy array, Abbott *et al.* (2001) have observed significant biological variability associated with mesoscale upwelling associated with the edges of eddies in the Southern Ocean.

The horizontal transport of nutrients associated with mesoscale eddies is locally of some significance in some areas, notably along the periphery of the oligotrophic ocean gyres (Oschlies and Garçon, 1998; Garçon *et al.*, 2001). It is unclear what the global significance of this transport is, but it likely scales with the distribution of eddy heat fluxes in the global ocean (Wunsch, 1999).

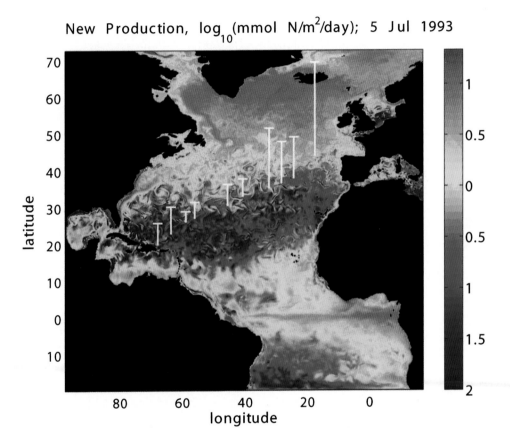

Plate 6.1. The total primary productivity measured by Steemann Nielsen during the *Galathea* cruise (Steemann Nielsen and Jensen, 1957–1959, white bars, relative scale) plotted on an image of the large-scale distribution of the rate of new production for a single day, derived from a high-resolution numerical model (0.1 degree, Los Alamos Parallel Ocean Program (POP); Smith *et al.*, 2000).

Consequences for the scales of biological variability

The mesoscale variability in the physical dynamics, and the consequences for variability in nutrient flux, are of profound importance for estimation of phyto-plankton productivity in the sea, and for estimation of some of the larger terms in the oceanic carbon cycle. Large-scale estimates of the productivity of the sea, either based on *in situ* data, satellite data or numerical models, at best largely averaged this variability, or at worst, are aliassed in time and space because the relevant processes and their scales are ignored. The attempt by Steemann Nielsen to estimate an average productivity rate for the open ocean based on three stations in the Sargasso Sea was courageous, but clearly naïve in the context of the mesoscale variability evident in Plate 6.1. Modern estimates are more robust, but given time-scales of the order of 10–20 days for both mesoscale dynamics and the biological response, the common practice of attempting to use monthly 'composite' satellite images of ocean colour to infer dynamics almost assuredly results in aliasing of the higher frequency and wavenumber variability. A major challenge for biological oceanographers, and those attempting to resolve large-scale oceanic biogeochemical cycles, is to appropriately accommodate mesoscale processes into the larger scale (Doney, 1999). This can be accomplished by explicit resolution of this variability with highly resolved (in both time and space) data sets or numerical models. However, this imposes a large cost in either intensive instrumentation arrays or in terms of com-puting resources, neither of which are likely to be sufficient in the near future to attack the longer-scale climate problems. A sensible means to parameterise this variability is required.

6.4 Parameterisation of mesoscale variability

Spectral estimation

A first step in the process would be to evaluate the resulting scales of variability in biological dynamics, as evidenced by variance in the concentration of the primary pigment, chlorophyll, in the upper ocean. Subsequent to Steemann Nielsen's pio-neering work, much was made of observations of large variability in plant biomass in the ocean. However, the root of the variability was only made clear with new high-resolution observation techniques (Herman and Platt, 1980) and the application of modern statistical approaches that decomposed this variance into the frequency and wavenumber domain (Platt and Denman, 1975). This approach continues to bear fruit (Denman and Abbott, 1988; Yoder *et al.*, 1993; Washburn *et al.*, 1998), and provides insight into the dominant physical processes driving variability in phyto-plankton dynamics.

Theoretical work in the wavenumber domain (Platt and Denman, 1975), based on simple models of plankton dynamics (Kierstead and Slobodkin, 1953), revealed sources of variability in phytoplankton which varied with respect to scale. At very high wavenumbers (equivalent to spatial scales less than 1 km), the turbulent processes that tend to dissipate spatial variance dominated; at lower frequencies, the biological dynamics were sufficiently strong to override the purely physical control, and the spatial variability is dominated by biological processes. It is of some interest to extend this work into the mesoscale domain to evaluate variability in biological properties at larger scales.

Gower *et al.* (1980) examined the wavenumber dependency of surface chlorophyll distributions from the (then) new ocean colour satellite over spatial scales of the order 100 km to 1 km. They found the spectral slope of the relationship between pigment variance and wavenumber to be close to -3, as predicted by theories of geostrophic turbulent mixing of passive tracers. Washburn *et al.* (1998) investigated spatial variability in phytoplankton in the North Atlantic, and found two distinct spectral regimes. At low wavenumbers, there was coherence between phytoplankton and non-conservative properties such as salinity, and at higher wavenumbers (length scales less than 25 km) spectral slopes changed, and coherence was lost. The conclusion is that physical processes dominated the distributions at the larger scales, and biological processes dominated at the smaller scales, a result in apparent contradiction with that of Platt and Denman (1975).

A resolution can be found in the dimensional analysis of Lewis and Platt (1982) which is presented in a modified form in what follows. Assume a spatial distribution of variance in pigment in the upper ocean which can be represented as a function of wavenumber as $E_\beta (k)$ ((mg chl m^{-3})^2 (cycle per metre)$^{-1}$). Furthermore, assume that at large scales, distributions of variance are governed solely by mesoscale velocity variance, $E_v (k)$, such as suggested in the data of Gower *et al.* (1980) and Washburn *et al.* (1998). Given the baroclinic nature of the mesoscale variance, an appropriate length scale would be the first mode internal Rossby radius defined above. On dimensional grounds, one can then write:

$$E_v \, \alpha f^2 k^{-3} F(R_0) \tag{6.1}$$

where F is an unspecified function of the Rossby radius, and the velocity variance goes as the -3 power of wavenumber, which also reflects $E_\beta (k)$ as in Gower *et al.* (1980). At higher wavenumbers, it can be assumed that the variance in pigment concentration is governed by biological processes that tend to increase variance, and dissipative processes (small-scale diffusion) that tend to erase it. Defining the rate of dissipation of turbulent kinetic energy, again at small scales, as ε, and the net growth rate of phytoplankton as r, then one predicts, again on dimensional grounds, that there exists a critical wavenumber, $k_{cr} (= r^3/\varepsilon)^{1/2}$, which defines the boundary between biological control and dissipation. At the smallest of scales, pigment behaves as a passive tracer, and at larger scales, biological processes dominate – a result consistent with that of Platt and Denman (1975).

A three-phase spectrum is then predicted. At the smallest of scales, mesoscale processes and biological dynamics are of little importance, and the distribution of pigment variance is predicted to go as the $-5/3$ power of wavenumber. At scales of the order of the Rossby radius, one might expect again that biological processes are of lesser importance, and that the variance would go as the -2 to -3 power of wavenumber. Finally, sandwiched in between in wavenumber space, is the 'biological window' where the growth dynamics of phytoplankton dominate the variance spectrum. This result reconciles the observations of Platt and Denman (1975), with those at the larger scale represented by Gower *et al.* (1980) and Washburn *et al.* (1998). Such a three-phase spectrum has been observed by Horwood (1981), with the apparent biological window spanning spatial scales from 10 to 25 km. The scales are consistent with those expected for this latitude.

This is an important result for model forecasts of mesoscale variability. For the larger scales, and to first order, pigment distributions can be treated as if they were passive tracers of the flow field. It also provides an interesting approach to estimating the net growth rate of phytoplankton through an examination of the spatial scales of variability along a meridional transect.

However, it is important to realise that the biological dynamics remain important for the larger scales. While it may not be possible to capture this high frequency and wave number variability explicitly because of limitations in computing power, if we simply average over long time and space scales, for say the ocean carbon cycle problem, the impact of these variations is eliminated. The problem is somewhat analogous to that for estimating the long time-scale response of climate to short-term weather.

Transfer functions

Consider an analogy to the problem faced by climate modellers who wish to examine the long-term variation in sea surface temperature as forced by higher-frequency weather fluctuations (Hasselmann, 1976; Frankignoul, 1979). Here, we are interested in the climate scale changes in surface pigment concentrations, B, with a simple model that looks like:

$$\frac{\partial B}{\partial t} = V(B, \boldsymbol{a}) \tag{6.2}$$

where V represents the 'high frequency' forcing function, and $\boldsymbol{a}$ is a set of variables representing cloudiness, nutrient input, etc. For this simple example, all advection and mixing of pigment is ignored by reference to the large scale under consideration. The coherent seasonal cycle is removed, and B redefined to represent departures from the long-term (many years) mean. Two time-scales for change are assumed: a fast one corresponding to those factors affecting biological productivity such as storms and mesoscale fluctuations (τ_a), and a slower time-scale for changes in the surface biomass (τ_B). Admittedly τ_a cannot be resolved but these scales must be parameterised if their importance is not to be lost for the larger-scale problem.

For time-scales short by comparison to τ_B, a stochastic forcing is assumed, and the variance in B can be written as:

$$\langle B^2 \rangle = \int_0^t \int_{-\tau}^{t-\tau} \langle V(\tau)V(\tau + \delta)\rangle d\delta d\tau \tag{6.3}$$

or

$$\langle B^2 \rangle = \langle [V(\tau)]^2 \rangle \tau_o t \tag{6.4}$$

where the angle brackets refer to an ensemble average, $[V(\tau)]^2$ is the variance in the higher frequency forcing at zero lag, and τ_o is twice the decorrelation time-scale for the higher frequency forcing. Variance grows with time. As a simple example, assume $V = V_o \cos\omega t$, then:

$$\langle B^2 \rangle = V_o^2/2\omega^2 \tag{6.5}$$

and the variance at these time-scales goes as the inverse square of the frequency of forcing. Expanding to include a spectrum of variability in the forcing, the transfer function between variability in the biomass and that in the forcing is then $\Gamma_B(\omega) = \Gamma_V(\omega)/\omega^2$ where Γ represents the respective power or variance spectra. If the spectrum of forcing is white at lower frequencies, then the spectrum of the pigment variance, as above, goes as the inverse square of the frequency.

Clearly, this cannot hold at very long time-scales, since the variance in pigment would grow without bound. A sensible steady state can be introduced by adding a negative feedback on the biomass such that:

$$\frac{\partial B}{\partial t} = V(B, a) - \lambda B \tag{6.6}$$

The feedback term (λ) perhaps can be considered to be a grazing rate. As above then, the transfer function becomes $\Gamma_B(\omega) = \Gamma_V(\omega)/(\omega^2 + \lambda^2)$, and furthermore, there is a time lag introduced of the magnitude of $\tan^{-1}(\omega/\lambda)$.

So for the largest of scales, the biomass largely tracks the longer scale fluctuations in the forcing terms, albeit with a lag of $\tan^{-1}(\omega/\lambda)$. At higher frequencies, variance in the forcing is smoothed in the resulting variation in the pigment biomass, and the fluid dissipation increasingly plays a role as described above.

Parameterisation of mesoscale processes for large-scale estimation of productivity

Such a stochastic model as developed above has value in that it permits the introduction of higher frequency variability into larger-scale models without reference to specific dynamical processes. Alternatively, one could take a more mechanistic approach to the parameterisation of unresolved mesoscale variance, assuming that the eddy-pumping mechanism described above is the dominant means by which eddies enhance the vertical nutrient flux and stimulate increased productivity. Siegel *et al.* (1999) have indicated that a perhaps suitable parameterisation of this process for larger-scale models might be cast in a mixing length form, which would consider the variance in the surface isopycnal displacement and an appropriate Lagrangian time-scale to define an eddy diffusion coefficient. This could then be used in conjunction with the mean nutrient profile to estimate the vertical nutrient flux associated with this mechanism. This is an insightful suggestion and it is interesting to see how it might be formulated.

The surface isopycnal displacement is related to changes in the sea-surface height as described above. Using the gain values from Siegel *et al.* (1999) for shallower depths, and those of Turk *et al.* (2001), 1 m of sea-surface height variation translates into approximately 200 m of subsurface isopycnal displacement. Referring to the global analysis of Stammer (1997) to estimate the surface displacement variability, and for the region considered by Siegel *et al.* (1999), a value of 0.01 m^2 can be taken as an estimate of the variance in surface height associated with the first baroclinic mode. Using the gain function above results in 400 m^2 as the variance in isopycnal displacement associated with the observed variance in surface height (equivalent to an rms isopycnal excursion of 20 m).

It remains to determine an appropriate time-scale. A useful estimate of the baroclinic time scale, T_{bc}, is $(Ri)^{1/2}/f$, where Ri is the Richardson number and f is the Coriolis parameter. The inverse of this number is the Eady growth rate for baroclinic eddies. Stammer (1998) has mapped global fields of this time-scale, and compared it to eddy time-scales inferred from the temporal auto-correlation function of surface height. For the Bermuda region, the mesoscale baroclinic time-scale is of the order of 20 days.

Dividing the variance in the subsurface displacement by the eddy time-scale provides a mixing coefficient, K_{bc}, a value estimated from the above at approximately 2.3×10^{-4} m^2 s^{-1}. Referring to the mean nutrient profile given in Siegel *et al.* (1999), the vertical gradient in nitrate below the impoverished region is ~ 0.02 mmol N m^{-4}. Finally, the flux is computed as the product of the derived mixing coefficient and the mean gradient. The resulting value is 4.6×10^{-6} mmol N m^{-2} s^{-1}, or if averaged over a year, 0.15 mol N m^{-2} y^{-1}, a value that is identical to that calculated by Siegel *et al.* (1999) from three years of time series data from the waters near Bermuda. This result, while perhaps fortuitous, lends credence to the approach first suggested by Siegel *et al.* (1999), to sensibly parameterise mesoscale variability in coarser-scale climate models.

Relationship to the historical context

With respect to the role of mesoscale eddies in the productivity of the global ocean, Steemann Nielsen perhaps wrote the preamble to the script on *Galathea*, but the play continues. The clear spatial and temporal heterogeneity that results from mesoscale variability occludes many of the arguments that have been presented with respect to the magnitude of the total and new production in the open ocean. While recognition of the importance of mesoscale variability with respect to the problem that has faced biological oceanographers for the last 50 years is a necessary first step (Platt *et al.*, 1989; Doney *et al.*, 2001), recognition alone does not solve it. The sampling problem is intense, as basic theory would insist on spatial sampling scales of 20 km or so, temporal resolution of approximately five days, and furthermore that the sampling take place over long (multi-year) time-scales. The appropriate para-meterisation of mesoscale (and finer) scale variability continues to elude us, and requires a fuller understanding of the fundamental mechanics that govern mesoscale eddy dynamics and their role in setting the scale for biological productivity in the open ocean. The impetus for this effort, as discussed above, can easily be seen in the fact that these open ocean areas occupy approximately 90% of the world's oceans, and continued uncertainty in the level of exported production in these areas pro-pagates directly into global scale estimates of the role of the oceans in the global carbon cycle.

6.5   Directions for future work

As somewhat of an addendum, it is useful to consider an experiment that might answer some of the persistent questions regarding the role of mesoscale variations, in a general sense, in present and future oceanic carbon cycles.

Clearly, as seen above, synopticity is a desirable feature, which would imply a reliance on satellite observations (Platt *et al.*, 1989). However, the satellite mea-surements are restricted to the upper 1's to 10's of metres, and it is only in resolving the vertical dimension that we can expect to advance our understanding.

One might picture a focused effort (Mesoscale Intensive Experiment or MIX) to solve the problem. As in Siegel *et al.* (1999), satellite observations, particularly altimetric observations of sea-surface height anomalies, would play a role, as would four-dimensional assimilative models as discussed in McGillicuddy *et al.* (1999). To fully resolve the mesoscale variability however, a rather intensive data collection effort would be required, which could only be accomplished with an ambitious autonomous mooring array.

If, as indicated above, the dominant horizontal spatial scale is ∼ 100 km at mid latitudes, the temporal scale is ∼ 20 days, and the vertical scale (typical excursion distance) is ∼ 20 m, then an array is required of 25 moorings, spaced on a 5 × 5 grid with a separation distance of 40–50 km between the moorings. Each array would need to resolve at minimum the vertical velocity structure, and the vertical density and nitrate concentration at scales of 10 m over the upper 200 m. Temporal sampling resolution could be as long as five to ten days which improves constraints on the data volume needed to be transmitted. Of course, if such an array could be implemented, other sensor systems to measure the pigment biomass variability, the vertical particle flux, and gas concentrations would be desirable. Such an array deployment would surely be ambitious, and somewhat expensive ($30–40 million), but would permit rigorous resolution of the mesoscale dynamics that are thought to dominate the biological signal in the open ocean, and perhaps bring to an end the debate that has energised the field over the last 50 years.

Acknowledgements

This effort has been partially funded by the Natural Sciences and Engineering Research Council (Canada), the National Aeronautics and Space Administration, and the US Office of Naval Research. I would like to thank Scott Doney, Dennis McGillicuddy and David Siegel for inspiration and insight.

References

Abbott, M.R., Richman, J.G., Nahorniak, J.S. & Barksdale, B.S. (2001) Meanders in the Antarctic polar frontal zone and their impact on phytoplankton. *Deep-Sea Research* (in press).

Christian, J.R., Lewis, M.R. & Karl, D. (1997) Vertical fluxes of carbon, nitrogen, and phosphorus in the North Pacific subtropical gyre near Hawaii. *Journal of Geophysical Research* **102**, 15667–77.

Denman, K.L. & Abbott, M.R. (1988) Time evolution of surface chlorophyll patterns from cross-spectrum analysis of satellite color images. *Journal of Geophysical Research* **93**, 6789–98.

Dickey, T., Frye, D., Jannasch, H., *et al.* (1998) Initial results from the Bermuda testbed mooring program. *Deep-Sea Research I* **45**, 771–94.

Doney, S.C. (1999) Major challenges confronting marine biogeochemical modeling. *Global Biogeochemical Cycles* **13**, 705–14.

Doney, S.C., Lindsay, K. & Moore, J.K. (2001) Global ocean carbon cycle modeling. In: *JGOFS/IGBP Synthesis* (ed. M.J.R. Fasham). In press.

Eppley, R.W. & Peterson, B.W. (1979) Particulate organic matter flux and planktonic new production in the deep ocean. *Nature* **282**, 677–80.

Falkowski, P.G., Ziemann, D., Kolber, Z. & Bienfang, P.K. (1991) Role of eddy pumping in enhancing primary production in the ocean. *Nature* **352**, 55–8.

Frankignoul, C. (1979) Stochastic forcing models of climate variability. *Dynamics of Atmospheres and Oceans* **3**, 465–79.

Garçon, V.C., Oschlies, A., Doney, S.C., McGillicuddy, D. & Waniek, J. (2001) The role of mesoscale variability on plankton dynamics in the North Atlantic. *Deep-Sea Research I* **48**, 2199–226.

Gill, A.E. (1982) *Atmosphere–Ocean Dynamics*. Academic Press, San Diego, California.

Gower, J.F.R., Denman, K.L. & Holyer, R.J. (1980) Phytoplankton patchiness indicates the fluctuation spectrum of mesoscale oceanic structure. *Nature* **288**, 157–9.

Hasselmann, K. (1976) Stochastic climate models. *Tellus* **28**, 473–85.

Herman, A. & Platt, T. (1980) Meso-scale spatial distribution of plankton: co-evolution of concepts and instrumentation. In: *Oceanography, The Past* (eds M. Sears & D. Merriman), pp. 204–25. Springer-Verlag, New York.

Holladay, C.G. & O'Brien, J.J. (1975) Mesoscale variability of sea surface temperatures. *Journal of Physical Oceanography* **5**, 761–72.

Horwood, J. (1981) Variation of fluorescence, particle-size groups and environmental parameters in the southern North Sea. *Journal du Conseil pour l'Exploration de la Mer* **39**, 261–70.

Jenkins, W.J. (1982) Oxygen utilization rates in the North Atlantic subtropical gyre and primary production in oligotrophic systems. *Nature* **300**, 246–8.

Jenkins, W.J. (1988) Nitrate flux into the euphotic zone near Bermuda. *Nature* **331**, 521–3.

Kierstead, H. & Slobodkin, L.B. (1953) The size of water masses containing plankton blooms. *Journal of Marine Research* **12**, 141–7.

Ledwell, J.R., Watson, A.J. & Law, C.S. (1993) Evidence of slow mixing across the pycnocline from an open ocean tracer-release experiment. *Nature* **364**, 701–703.

Le Traon, P.Y. (1991) Time scales of mesoscale variability and their relationship with space scales in the North Atlantic. *Journal of Marine Research* **49**, 467–92.

Lewis, M.R. & Platt, T. (1982) Scales of variability in estuarine ecosystems. In: *Estuarine Comparisons* (ed. V. Kennedy), pp. 3–20. Academic Press, New York.

Lewis, M.R., Harrison, W.G., Oakey, N.S., Hebert, D. & Platt, T. (1986) Vertical nitrate fluxes in the oligotrophic ocean. *Science* **234**, 870–73.

Lohrenz, S.E., Wiesenburg, D.A., DePalma, I.P., Johnson, K.S. & Gustafson, D.E., Jr (1988) Interrelationships among primary production, chlorophyll, and environmental conditions in frontal regions of the western Mediterranean Sea. *Deep-Sea Research* **35**, 793–810.

Lohrenz, S.E., Cullen, J.J., Phinney, D.A., Olson, D.B. & Yentsch, C.S. (1993) Distributions of pigments and primary production in a Gulf Stream meander. *Journal of Geophysical Research* **98**, 14545–60.

McGillicuddy, D.J., Jr., Johnson, R., Siegel, D.A., Michaels, A.F., Bates, N.R. & Knap, A.H. (1999) Mesoscale variations of biogeochemical properties in the Sargasso Sea. *Journal of Geophysical Research* **104**, 13381–94.

McGillicuddy, D.J., Jr & Robinson, A.R. (1997) Eddy-induced nutrient supply and new production. *Deep-Sea Research I* **44**, 1427–50.

McGillicuddy, D.J., Jr., Robinson, A.R., Siegel, D.A., *et al.* (1998) Influence of mesoscale eddies on new production in the Sargasso Sea. *Nature* **394**, 263–6.

McNeil, J.D., Jannasch, H.W., Dickey, T. McGillicuddy, D.J., Jr, Brzezinski, M. & Sakamoto, C.M. (1999) New chemical, bio-optical and physical observations of upper ocean response to the passage of a mesoscale eddy off Bermuda. *Journal of Geophysical Research* **104**, 15537–48.

Oschlies, A. & Garçon, V. (1998) Eddy-induced enhancement of primary production in a model of the North Atlantic ocean. *Nature* **394**, 266–9.

Peterson, B.J. (1980) Aquatic primary productivity and the $^{14}C–CO_2$ method: A history of the productivity problem. *Annual Review Ecology and Systematics* **11**, 359–85.

Platt, T. & Denman, K.L. (1975) Spectral analysis in ecology. *Annual Review Ecology and Systematics* **6**, 189–210.

Platt, T. & Harrison, W.G. (1985) Biogenic fluxes of carbon and oxygen in the ocean. *Nature* **318**, 55–58.

Platt, T., Harrison, W.G., Lewis, M.R., Li, W.K.W., Sathyendranath. S., Smith, R.E. &

Vezina, A.F. (1989) Biological production of the oceans: the case for a consensus. *Marine Ecology Progress Series* **52**, 77–88.

Schulenberger, E.L. & Reid, J.L. (1981) The Pacific shallow oxygen maximum, deep chlorophyll maximum, and primary productivity, reconsidered. *Deep-Sea Research* **28**, 901–19.

Siegel, D.A., McGillicuddy, D.J., Jr. & Fields, E.A. (1999) Mesoscale eddies, satellite altimetry, and new production in the Sargasso Sea. *Journal of Geophysical Research* **104**, 13359–79.

Smith, R.D., Maltrud, M.E., Bryan, F.O. & Hecht, M.W. (2000) Numerical simulation of the North Atlantic Ocean at 1/10 degrees. *Journal of Physical Oceanography* **30**, 1532–61.

Stammer, D. (1997) Global characteristics of ocean variability estimated from regional TOPEX/POSEIDON altimeter measurements. *Journal of Physical Oceanography* **27**, 1743–69.

Stammer, D. (1998) On eddy characteristics, eddy transports, and mean flow properties. *Journal of Physical Oceanography* **28**, 727–39.

Stammer, D. & Wunsch, C. (1999) Temporal changes in eddy energy of the oceans. *Deep-Sea Research* **46**, 77–108.

Steemann Nielsen, E. (1951) Measurement of production of organic matter in the sea by means of Carbon-14. *Nature* **167**, 684–5.

Steemann Nielsen, E. (1952a) Production of organic matter in the sea. *Nature* **169**, 956–7.

Steemann Nielsen, E. (1952b) The use of radioactive carbon (C^{14}) for measuring organic production in the sea. *Journal du Conseil International pour l'Exploration de la Mer* **18**, 117–40.

Steemann Nielsen, E. & Jensen, E.A. (1957–1959) Primary oceanic production. The autotrophic production of organic matter in the oceans. In: *Galathea Report*, Vol. 1, (eds A.F. Bruun, S.V. Greve & R. Spärck), pp 49–136. The Galathea Committee, Copenhagen.

Turk, D., McPhaden, M.J., Busalacchi, A.J. & Lewis, M.R. (2001) Remotely-sensed biological production in the tropical Pacific during 1992–1999 El Niño and La Niña. *Science* **293**, 471–4.

Uz, B.M., Yoder, J.A. & Osychny, V. (2001) Pumping of nutrients to ocean surface waters by the action of propagating planetry waves. *Nature* **409**, 597–600.

Washburn, L., Emery, B.M., Jones, B.H. & Ondercin, D.G. (1998) Eddy stirring and phytoplankton patchiness in the subarctic North Atlantic in late summer. *Deep Sea Research I*, **45**, 1411–39.

Woods, J.D. (1988) Mesoscale upwelling and primary production. In: *Towards a Theory on Biological–Physical Interactions in the World Ocean* (ed. B.J. Rothschild), pp. 7–38. Kluwer Academic Press, NATO ASI Series.

Wunsch, C. (1999) Where do eddy heat fluxes matter? *Journal of Geophysical Research* **104**, 13235–49.

Wyrtki, K., Magaard, L. & Hager, J. (1976) Eddy energy in the oceans. *Journal of Geophysical Research* **81**, 2641–6.

Yoder, J.A., Aiken, J., Swift, R.N., Hoge, F.E. & Stegmann, P.M. (1993) Spatial variability in near-surface chlorophyll *a* fluorescence measured by the Airborne Oceanographic Lidar (AOL). *Deep Sea Research* **40**, 37–53.

Chapter 7
Assessment of Primary Production at the Global Scale

Michael J. Behrenfeld, Wayne E. Esaias and Kevin R. Turpie

7.1 Introduction

Since the evolution of the first photosynthetic organisms some 3.8 billion years before present (Schopf, 1983), photoautotrophic organisms and the communities they support have continuously altered the chemical composition of the oceans and, through exchange across the air–sea interface, influenced the composition of the overlying atmosphere. Variations in the concentration of radiatively sensitive gases in the atmosphere (e.g. CO_2, CH_4), in turn, influence global climate and consequently ocean circulation, stratification, and the transport of dust to remote ocean regions. These physical forcings govern spatio–temporal variability in phytoplankton distributions through the direct effect of temperature on growth and through their secondary influence on factors such as mixed-layer light availability and the distribution of macro- and micro-nutrients. Perhaps beyond any other observations, satellite measurements of global phytoplankton pigment concentrations have most clearly demonstrated this dependence of ocean productivity on physical processes.

Biogeochemical cycles in the oceans are clearly not in steady state (Falkowski *et al.*, 1998). Biological responses to global perturbations in the physical environment are delayed by the buffering effects of ecosystem complexity. An urgency to characterise such physical–biological feedbacks developed during the final decades of the twentieth century due to escalating public and scientific concerns that environmental impacts of human activities were transitioning from the local to the global scale. At the forefront of these emergent global issues is the potential for a change in climate resulting from rising atmospheric concentrations of carbon dioxide and other 'greenhouse' gases. Unquestionably, sequestration of CO_2 by the photosynthetic biosphere (land and oceans) will play a critical role in future climate trends, but quantifying this CO_2 exchange and its temporal sign remains an uncertainty in global climate models (GCMs).

The most accurately constrained carbon fluxes in GCMs are the release from fossil fuel combustion (presently, $5.5 \pm 0.5 \times 10^{15}$ g C y^{-1} = 5.5 petagrams (Pg) y^{-1}) and atmospheric CO_2 accumulation (3.3 ± 0.2 Pg C y^{-1}). Less well quantified is the terrestrial carbon source from land-use change and deforestation, estimated at 1.6 ± 1.0 Pg C y^{-1} (Sarmiento and Wofsy, 1999). Thus, of the 7.1 Pg C released annually, 3.3 Pg C y^{-1} are retained in the atmosphere and the remaining 3.8 Pg C are removed both abiotically and through photosynthetic fixation. Based on ≈ 2 million measurements of the partial pressure of CO_2 over the oceans (pCO_2) collected across 25 years, the annual oceanic sink for CO_2 has been estimated at 2.0 ± 0.8 Pg C (Sarmiento and Wofsy, 1999). Balancing the global CO_2 budget thus requires an additional sink of 1.8 ± 1.6 Pg C y^{-1}. This 'missing sink' is assumed to involve the biosphere and likely entails both oceanic and terrestrial components. Partitioning of the unaccounted carbon between the land and oceans is difficult, however, because it represents less than 2% of biospheric net primary production, estimated at 111–117 Pg C y^{-1} (Behrenfeld *et al.* 2001a).

Assuming a historically balanced global carbon budget, perturbation by increased emissions during the industrial revolution implies that the unidentified carbon sink(s) have been stimulated, either directly or indirectly, by increased atmospheric CO_2 concentrations. A variety of approaches have been employed to identify where these changes have occurred, including the analysis of north–south atmospheric CO_2 gradients (Tans *et al.*, 1990) and changes in $^{13}C{:}^{12}C$ and $O_2{:}N_2$ ratios (Ciais *et al.*, 1995; Keeling *et al.*, 1996). Biologically mediated carbon fluxes over land and in the oceans can also be investigated through quantitative models of net primary production (NPP = photosynthetically fixed carbon that is not respired by the plant biomass itself and thus is available to the first heterotrophic level (Lindeman, 1942)). Assessing the fractional loss of carbon to long-term storage pools from model estimates of NPP is complicated by the complexity and variable time-scales of carbon remineralisation pathways. Nevertheless, NPP models can provide critical information on whether the total and/or spatial distributions of oceanic and terrestrial primary production are in steady state or changing, provided such estimates are based on temporally resolved global scale measurements of plant biomass (i.e. not climatologies).

The potential to detect changes in biospheric photosynthesis can largely be attributed to the availability of global satellite imagery, without which errors resulting from the spatial extrapolation of local-scale, process-oriented models would be prohibitive. For terrestrial systems, NPP can be calculated as the product of light absorption by the plant canopy (APAR – absorbed photosynthetically active radiation) and an average light utilisation efficiency (ε) (Field *et al.*, 1998). For such calculations, APAR is typically determined from satellite-based estimates of vegetation greenness (e.g. the normalised difference vegetation index (NDVI)) and modelled as an empirical function of environmental factors, such as water stress and temperature (Potter *et al.*, 1993). A greater than 20-year continuous record of global satellite NDVI measurements has allowed characterisation of regional interannual variability (Tucker and Nicholson, 1999) and the detection of increasing high-latitude growing seasons in the northern hemisphere (Myneni *et al.*, 1998).

For aquatic systems, the fundamental approach to modelling NPP at the global scale is very similar to terrestrial APAR models. Satellite measurements of water-leaving radiance (L_w) are used to derive global fields of near-surface phytoplankton chlorophyll concentrations (C_{sat}) and then NPP related to the product of C_{sat} and an empirically derived estimate of ε. Estimating NPP in aquatic systems requires explicit models for, or assumptions regarding, the depth-dependent distribution of plant biomass through the water column (since satellite L_w data only reflect variability in the optical properties of the upper portion of the water column). Once phytoplankton biomass is assessed, a variety of model formulations can be employed to estimate NPP (Box 7.1). Depth-integrated models (DIMs) are the simplest and do not explicitly resolve any vertically varying characteristics of the water column (Behrenfeld and Falkowski, 1997a). Wavelength-resolved models (WRMs), in contrast, are the most fully expanded and entail (1) spectrally resolving the underwater light field to calculate depth-dependent absorption of light by phytoplankton, (2) applying an empirical relationship between absorbed light and photosynthesis, and then (3) integrating NPP over time and depth (Morel, 1991; Platt and Sathyendranath, 1988; Antoine *et al.*, 1996).

Irrespective of the specific approach employed, the foundation of aquatic NPP modelling is deeply rooted in basic concepts of phytoplankton photosynthesis as a function of irradiance. Introduction of the ^{14}C method by Steemann Nielsen (1952) made a critical contribution towards quantifying global NPP because this rapid, sensitive technique for assessing water column carbon fixation has since yielded thousands of NPP measurements for model parameterisation, spanning the entire range of naturally occurring phytoplankton concentrations. Notwithstanding, the underlying depth-dependent relationship between submarine irradiance and photosynthesis was well understood from O_2 measurements (largely from freshwater lakes) long before these large ^{14}C-based data sets were available.

Since the earliest descriptions of water column NPP (Ryther, 1956; Ryther and Yentsch, 1957; Talling, 1957; Rodhe *et al.*, 1958; Wright, 1959; Vollenweider, 1966), the evolution of productivity modelling has generally progressed from simple DIMs and time-integrated models (TIMs) toward WRMs (Behrenfeld and Falkowski, 1997a). Our understanding of the mechanistic basis for the photosynthesis–irradiance relationship has also developed tremendously during this period. In the following section, we begin our discussion of global productivity modelling by reversing the direction of this evolutionary process and start with a physiological description of the photosynthesis–irradiance relationship, followed by a WRM-type calculation of time- and depth-dependent NPP for a model water column, and then integrate these results over time (and depth) to arrive back at the relationships between surface irradiance and photosynthesis originally described in the first DIMs and TIMs.

Box 7.1   Classification of aquatic primary production models.

A variety of formulations have been described that relate measures of phytoplankton biomass (most commonly, chlorophyll concentration) to rates of photosynthesis. These models are often delineated into empirical, semi-analytical and analytical categories, but the distinction between these groups is ambiguous. In truth, all NPP models developed to date are, to some degree, dependent on empirical parameterisations that describe, for example, the relationship between photosynthesis and light or between C_{sat} and the vertical distribution of chlorophyll. In this chapter, we have employed the alternative classification system proposed by Behrenfeld and Falkowski (1997a), which delineates model groups according to the level of integration over wavelength, time and depth. Their four model categories are as follows.

I.   *Depth-integrated models (DIMs).*   These are the simplest models of daily, water-column net primary production (ΣNPP). DIMs calculate ΣNPP as the product of C_{sat}, surface photosynthetically active radiation (PAR), the depth of the euphotic layer (Z_{eu} = penetration depth of 1% surface PAR), and the maximum chlorophyll-normalised daily rate of carbon fixation (P^b_{opt}).

II. *Time-integrated models (TIMs).*   Like DIMs, these models do not explicitly resolve changes in photosynthesis over the course of a photoperiod. However, TIMs describe PAR, chlorophyll concentration, and chlorophyll-normalised carbon fixation rates as a function of depth. ΣNPP is then calculated by integrating the product of these variables from the surface to Z_{eu}.

III. *Wavelength-integrated models (WIMs).*   These models describe depth-dependent changes in photosynthesis as a time-resolved function of subsurface PAR. Unlike DIMs and TIMs which are parameterised using ^{14}C-uptake results from 6 to 24 h incubations under ambient sunlight, WIMs are parameterised with photosynthesis–irradiance data from short-term (0.5–2 h) incubations that employ a range of artificial light intensities.

IV. *Wavelength-resolved models (WRMs).*   Subsurface irradiance is spectrally resolved in WRMs. This treatment has two primary objectives: (1) to more accurately describe the attenuation of light through the water column, and (2) to permit the description of photosynthesis as a function of absorbed light, rather than incident PAR. Parameterisation of WRMs thus requires field data on variations in algal absorption characteristics, as well as results from photosynthesis–irradiance measurements. WRMs are the most fully expanded productivity models and explicitly describe time-, depth- and wavelength-dependent changes in photosynthesis.

7.2   Modelling water column primary production

Depth-dependent changes in photosynthesis (P_z) measured during standard 6 h to 24 h ^{14}C incubations exhibit a clear dependence on the vertical attenuation of light. This relationship between P_z and subsurface irradiance has been recognised for 50 years and its description remains a primary focus of productivity model development. The relationship originates from light-dependent changes in photosynthesis that occur within the water column over the course of a photoperiod. A brief description of the physiological processes involved in the photosynthesis–irradiance relationship is therefore justified to better understand the results and assumptions of daily primary production models.

Photosynthesis involves both light-dependent and light-independent ('dark') reactions. The 'light reactions' include the harvesting of photon energy by photosystems I and II (PSI, PSII), electron transport, and ATP and NADPH generation. The 'dark reactions' of the Calvin cycle subsequently utilise the products of the light reactions to fix CO_2 into organic carbon products. At low light, photosynthesis is limited by the light-harvesting capacity of PSII, such that the slope of the photosynthesis–irradiance relationship (α) can be expressed as the product of the functional absorption cross-section of PSII (σ_{PSII}) and the number of functional PSII reaction centres (n) (Ley and Mauzerall, 1982; Dubinsky *et al.*, 1986; Falkowski and Raven, 1997) (Fig. 7.1). At saturating light, photosynthesis becomes limited by the carbon fixing *capacity* of the dark reactions (P_{max}) (Stitt, 1986; Sukenik *et al.*, 1987), which is a product of the concentration of Calvin cycle enzymes and their

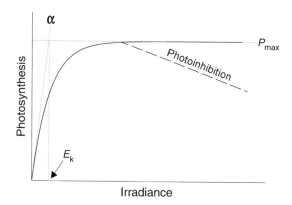

Fig. 7.1   Photosynthesis as a saturating function of incident irradiance. At very low light, photosynthesis is a linear function of irradiance with a slope (α) dependent on the number of functional PSII reaction centres and their average functional absorption cross-section. At light saturation, the photosynthetic rate is set by the capacity of Calvin cycle reactions (P_{max}). The ratio, $P_{max}{:}\alpha$, provides a useful modelling index (E_k) for the onset of light saturation. Under certain experimental conditions, a photoinhibitory decrease in photosynthesis may be observed at very high light (dashed line), which was modelled here as a linear function of irradiance with an onset only after light saturation (equation (7.1)).

activity (Fig. 7.1). A useful index for modelling primary production is the onset of light saturation (E_k), calculated as: $E_k = P_{max} \times \alpha^{-1}$ (Talling, 1957) (Box 7.2). Finally, a decrease from P_{max} at very high light (i.e. 'photoinhibition') may also be observed under certain experimental conditions (Fig. 7.1). Inclusion or exclusion of photoinhibition is a source of divergence between NPP models and will be discussed more fully in subsequent sections.

Wavelength-resolved (or 'bio-optical') models employ empirical parameterisations for the photosynthesis–irradiance variables (Fig. 7.1) and then calculate carbon fixation as a function of time- and depth-dependent changes in absorbed light. To illustrate the product of such a model, we calculated photosynthetic rates for a hypothetical, low-latitude water column in an oligotrophic open ocean region. Surface photosynthetically active radiation (PAR) was taken from measurements made under generally clear skies at 0°, 140°W. The underwater light field was calculated from surface PAR values (assuming negligible surface reflection) and spectral attenuation coefficients measured in Case I (Morel and Prieur, 1977) waters at 1° N, 26° W. For each depth and time interval, algal light absorption was derived by multiplying the spectral PAR values by an absorption spectrum (a_{ph}) measured for a marine strain of *Synechococcus*. The ratio of chlorophyll to a_{ph} was assumed constant. Thus, to simplify notations in this chapter, photosynthesis was expressed per unit of chlorophyll ($P^b_{z,t}$, where the superscript b denotes normalisation to chlorophyll) and described as a function of light using the relationship:

$$P^b_{z,t} = P^b_{max} \left(1 - e^{-Ez,t/Ek}\right) f(E_{z,t}) \tag{7.1}$$

where P^b_{max} = chlorophyll-normalised, light-saturated photosynthesis and $E_{z,t}$ = irradiance at time t and depth z. The term $f(E_{z,t})$ provides an optional correction for the decrease in $P^b_{z,t}$ at very high light (i.e. photoinhibition) and is assigned a value of unity when photoinhibition is assumed negligible. When included, photoinhibition was modelled as a linear function of $E_{z,t}$, with an onset only after light saturation of photosynthesis (as illustrated in Fig. 7.1).

Results for our modelled water column are shown in Fig. 7.2 and closely resemble results from similar calculations reported 46 years ago by Ryther (1956). When photoinhibition was assumed significant, modelled photosynthesis exhibited a clear midday depression to 1.5 optical depths (i.e. optical depth (o.d.) $= k_d \times Z_{eu}$, where k_d = the average attenuation coefficient for PAR and Z_{eu} = penetration depth for 1% surface irradiance (Talling, 1957)) and then remained light-limited at greater depths (Fig. 7.2(a)). When photoinhibition was excluded, photosynthesis near the surface remained at P_{max} for nearly the duration of the photoperiod (Fig. 7.2(b)). For these calculations, diurnal variability in the photosynthesis–irradiance relationship was neglected.

The depth-dependent, diurnal patterns in $P^b_{z,t}$ illustrated in Fig. 7.2 represent a typical result from a WRM or wavelength-integrated model (WIM) (Behrenfeld and Falkowski, 1997a). Summing $P^b_{z,t}$ over the photoperiod results in a vertical

Box 7.2   Nomenclature.

Light

PAR	Photosynthetically active radiation
E_0	Surface downwelling PAR
DL	Daylength
MLD	Mixed layer depth
k_d	Mean attenuation coefficient for PAR for the water column
Z_{eu}	Penetration depth of 1% of surface PAR
I_g	Light level to which phytoplankton are photoacclimated

Phytoplankton biomass

C_z	Depth-dependent distribution of phytoplankton chlorophyll
C_{sat}	Satellite-derived, near-surface chlorophyll concentration

Physiological variables

PSI	Photosystem I
PSII	Photosystem II; the oxygen evolving complex
n	Number of functional PSII reaction centres
σ_{PSII}	Functional absorption cross-section of PSII
a_{ph}	Phytoplankton absorption spectrum for PAR
CC_{cap}	Carbon fixing capacity of the Calvin cycle reactions

Photosynthesis–irradiance relationships

α	Chlorophyll-normalised, initial slope of the photosynthesis–irradiance relationship
P^b_{max}	Chlorophyll-normalised, light-saturated rate of photosynthesis
E_k	Light saturation index; calculated as the ratio, $P^b_{max}{:}\alpha^b$

Water column primary production

P^b_z	Depth-dependent, chlorophyll-normalised carbon fixation
P^b_{opt}	Maximum value of P^b_z within a water column
α^*	Light-limited slope of the P^b_z profile
E^*_k	Light-saturation index for the P^b_z profile; calculated as the ratio, $P^b_{opt}{:}\alpha^*$
F	Dependence of ΓNPP on E_0 resulting from a change in the depth of light-saturated photosynthesis
NPP	Net primary production
ΣNPP	Net primary production integrated from the surface to Z_{eu}

profile of daily NPP (P^b_z) that exhibits a maximum (P^b_{opt}) which is always less than the light-saturated rate, P^b_{max} (Fig. 7.3(a)). This difference between P^b_{opt} and P^b_{max} increases with the assumed severity of photoinhibition and results because photosynthesis is not maintained at P^b_{max} throughout the photoperiod, but rather varies between light-limited, light-saturated, and photoinhibited rates (Fig. 7.2). The slope

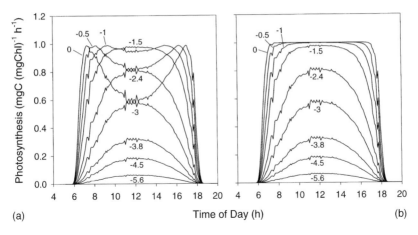

(a)   Time of Day (h)   (b)

Fig. 7.2 Temporal patterns in photosynthesis at specific optical depths within a hypothetical, low-latitude water column. Photosynthesis at a give time and depth ($P^b_{z,t}$) was calculated using equation (7.1) and a light-saturated photosynthetic rate of $P_{max} = 1\,mgC\,(mgChl)^{-1}\,h^{-1}$. Photoinhibition was assumed significant in (a) and neglected in (b). Optical depths are indicated by numbers adjacent to each curve. These results exemplify the output of a wavelength resolved model (WRM).

of the light-limited portion of the P^b_z profile (α^*) varies only if the chlorophyll-normalised initial slope of the photosynthesis–irradiance relationship (α^b) changes as function of depth in the water column. As with the photosynthesis–irradiance variable, E_k, the onset of light saturation for the P^b_z profile can be represented by the variable, $E^*_k = P^b_{opt} \times (\alpha^*)^{-1}$.

Differences between the P^b_z variables (P^b_{opt}, α^*, and E^*_k) and the photosynthesis–irradiance variables (P^b_{max}, α^b, and E_k) are not critical to the parameterisation of global productivity models, but their distinction was recommended by Behrenfeld and Falkowski (1997b) simply as a notational recognition of the intrinsic time-integration of variable photosynthetic rates when carbon fixation is measured under natural light conditions during prolonged incubations. From a practical standpoint, variability in P^b_{opt} is almost entirely due to changes in P^b_{max} and these two variables have nearly identical values when photoinhibition is negligible (Fig. 7.3(a)).

Time-integrated productivity models (TIMs) attempt to describe variability in P^b_z as a function of daily surface PAR (E_0). By retaining an explicit description of depth-dependent changes in P^b_z, TIMs can incorporate empirical models that derive vertical profiles of phytoplankton biomass (C_z) from given values of C_{sat} (e.g. Platt and Sathyendranath, 1988; Morel and Berthon, 1989). A real daily primary production (ΣNPP: $mgC\,m^{-2}\,d^{-1}$) is then calculated by integrating over depth the product, $P^b_z \times C_z$. If, however, phytoplankton biomass is assumed to be uniformly distributed through the water column, TIMs can be collapsed further into DIMs, the simplest form of daily primary productivity models. This assumption of uniform C_z can result in substantial errors (of the order of 20%) if a chlorophyll maximum exists

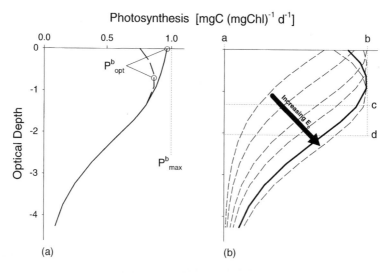

Fig. 7.3 Photosynthesis integrated over a photoperiod (P^b_z) as a function of optical depth. Results in (a) represent the time-integration of data illustrated in Fig. 7.2a (dashed line) and Figure 7.2b (solid line). The maximum daily photosynthetic rate (P^b_{opt}) is less than the light-saturated rate (P^b_{max}) due to light limitation and photoinhibition during the photoperiod. (b) An increase in surface irradiance (E_0) (indicated by the large black arrow) causes a deepening of the light-saturated portion of the P^b_z profile (dashed lines). The area encompassed by a given P^b_z profile represents water column net primary production (ΣNPP) and can be related to rectangular equivalents. For the P^b_z profile indicated by the heavy solid line, the light-saturated portion of the water column can be equated to the rectangle, R_1, with dimensions $ab \times bc$. The depth of c corresponds to the depth of the light saturation parameter, E^*_k. The light-limited portion of the water column can be equated to the rectangle, R_2, with dimensions $ab \times cd$.

between approximately 1 and 2 optical depths (Platt and Sathyendranath, 1988), but otherwise the resultant errors are relatively small.

Conversion of a TIM to a DIM is perhaps best envisioned by equating the area bound by the P^b_z profile to the sum of areas for two rectangles, R_1 and R_2 (Behrenfeld and Falkowski, 1997a) (Fig. 7.3(b)). In this treatment, R_1 encompasses the light-saturated portion of the P^b_z profile and is defined by the horizontal dimension, ab, and vertical dimension, bc (Fig. 7.3(b)). If P^b_{opt} has units of mgC (mgchl)$^{-1}$ h^{-1}, then $ab = P^b_{opt} \times C_{sat} \times DL$, where DL is the duration of the photoperiod in hours. The vertical dimension, bc, corresponds to the depth of E^*_k and is thus a function of E_0. At low light, P^b_z exhibits a light-limited exponential decrease from the surface to Z_{eu} and $bc = 0$. As E_0 increases, the light-saturated fraction of the euphotic zone increases (Fig. 7.3(b)) and bc varies according to:

$$bc = \ln(E_0/E^*_k) \times k_d^{-1} \qquad (7.2)$$

Obviously, the area of R_1 slightly overestimates P^b_z in the light-saturated portion of the water column when photoinhibition is substantial, but provides a better approximation when photoinhibition is negligible (Fig. 7.3).

Rectangle R_2 is defined by the dimensions *ab* and *cd* and has an area corresponding to that bounded by the light-limited portion of the $P^b{}_z$ profile. R_2 is essentially independent of E_0 for $E_0 > E^*{}_k$ and its area can be expressed as the product: $R_2 = P^b{}_{opt} \times C_{sat} \times DL \times i \times k_d{}^{-1}$. The value of i varies according to the presumed kinetics for light saturation, but generally is of the order of 0.7–0.8 (Behrenfeld and Falkowski, 1997a).

From the above discussion, it should be clear that variability in E_0 influences ΣNPP by changing the dimension *bd* according to $\ln(E_0/E^*{}_k)$. By defining the variable $F = bd \times Z_{eu}{}^{-1}$, equations for rectangles R_1 and R_2 (Fig. 7.3(b)) can be combined to yield the standard formulation of a DIM:

$$\Sigma NPP = P^b{}_{opt} \times C_{sat} \times DL \times Z_{eu} \times F \qquad (7.3)$$

Equation 7.3 has been published in various forms and notations approximately once every two years for the past 50 years, with the primary difference between DIMs being the description of F (Behrenfeld and Falkowski, 1997a).

In this section, we have described relationships between various productivity modelling approaches by beginning with a photosynthesis–irradiance relationship, applying this to a temporally varying underwater light field, and then integrating over time and depth to arrive at a simple relationship between ΣNPP and light, phytoplankton biomass, and ε (i.e. $P^b{}_{opt}$), essentially a 'reductionist to generalist' approach. Each level of simplification in this process requires certain assumptions regarding the importance of variability in the biological and physical characteristics of the water column. Critical assumptions for converting a WRM to a WIM are that an average attenuation coefficient for PAR is sufficient for describing submarine light and that variability in algal absorption spectra does not significantly influence ΣNPP. Conversion of a WIM to a TIM requires the additional assumption that diurnal variability in the physiological characteristics of phytoplankton has little influence on ΣNPP. Finally, DIMs assume that reasonable estimates of ΣNPP can be achieved without accounting for the depth-dependent distribution of phytoplankton biomass.

In truth, the evolution of productivity modelling has generally followed a DIM to WRM direction – in other words, a 'generalist to reductionist' approach. Arguably, the greatest advances in this procession have been in the description of the underwater light field (Morel, 1991; Platt and Sathyendranath, 1988; Antoine *et al.*, 1996). Unfortunately, the benefits of such models have not yet been fully realised due to slower progress in effectively modelling physiological variability. In the following section, we investigate the relative influence of $P^b{}_{max}$ and α^b on variability in $P^b{}_z$ and ΣNPP.

7.3   Importance of physiological variability

Light-limited versus light-saturated photosynthesis

A change in the light-limited slope of the photosynthesis–irradiance relationship (α^b) alters the profile of P^b_z in a similar manner as would a change in surface PAR (E_0). To illustrate, we employed our wavelength-resolved model for the hypothetical water column described above and varied the assumed value for α^b by a factor of 5. For this calculation, E_0 was held constant and P^b_{max} assigned a value of $6\,mgC$ $(mgChl)^{-1}\,h^{-1}$, which might be considered on the higher end of the average value for the global oceans. Photoinhibition was assumed to increase with α^b, as might be expected for low-light acclimation in phytoplankton.

Rapid attenuation of light through the water column desensitises ΣNPP to variability in α^b, such that the modelled factor of 5 increase in α^b only yielded a 50% increase in ΣNPP (Fig. 7.4). As with an increase in E_0, an increase in α^b results in a deepening of the light-saturated fraction of the euphotic zone, due to the associated decrease in E_k (Fig. 7.4). If α^b is assumed constant throughout the water column, the light-limited portion of the P^b_z profile (α^*) is unaltered by the factor of 5 increase in α^b. An increase in α^b from the surface to Z_{eu}, however, would result in a slight increase in α^* and a P^b_z profile intermediate to those shown in Fig. 7.4. This relative stability of α^* has long been recognised (Talling, 1957; Rodhe, 1966; Vollenweider, 1966), but misrepresents the influence of α^b on ΣNPP.

The influence of P^b_{max} on ΣNPP can be illustrated in a similar manner. Variability in P^b_{max} is considerably greater than photoacclimation-related changes in α^b. P^b_{max}

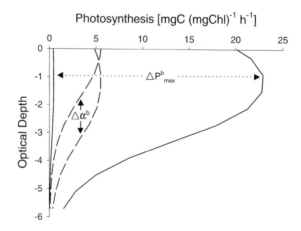

Fig. 7.4   Sensitivity of depth-dependent daily photosynthesis (P^b_z) to variability in the light-limited (α^b) and light-saturated (P^b_{max}) photosynthetic rate. Dashed lines illustrate the change in P^b_z when P^b_{max} is held constant at $6\,mgC$ $(mgChl)^{-1}\,h^{-1}$ and α^b is changed (α^b) by a factor of 5. Solid lines illustrate the change in P^b_z when α^b is held constant and P^b_{max} is changed (ΔP^b_{max}) from 0.5 to $25\,mgC$ $(mgChl)^{-1}\,h^{-1}$.

is minimal ($\approx 0.5\,\mathrm{mgC}\ (\mathrm{mgChl})^{-1}\ \mathrm{h}^{-1}$] under low-light, low-temperature conditions (e.g. Dierssen *et al.*, 2000) and maximal ($\approx 25\,\mathrm{mgC}\ (\mathrm{mgChl})^{-1}\ \mathrm{h}^{-1}$) under ideal growth conditions of high-light and replete nutrients (Falkowski, 1981; Behrenfeld *et al.*, 2001b). Applying this range in P^b_{max} to our modelled water column and assuming the same E_0 as in the previous calculations yielded a range in P^b_{opt} of 0.46–22.9 mgC $(\mathrm{mgChl})^{-1}\ \mathrm{h}^{-1}$ and a factor of 50 variability in ΣNPP (Fig. 7.4). This result clearly demonstrates the overwhelming importance of accurately modelling P^b_{opt} relative to α^*.

The influence of P^b_{opt} on ΣNPP can also be demonstrated using ^{14}C-based field data. Here, we have employed the productivity database of Behrenfeld and Falkowski (1997b) and the DIM version of their vertically generalised production model (VGPM):

$$\Sigma\mathrm{NPP} = 0.66125 \times P^b_{opt} \times C_{sat} \times DL \times Z_{eu} \times [E_0/(E_0 + 4.1)] \qquad (7.4)$$

For the $\approx 1700\,P^b_z$ profiles included in this data set, equation (7.4) explained 38% of the observed variance in ΣNPP when P^b_{opt} was assumed constant (Fig. 7.5(a)), which largely reflects the correlation between phytoplankton biomass ($C_{sat} \times Z_{eu}$) and daily primary production. In contrast, when measured values of P^b_{opt} are included in equation (7.4), the model explained 86% of the observed variance in ΣNPP (Fig. 7.5(b)).

It is tempting to conclude from the above analysis that, if a simple DIM can account for nearly all the variability in ΣNPP, then the additional computational overhead associated with more complicated TIMs, WIMs, and WRMs is unnecessary, since it can at best only account for the remaining 14% unexplained variance.

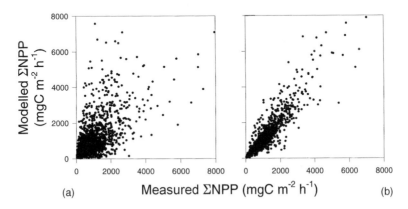

Fig. 7.5 Measured depth-integrated net primary production (ΣNPP: mgC m^{-2} d^{-1}) versus modelled ΣNPP for the data set of Behrenfeld and Falkowski (1997b) ($n = 1698$). ΣNPP was modelled using the vertically generalised production model (VGPM) of Behrenfeld and Falkowski (1997b) and measured surface chlorophyll, euphotic depths, photoperiods, and incident PAR. (a) ΣNPP modelled using a constant value for P^b_{opt} of 4.54 mgC $(\mathrm{mgChl})^{-1}\ \mathrm{h}^{-1}$ (i.e. the observed mean for the data set) ($r^2 = 0.38$). (b) ΣNPP modelled using measured values of P^b_{opt} ($r^2 = 0.86$).

As will be expounded upon later, such a conclusion is premature if P^b_{max} varies as a function of the underwater light field. But before addressing this issue, we review in the following subsections the various models described for P^b_{max} and P^b_{opt} and then compare global NPP distributions based on two of these models.

Modelling P^b_{opt}

All DIMs and TIMs require an estimate of P^b_{opt} to convert the chosen measure of phytoplankton biomass (e.g. C_{sat}) into NPP (a rate). Likewise, WIMs and WRMs employ empirical parameterisations for P^b_{max} to calculate photosynthesis. As discussed above, P^b_{opt} is primarily a function of P^b_{max}. Thus, an effective model for either should provide robust estimates of both.

Ryther and Yentsch (1957) assigned P^b_{opt} a constant value of 3.7 mgC (mgChl)$^{-1}$ h^{-1}, but acknowledged that this assumption was potentially the weakest aspect of their model. The first variable model for P^b_{opt} was described by Megard (1972) and based on ^{14}C-uptake measured in phytoplankton sampled from Lake Minnetonka, Minnesota. Megard (1972) described P^b_{opt} as a linear function of temperature (T^o): $P^b_{opt} = 17.1 + 1.61 \times T^o$, with P^b_{opt} expressed as a *daily* rate. Temperature has since been the most common environmental variable used for modelling P^b_{opt} and P^b_{max}. Morel (1991) described P^b_{max} as an exponential function of temperature with a Q_{10} of 1.88 (from Eppley, 1972). In contrast, Balch et al. (1992) described P^b_{opt} as an inverse function of T^o, while Balch and Byrne (1994) adopted a biphasic approach that attributed variability in P^b_{opt} to both T^o and nutrient limitation. Behrenfeld and Falkowski (1997b) described P^b_{opt} as increasing with temperature up to 20°C and then decreasing at higher temperatures.

The tremendous variability in temperature-dependent P^b_{opt} models described above results because temperature only serves as a proxy for the environmental factors directly influencing P^b_{max}. Temperature likely only imparts a direct physiological limitation on P^b_{max} below about 5°C, while above 5°C it is weakly correlated with other directly limiting factors, such as light and nutrient availability (Behrenfeld et al., 2001b). A general insensitivity of P^b_{max} to temperature is anticipated because:

(1) taxonomic adjustments in the phytoplankton assemblage select for species adapted to the ambient temperature (Yentsch, 1974; Kirk, 1983),

(2) temperature-dependent changes in Calvin cycle enzyme activity can be compensated for by increased enzyme concentrations (Steemann Nielsen and Hansen, 1959; Steemann Nielsen and Jørgensen, 1968; Geider et al., 1985; Geider, 1987), and

(3) changes in the Calvin cycle capacity can be paralleled by changes in chlorophyll (e.g. Durbin, 1974; Yoder, 1979; Verity, 1981; Lapointe et al., 1984), such that normalisation of P_{max} to chlorophyll (i.e. P^b_{max}) masks temperature dependence.

Variability in P^b_{opt} models therefore arises when the correlation between temperature and the physiologically limiting factors differs between data sets used for model parameterisation.

When tested against large observational data sets, temperature-dependent models typically explain only a small fraction of P^b_{opt} variability (about 20% *at best*) (Behrenfeld and Falkowski, 1997a; Siegel *et al.*, 2001; Behrenfeld *et al.*, 2001b). Application of such models to global estimates of NPP thus assumes that model error is unbiased and, when averaged over large space and time-scales, yields accurate mean values. An alternative modelling approach is to define seasonally and regionally varying climatological mean P^b_{max} values (Platt and Sathyendranath, 1988; Platt *et al.*, 1991; Longhurst *et al.*, 1995; Longhurst, 1995). For this method, geographical partitioning of the ocean is effectuated at either the 'domain' level, based on characteristic seasonal cycles in stability, nutrients, and light, or at the 'province' level, based on ocean currents, fronts, topology, and phytoplankton biomass features. The 'biogeographical province' approach thus has a stronger ecological and oceanographic basis than the simpler temperature-dependent models, but again relies on the assumption of an unbiased error around the assigned provincial mean P^b_{max} values. As would be anticipated, these mean values do not account for the much finer scale physiological variability in P^b_{max} observed within a particular province (e.g. Marañón and Holligan, 1999). Thus, like the temperature-dependent models, biogeographical province-based estimates of global NPP are difficult to verify.

Influence of physiological models at the global scale

If modelled distributions of global primary production were insensitive to variability in P^b_{opt}, then choosing between the afore-mentioned models would be irrelevant. Unfortunately, this is not the case. To exemplify the potential divergence between modelled NPP distributions, we calculated global annual average NPP using the VGPM (Behrenfeld and Falkowski, 1997b) and the temperature-dependent P^b_{opt} models of Morel (1991) and Behrenfeld and Falkowski (1997b). The primary difference between these two models is that the former assumes a monotonic increase in P^b_{opt} with temperature, while the latter assumes a decrease above 20°C.

Global NPP was estimated using monthly mean C_{sat} fields measured between November 1997 and October 1999 with the sea-viewing wide field-of-view sensor (SeaWiFS). Sea surface temperature (SST) fields were provided by the advanced very high resolution radiometer (AVHRR) (Reynolds and Smith, 1994) and PAR obtained at 9 km resolution from five SeaWiFS visible wave bands using an algorithm adapted from Frouin and Chertock (1992). Bathymetry was not considered, so NPP was integrated from the surface to the 1% light level regardless of location. Mean attenuation coefficients for PAR were derived from C_{sat} following Morel and Berthon (1989). Weekly averaged, smoothed SST data (Reynolds and Smith, 1994) and mixed layer depth (MLD) fields were averaged over monthly time intervals and latitude–longitude grid resolution changed to 9 km (2048 × 1024) using bilinear interpolation. Monthly SeaWiFS C_{sat} data were reconstituted from log-scaled, 8-bit digital values (which introduces an uncertainty of a few per cent due to truncation of concentration values). Anomalously high C_{sat} values (> 10 mg m^{-3}) were also removed, which decreased global NPP by less than 1%.

The VGPM estimate for annual global NPP was 59 Pg C y^{-1} using the exponential P^b_{opt} model of Morel (1991) (Plate 7.1a). Generally, NPP for this model only exceeded 400 gC m^{-2} y^{-1} over shelf regions, along the eastern-margin upwelling centres, and in the monsoon-driven upwelling region of the Indian ocean. In the open ocean, NPP was moderately high (approximately 200–300 gC m^{-2} y^{-1}) across broad bands of the equatorial region (reflecting modest levels of phytoplankton biomass and high P^b_{opt} values associated with elevated sea surface temperatures) and in the North Atlantic (due to much higher C_{sat} values compensating for relatively low P^b_{opt} values) (Plate 7.1a).

Applying the polynomial P^b_{opt} model of Behrenfeld and Falkowski (1997b) yielded a slightly lower estimate for annual global NPP of 56 Pg C y^{-1} and a very different distribution of productivity than the exponential model (compare Plate 7.1a with 7.1b). For the polynomial model, equatorial NPP was greater than 200 gC m^{-2} y^{-1} only near the upwelling centres of the Pacific and Atlantic Oceans. Indian Ocean NPP was greatly reduced both in the north-western quadrant and the central basin (Plate 7.1b). High sea surface temperatures in the central ocean gyres were associated with reduced values of P^b_{opt} and consequently lower NPP. High-latitude open-ocean NPP, on the other hand, was considerably enhanced and frequently exceeded 300 gC m^{-2} y^{-1} in the North Atlantic and Southern hemisphere subtropical convergence (Plate 7.1b).

The two models chosen for this comparison yield P^b_{opt} values that only range from approximately 1 to 9 mgC (mgChl)$^{-1}$ h^{-1} for the global temperature range of $-2°C$ to $30°C$. Estimates of global annual NPP differed by a mere 5% because higher P^b_{opt} values for the polynomial model at mid-temperatures were compensated by lower values at high temperatures, relative to the exponential model. However, the spatial distribution of NPP for the two P^b_{opt} models was strikingly divergent, with regional productivity occasionally differing by > 100% (Plate 7.1). Achieving a reasonable representation of the spatial patterns in NPP is a critical aspect of productivity modelling because the fraction of NPP exported from the upper water column (i.e. 'new production') versus that which is rapidly recycled by the near-surface marine food web exhibits strong regional dependence.

Unfortunately, the reliance of temperature-dependent P^b_{opt} models on achieving accurate estimates through spatial and temporal averaging compromises our ability to decipher which, if any, of the resultant modelled NPP distributions is correct. This issue can be restated in the following manner: *The ability to assess model performance requires that measured and modelled NPP be comparable on the same time and space scales.* Meeting this requirement demands the development of models based on environmental factors directly controlling P^b_{max}, rather than weak proxies or regionally averaged climatologies.

Recapitulation

At this point, we have discussed the utility of NPP models for investigating changes in biospheric CO_2 uptake, reviewed the conceptual basis of water column productivity models and the relationship between different formulations, and evaluated principal physiological sources of variability in P^b_z. We have also demonstrated in the foregoing section the sensitivity of global productivity estimates to variability in P^b_{opt} models and argued that model performance must be testable at the local scale of field ^{14}C measurements. Earlier, we hinted that an advantage might be gained by employing a spectrally resolved model for NPP if variability in P^b_{max} was light-dependent. In the following section, we:

(1) describe why P^b_{max} should indeed vary as a strong function of light,
(2) present a light-nutrient model for P^b_{max}, and
(3) apply this model for the first time to generate revised estimates of global NPP and its distribution.

7.4   A revised assessment of ocean productivity

A light-nutrient model for P^b_{max}

Two principal benefits of developing a productivity model based on chlorophyll-normalised physiological variables are:

(1) satellite C_{sat} data can be readily employed for spatial and temporal extrapolations, and
(2) cellular chlorophyll is, to first order, a dependent function of the Calvin cycle capacity.

This later consideration is due to the fact that the primary role of light harvesting (i.e. pigments) is to support the ATP and NADPH demands of the Calvin cycle. Consequently, the ratio of the Calvin cycle capacity to chlorophyll (P^b_{max}) is somewhat insensitive to changes in growth conditions. What causes P^b_{max} to vary, then, is environmental factors that shift the balance between light harvesting and carbon fixation. From this vantage, effectively modelling P^b_{max} can be viewed as simply a matter of describing how specific growth conditions cause chlorophyll to change *relative* to the Calvin cycle capacity (CC_{cap}).

One of the primary factors causing chlorophyll to vary independently of CC_{cap} is light. Photoacclimation may be considered a physiological response aimed at minimising the influence of light variability on a cell's daily allocation of fixed carbon. This is achieved by increasing light absorption (chlorophyll) as growth irradiance decreases. The relationship between growth irradiance and chlorophyll has been repeatedly described from laboratory photoacclimation studies and generally follows the form:

$$\text{Chl} = c_1 + c_2 \times e^{-c^3 \times I_g} \tag{7.5}$$

where c_1, c_2, and c_3 are constants and I_g is growth irradiance (Behrenfeld *et al.*, 2001b).

Under natural conditions, three principal factors must be considered when assessing I_g:

(1) downwelling irradiance,
(2) the spectral attenuation characteristics of the water column (thus, phytoplankton biomass), and
(3) the depth of mixing.

It is this requirement to accurately characterise the underwater light field in order to estimate P^b_{max} that will impart the greatest advantage to using a spectrally resolved model (i.e. a WRM) for quantifying global NPP.

Modelling P^b_{max} by describing relative changes in chlorophyll and CC_{cap} was the approach adopted by Behrenfeld *et al.* (2001b) in developing their 'PhotoAcc' model. For this model, CC_{cap} was assigned a value of $1\,\text{mgC}\,\text{m}^{-3}\,\text{h}^{-1}$ for mixed-layer conditions of replete nutrients, high light, and temperature $>5°C$. Changes in chlorophyll and CC_{cap} under all other conditions were then described relative to this value. I_g was assessed as the average daily PAR either at the bottom of the mixed layer or at the depth of interest below the mixed layer. CC_{cap} was assumed to decrease at low I_g below the mixed layer. With this conceptual basis, equation (7.5) could be parameterised from the ratio of $CC_{cap}:P^b_{max}$ as a function of I_g using field data from two Atlantic Meridional Transect studies (AMT-2 and AMT-3) (Robins *et al.*, 1996; Marañón and Holligan, 1999; Marañón *et al.*, 2000; Behrenfeld *et al.*, 2001b).

In addition to light, the analysis of Behrenfeld *et al.* (2001b) also indicated a dependence of P^b_{max} on the history of the water column, which was interpreted as an effect of nutrient stress. This conclusion was based on the observation that their 'nutrient charged' model generally overestimated P^b_{max} under conditions where stratification had been maintained for a sufficient period to cause a separation of mixed layer and nutricline depths. A 'nutrient depleted' model was thus described that entailed a lower value to CC_{cap} and the same chlorophyll–irradiance relationship as applied to 'nutrient charged' conditions.

For phytoplankton in the mixed layer, the PhotoAcc model assigns CC_{cap} a value of $1\,\text{mg C}\,\text{m}^{-3}\,\text{h}^{-1}$ for 'nutrient charged' conditions and $0.4\,\text{mg C}\,\text{m}^{-3}\,\text{h}^{-1}$ for 'nutrient depleted' conditions, irrespective of light level. A single chlorophyll–irradiance relationship is used for all nutrient conditions and then P^b_{max} is calculated as the ratio of CC_{cap}:chlorophyll. Although the PhotoAcc model CC_{cap} and chlorophyll relationships cannot yet be directly tested in the field, they are consistent with laboratory results (Fig. 7.6).

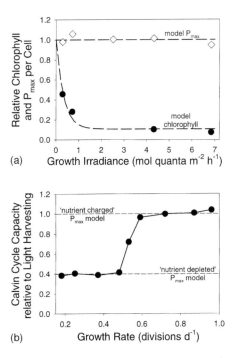

(a)

(b)

Fig. 7.6 The PhotoAcc model (Behrenfeld *et al.*, 2001b) estimates P^b_{max} by describing relative changes in cellular chlorophyll (Chl_{rel}) and carbon fixing capacities (P_{max}). In the field, Chl_{rel} and P_{max} are not routinely measured, but they can be determined in the laboratory. (a) Sukenik *et al.* (1987) reported changes in cellular chlorophyll (●) and P_{max} (◇) as a function of growth irradiance for the marine chlorophyte, *Dunaliella tertiolecta*. Their results correspond well with the predictions of the PhotoAcc model (dashed lines). (b) Falkowski *et al.* (1989) measured changes in photosynthetic electron transport (PET) components (PSII and PSI reaction centre proteins, cytochrome b_6f) and the Calvin cycle enzyme, ribulose 1,5-bisphophate carboxylase (RUBISCO), as a function of nitrogen-limited growth rate in the marine haptophyte, *Isochrysis galbana*. They reported a rapid decrease in the ratio of PET components to RUBISCO concentration. They also found cellular chlorophyll to vary in proportion to the concentration of PET components. If RUBISCO is assumed proportional to P_{max}, then the change in the balance between Chl_{rel} and P_{max} observed by Falkowski *et al.* (1989) (—●—) is consistent with the nutrient-dependent shift employed by the PhotoAcc model (dashed lines).

The full set of PhotoAcc model equations is provided in Table 7.1. Performance of the model was tested using the AMT data and a six-year light and primary production record from the US-JGOFS Bermuda Atlantic Time Series (BATS) and Bermuda BioOptics Programs (BBOP) (Michaels and Knap, 1996; Siegel *et al.*,

Table 7.1 Relationships of the PhotoAcc model (Behrenfeld *et al.*, 2001b) describing relative changes in the Calvin cycle capacity (CC_{cap}) and chlorophyll (Chl_{rel}). CC_{cap} is assigned a value of 1 mgC m^{-3} h^{-1} for mixed-layer phytoplankton under nutrient 'charged' conditions (i.e. conditions where growth is not strongly limited by nutrient availability). CC_{cap} and Chl_{rel} for all other conditions are expressed relative to this constant value and described as a function of growth irradiance (I_g). For the mixed layer, I_g is the average daily irradiance (mol quanta m^{-2} h^{-1}) at the depth of mixing (MLD). Below the MLD, I_g is the average daily irradiance at the depth of interest. A unique set of relationships for CC_{cap} and Chl_{rel} is applied for calculations of $P^b{}_{max}$ above the nitracline depth (NCD) under nutrient 'depleted' conditions. For all growth conditions, $P^b{}_{max} = CC_{cap} \times Chl_{rel}$. The final equation in the table is the correction to nutrient 'charged' CC_{cap} suggested by Behrenfeld *et al.* (2001b) for very low temperatures (<5°C), which was applied in our calculations of global NPP.

Growth condition	Depth (z)	CC_{cap}	Chl_{rel}
Nutrient 'charged'	$z < $ MLD	1	$0.036 + 0.3 \times e^{-3} \times I_g$
	$z > $ MLD	$0.1 + 0.9 \times [1 - (5 \times 10^{-9})^{I_g}]$	$0.036 + 0.3 \times e^{-3} \times I_g$
Nutrient 'depleted'	$z < $ MLD	0.4	$0.036 + 0.3 \times e^{-3} \times I_g$
	NCD $> z >$ MLD	$0.04 + 0.36 \times [1 - (5 \times 10^{-9})^{I_g}]$	$0.036 + 0.3 \times e^{-3} \times I_g$
Applied $T°$ correction		$0.4 + 0.6 \times [1 - \exp^{-0.6 \times (T+6) - 50}]$	

1995a, b, 2000). It is noteworthy that the AMT studies involved photosynthesis–irradiance measurements that yielded $P^b{}_{max}$ data, whereas BATS measurements entailed sunrise-to-sunset, light–dark *in situ* incubations that provided values of $P^b{}_{opt}$. Despite this difference, the PhotoAcc model accounted for 73–80% of the variance in $P^b{}_{max}$ and $P^b{}_{opt}$ for the combined data set, while previously described temperature-dependent models explained $<9\%$ of the variance (Behrenfeld *et al.*, 2001b).

The enhanced performance of the PhotoAcc model relative to temperature-dependent models clearly indicates that the PhotoAcc model captures important sources of variability in $P^b{}_{opt}$. We therefore extended the analysis of Behrenfeld *et al.* (2001b) by applying modelled $P^b{}_{opt}$ values for the BATS record to calculations of ΣNPP. Two comparisons were made between measured and modelled ΣNPP. First, $P^b{}_z$ profiles were normalised to measured $P^b{}_{opt}$ and then rescaled to modelled $P^b{}_{opt}$. ΣNPP was then calculated by multiplying modelled $P^b{}_z$ by measured Chl_z and integrating from the surface to Z_{eu}. In this case, differences between measured and modelled ΣNPP resulted solely from errors in modelled $P^b{}_{opt}$, since information on vertical variability was retained. For the second comparison, ΣNPP was modelled using equation (7.3) and PhotoAcc estimates of $P^b{}_{opt}$. C_{sat} was taken as the average chlorophyll concentration for the three surface-most sampling depths and F was modelled following Talling (1957), since photoinhibition was minimal in the BATS data (see section 7.2 above). Unexplained variance in measured ΣNPP for this second comparison thus resulted both from neglecting vertical variability in photosynthesis and chlorophyll and from errors in modelled $P^b{}_{opt}$.

For both sets of calculations, greater than 70% of the observed variance in ΣNPP was accounted for using the PhotoAcc estimates of $P^b{}_{opt}$ (Fig. 7.7). When ΣNPP was recalculated using the temperature-dependent functions of Megard (1972), Morel (1991) and Behrenfeld and Falkowski (1997b), $\leq 36\%$ of the variance in ΣNPP was explained (primarily reflecting the correlation between ΣNPP and the product, $C_{sat} \times Z_{eu}$). We therefore endeavoured to recalculate global NPP using the PhotoAcc model and SeaWiFS data collected between November 1997 and October 1999.

Implementing the PhotoAcc model globally

As in the previous calculations (Plate 7.1), SeaWiFS measurements provided C_{sat} and PAR fields. $P^b{}_{opt}$ was calculated as a function of I_g for the mixed layer. Monthly global fields of mixed layer depths were provided by the Fleet Numeric Meteorology and Oceanography Center (FNMOC) (7 Grace Hopper Avenue, Stop 1, Monterey, California, USA 93940). I_g at the bottom of the mixed layer was calculated from the monthly mean PAR data and mean attenuation coefficients derived from C_{sat} following Morel and Berthon (1989). Application of the PhotoAcc model required an environmental index for assessing when to model $P^b{}_{opt}$ as either nutrient 'charged' or 'depleted' (Table 7.1). Behrenfeld *et al.* (2001b) used the

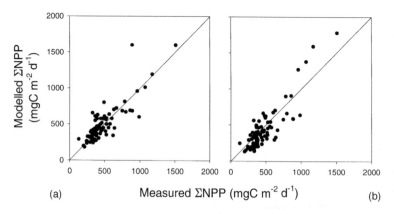

Fig. 7.7 Measured versus modelled depth-integrated net primary production (ΣNPP: mgC m^{-2} d^{-1}) for the six-year primary production record from the US-JGOFS Bermuda Atlantic Time Series (BATS) programme. (a) ΣNPP modelled using measured P^b_z profiles normalised to measured P^b_{opt} and then rescaled using P^b_{opt} values from the light-nutrient PhotoAcc model of Behrenfeld *et al.* (2001b) ($n = 81$; $r^2 = 0.71$). (b) ΣNPP modelled using the vertically generalised production model (VGPM) of Behrenfeld and Falkowski (1997b), measured surface chlorophyll concentrations, incident PAR, and euphotic depths, and PhotoAcc model estimates of P^b_{opt} ($n = 81$; $r^2 = 0.76$).

relationship between mixed layer and nutricline depths as their criterion, but such information is not yet available on the global scale. We therefore employed monthly nutrient depletion maps (Carder *et al.*, 1999) based on SeaWiFS-coincident surface temperature fields (Reynolds and Smith, 1994) and climatological nutrient–temperature relationships (Kamykowski and Zentara, 1986). The nutrient 'depleted' PhotoAcc equations were applied where surface temperatures exceeded the regionally dependent depletion temperature. In all other cases, the nutrient 'charged' PhotoAcc equations were used. This approach is clearly 'first order' and should be revisited in future applications of the model. Finally, the original data set used to parameterise the PhotoAcc model was not globally representative in its sea surface temperature range. We therefore applied the correction suggested by Behrenfeld *et al.* (2001b) for temperatures less than 5°C (Table 7.1).

Global application of the PhotoAcc model resulted in far greater spatial and seasonal variations in P^b_{opt} than generated using the temperature-dependent models of Morel (1991) and Behrenfeld and Falkowski (1997b). For the PhotoAcc model, P^b_{opt} varies as a function of MLD, surface PAR, and C_{sat} (due to associated changes in light attenuation (k_d)), as well as nutrient status. Thus, for a given temperature, PhotoAcc estimates of P^b_{opt} were lower during deep winter mixing than for the two temperature-dependent functions, but higher during springtime stratification when phytoplankton biomass was moderately low. During later stages of a phytoplankton bloom, P^b_{opt} estimates for the PhotoAcc model once again decreased due to the higher associated k_d values. In the central ocean gyres, P^b_{opt} was generally lower than estimated using the temperature-dependent models, despite high PAR and low k_d, because these regions were classified as nutrient 'depleted'.

The diverse spatial patterns in P^b_{opt} generated using the PhotoAcc model resulted in a similar degree of heterogeneity in global NPP estimates. Seasonal distributions of NPP exhibited sharp gradations between low and high production regions (Plate 7.2). During austral summer (December to February), NPP was high along the south-eastern margins of Africa and South America, with enhanced productivity in the eastern equatorial Pacific extending nearly to the Gulf of California (Plate 7.2a). NPP was also generally elevated in the southern half of the Indian Ocean (note: the exceptionally high NPP south-east of Madagascar is likely an overestimate resulting from extrapolation artefacts in the FNMOC MLD fields), around New Zealand, and south of about 40° S (Plate 7.2a). During boreal summer (June to August), the equatorial Pacific band of enhanced NPP was latitudinally restricted, but reached further west to about 130° W (Plate 7.2b). Regions of enhanced NPP in the Northern hemisphere extended to lower latitudes for the PhotoAcc results than for the temperature-dependent models, and NPP was considerably elevated at high northern latitudes along the western margins of the Atlantic and Pacific (Plate 7.2b). Lower NPP values in the central North Atlantic, relative to values to the west and north-east, reflected decreases in P^b_{opt} resulting from bloom-associated decreases in k_d and seasonal nutrient depletion in this region (Plate 7.2b).

The VGPM–PhotoAcc estimate of annual global NPP for the 1997–1999 period was 41 Pg C y^{-1}, which is 15 Pg C y^{-1} lower than the polynomial temperature-dependent model. This difference largely reflects a decrease in North Atlantic NPP for the PhotoAcc model, as well as general decreases in NPP in mid- and low-latitude open ocean regions and the subtropical convergence of the Southern hemisphere (see Plates 7.1b and 7.3a). The bivariate histogram of NPP (Esaias *et al.*, 1999) for the two models reveals a bimodal distribution for the PhotoAcc model that reflects, in part, differences in P^b_{opt} estimates for nutrient 'charged' and 'depleted' conditions (Plate 7.3b).

The range of NPP values in the bivariate histogram (Plate 7.3b) primarily reflects variability in C_{sat} and P^b_{opt}. Since the same SeaWiFS C_{sat} data were used for both estimates of global productivity, spatial differences in modelled P^b_{opt} can be illustrated by separating the bivariate histogram into sections (A to F in Plate 7.3b) and plotting the global distribution of these segments. Differences in model estimates of P^b_{opt} for January and July of 1999 (Plate 7.4) clearly illustrate the influence of seasonal changes in MLDs, phytoplankton biomass and nutrients. During austral summer, the PhotoAcc model yields enhanced P^b_{opt} values in stratified regions of the Southern hemisphere (cool colours) and low values in deeply mixed regions of the Northern hemisphere (warm colours) (Plate 7.4a). At mid- and low-latitudes of the Southern hemisphere, decreases in P^b_{opt} relative to the polynomial model resulted from regional changes in nutrient distributions (Plate 7.4a). In a similar manner, P^b_{opt} values in the Northern hemisphere were generally elevated over the polynomial estimates during the boreal summer, whereas values in the Southern hemisphere were generally diminished (Plate 7.4b).

7.5   Directions for model development

As stated in the introduction to this chapter, one application for quantitative NPP models is to detect changes in the magnitude and distribution of biospheric primary production. At present, we are not even certain of the sign for long-term temporal trends in ocean NPP, largely because a consistent time series of C_{sat} data only began with the launch of SeaWiFS in 1997. However, it is safe to assume that if ocean NPP is trending at the decadal scale, its rate is <1 Pg C per year. By comparison, global NPP estimates presented here and based on the same PAR, SST, and C_{sat} input fields varied over a range of 18 Pg C y^{-1} (Plates 7.1 and 7.2). Decreasing this uncertainty hinges critically on improving estimates of spatial and temporal variability in P^b_{opt}. We propose that the PhotoAcc model provides a useful foundation for building a more complete P^b_{opt} model because:

(1)   it is based on environmental factors directly influencing P^b_{opt} and
(2)   its division into separate expressions for CC_{cap} and chlorophyll imparts considerable flexibility for incorporating new formulations describing additional environmental constraints.

We have made an initial attempt at applying the PhotoAcc model globally. The resultant P^b_{opt} (and NPP) distributions deviated substantially from the much smoother spatial gradations generated by the temperature-dependent functions (Plates 7.1, 7.3a, 7.4), with results for the PhotoAcc model being consistent with oceanographic patterns in seasonal mixing and stratification (Plate 7.4). A weakness in the current application is the employment of nutrient depletion fields for choosing between nutrient 'charged' and 'depleted' PhotoAcc equations. The approach leads to clear delineations in NPP fields at both the seasonal and annual scale (Plates 7.2 and 7.3a). In the future, improved treatment of nutrient limitation effects on P^b_{opt} may be achieved by considering temporal patterns in mixing, which would also be more consistent with the results of Behrenfeld *et al.* (2001b). Improvements in the monthly global mixed layer depth fields are also needed. Research is now under way to investigate potential lidar-based techniques for remotely measuring surface mixing depths, but an operational mission will not be realised for years. During the interim, advanced modelling approaches should be pursued.

Careful assessment of uncertainties in global satellite-derived NPP estimates is also necessary, but difficult to conduct. Regardless of the model employed, global estimates of NPP are sensitive to errors in C_{sat}. Stringent point-by-point comparisons of satellite-derived C_{sat} values with field measurements of near-surface

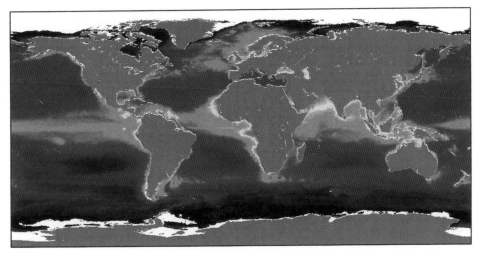

(a)

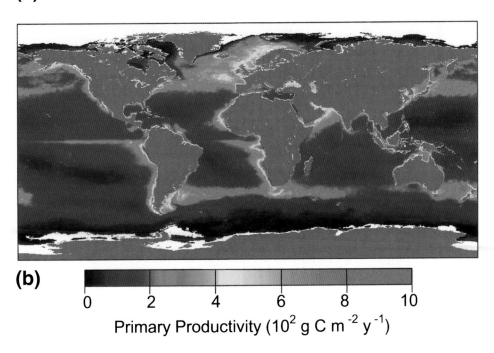

(b)

0   2   4   6   8   10

Primary Productivity (10^2 g C m^{-2} y^{-1})

Plate 7.1. Global annual net primary production (ΣNPP: $g\,Cm^{-2}y^{-1}$) for November 1997 to October 1999 calculated using monthly SeaWiFS satellite measurements of near-surface chlorophyll and the vertically generalised production model (VGPM) of Behrenfeld and Falkowski (1997b). (a) P^b_{opt} estimated using the exponential function of Morel (1991). (b) P^b_{opt} estimated using the polynomial function of Behrenfeld and Falkowski (1997b). Tan colours indicate land; white indicates ice. Annual NPP values for each pixel represent the average for the two year period. Zero productivity in areas of seasonal darkness or sea ice was included in these averages, but all other zero values were treated as missing data and not counted. Total global production was calculated from the average NPP values weighted by the area (m^2) of each pixel.

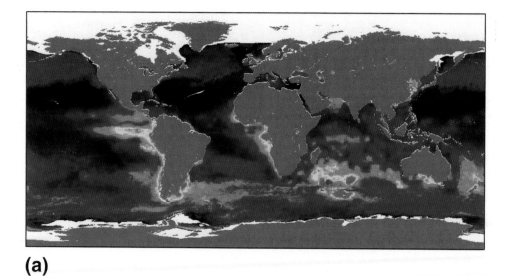

(a)

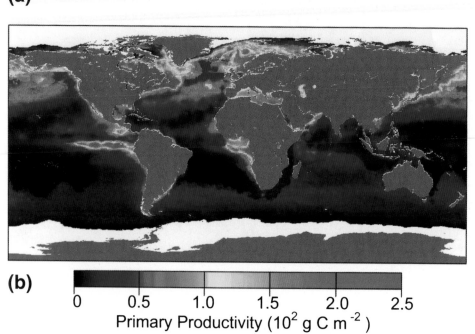

(b)

| 0 | 0.5 | 1.0 | 1.5 | 2.0 | 2.5 |

Primary Productivity (10^2 g C m^{-2})

Plate 7.2. Seasonal total net primary production (ΣNPP: g C m^{-2}) for (a) austral summer (December to February) and (b) boreal summer (June to August). ΣNPP was calculated using (1) SeaWiFS chlorophyll data collected between November 1997 and October 1999, (2) the vertically generalised production model (VGPM) of Behrenfeld and Falkowski (1997b), and (3) the PhotoAcc model for P^b_{opt} (Behrenfeld *et al.*, 2001b). ΣNPP values represent the total carbon fixed during each three month season. Note that the colour scale is different from Plates 7.1 and 7.3a.

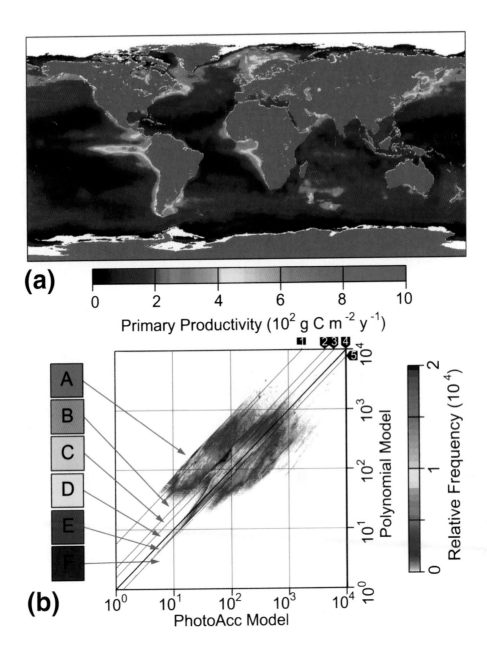

Plate 7.3. (a) Global annual net primary production (ΣNPP: $gCm^{-2}y^{-1}$) estimated using the VGPM and PhotoAcc models. ΣNPP values represent averages for the two year period of November 1997 to October 1999. (b) Scatter plot of monthly average ΣNPP ($gCm^{-2}y^{-1}$) for January and July 1999 obtained with (x-axis) the PhotoAcc model for P^b_{opt} versus (y-axis) the polynomial P^b_{opt} model of Behrenfeld and Falkowski (1997b). The colour scale indicates the frequency of points within bins of 0.01 log units. The heavy line indicates a 1:1 relationship between NPP values for the two P^b_{opt} models. The distribution of points in this bivariate histogram is divided into six clusters (labelled from A to F) that exhibited strong geographical coherence (Plate 7.4). The clusters are delineated here by the five thin solid lines (numbered at the top right-hand corner) associated with features aligned parallel to the 1:1 line.

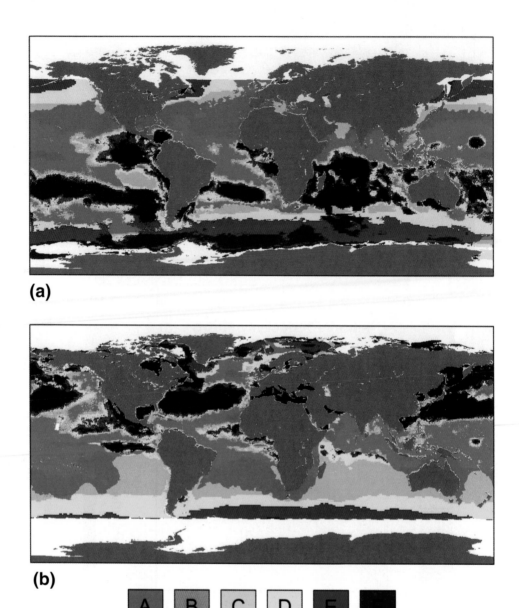

Plate 7.4. Geographic distribution of the six difference clusters (A through F) identified in Plate 7.3b. (a) January 1999. (b) July 1999. Warm colours correspond to regions where NPP estimates for the PhotoAcc model were lower than the polynomial model of Behrenfeld and Falkowski (1997b). Cool colours indicated higher NPP values for the PhotoAcc model than the polynomial model.

chlorophyll concentrations indicate a minimum uncertainty in C_{sat} of $\approx 35\%$, which is likely an underestimate for global data sets. Errors in C_{sat} are common to all NPP estimates and thus not a source of divergence between model results. These errors do, however, contribute to the overall accuracy of the modelled NPP values. Uncertainties in global productivity estimates also arise from uncertainties in model formulations and coefficients, as well as from ancillary data sets (e.g. PAR, SST, nutrient depletion criteria, MLD). Different methodological approaches to spatial and temporal averaging, treatment of missing data, and application of monthly averages also contribute a degree of uncertainty to global annual NPP estimates. Uncertainties also exist in the field ^{14}C-uptake data used for model parameterisation and testing.

In addition to the rather practical issues identified above, future applications of the PhotoAcc model will require modifications to the basic equations involved. Most critical will be a revised treatment of nutrient limitation. In the current formulation, the PhotoAcc model simply switches between nutrient 'charged' and 'depleted' equations (Table 7.1). Although such a 'switch' in the ratio of light harvesting to CC_{cap} has been observed in the laboratory (Falkowski *et al.*, 1989) (Fig. 7.6b), it is likely not a general phenomenon. More probable is that this ratio changes in proportion to the severity of nutrient limitation. Developing a robust model of nutrient effects will require additional laboratory studies focused on identifying the physiological basis for why nutrient limitation increases the demand for light reaction products relative to carbon fixing capacity. Behrenfeld *et al.* (2001b) suggested the adjustment reflects an additional ATP requirement for extracting nutrients from the surrounding medium. Whatever the mechanism, results from various field photosynthesis–irradiance studies appear to advocate employing a gradual transition between nutrient-sufficient and nutrient-depleted P^b_{max} values, rather than an abrupt shift. Specifically, Platt and Jassby (1976), Harding *et al.* (1987), Claustre *et al.* (1997) and Moline *et al.* (1998) all reported a positive correlation between α^b and P^b_{max}. Such a relationship can not be accounted for by classic photoacclimation theory (which tends to produce a negative correlation between α^b and P^b_{max}) and is likely related to nutrient availability. Resolving the physiological basis for this positive correlation could very well make one of the most significant contributions towards improving global NPP estimates.

An additional alteration to the PhotoAcc model that requires attention is the inclusion of separate expressions for iron-limiting growth conditions. Iron limitation causes unique stoichiometeric changes in the composition of the photosystems (Guikema and Sherman, 1983; Sandmann, 1985; Greene *et al.*, 1992; Straus, 1994; Vassiliev *et al.*, 1995; Behrenfeld and Kolber, 1999). A typical response to iron limitation is an increase in the ratio of PSII:PSI, which consequently should decrease chlorophyll-specific efficiencies for NADPH production and thus lower P^b_{max} relative to similar levels of growth limitation by nitrogen or phosphate. Application of iron-specific PhotoAcc equations would require defining the global distribution of iron-limited phytoplankton populations.

Finally, the prevalence of photoinhibition *in situ* requires re-evaluation. Although a near-surface decrease in ^{14}C-uptake is a common feature of field P^b_z profiles, it is not clear whether such measurements provide an adequate representation of photoinhibition. The problem lies in the discrepancy between the duration of high light exposure in incubation bottles and the actual exposure experienced by phytoplankton in an actively mixing surface layer. If a decrease in carbon fixation is a common phenomenon *in situ*, then P^b_{max} should exhibit midday minima when PAR is highest. In contrast, phytoplankton sampled at the surface often exhibit midday maxima in P^b_{max} (Doty and Oguri, 1957; Yentsch and Ryther, 1957; Sournia, 1974; MacCaull and Platt, 1977; Gargas *et al.*, 1979; Malone *et al.*, 1980; Cullen *et al.*, 1992; Behrenfeld *et al.*, 1998). In fact, when conducted in parallel using samples from the same phytoplankton population, photosynthesis-irradiance measurements can indicate a midday maximum in P^b_{max} while sunrise–sunset *in situ* incubations exhibit significant reductions in surface carbon fixation (Cullen *et al.*, 1992).

The physiological basis for the afore-mentioned discrepancy is that photoinhibition is time-dependent and results from the overexcitation and subsequent inactivation of PSII reaction centres. A decrease in the number of function PSII centres (n), however, is not necessarily coupled to a decrease in carbon fixation. At light saturation, carbon fixation is limited by the Calvin cycle reactions and a decrease in n simply results in an increase in the electron turn-over rate of the remaining functional PSII centres (Behrenfeld *et al.*, 1998). The extent of this 'excess capacity' of the light reactions over the dark reactions is generally on the order of 40–65%, but varies with photoacclimation state and species-specific differences in photoacclimation strategies (i.e. a preferential increase in n or σ_{PSII} with decreasing light) (Kok, 1956; Weinbaum *et al.*, 1979; Heber *et al.*, 1988; Leverenz *et al.*, 1990; Behrenfeld *et al.*, 1998). In other words, approximately half of the functional PSII reaction centres must typically be inactivated before a reduction in light saturated carbon fixation occurs. If phytoplankton near the surface are transported by physical mixing away from photoinhibiting light before such large decreases in n develop, the daily allocation of fixed carbon will be unaffected. In a static bottle incubation, this reprieve from photoinhibition is impossible.

7.6   Summary

Primary productivity models, when coupled with time-varying global measurements of phytoplankton biomass, provide critical information for detecting changes in oceanic carbon fluxes. Such models are also beneficial for assessing ecological responses to regional changes in nutrient loading and may provide estimates of biologically mediated long-term carbon sinks if linked to secondary models of export production. The foundation of both aquatic and terrestrial NPP models is similar and, in the simplest terms, can be described as equating NPP to the product of plant biomass and light utilisation efficiency. From this basic relationship, a wide variety of phytoplankton productivity models have developed. At the categorical level, these models differ with respect to the presumed importance of resolving the time, depth, and spectral dependence of photosynthesis. Within each model category, differences largely centre around the specific treatment of the photosynthesis–irradiance relationship and in the characterisation of the underwater light field.

In this chapter, we have attempted to describe the relationship between various NPP models and the assumptions required to progress from fully expanded WRMs to the very simple DIMs. We have argued that all global estimates of NPP are critically dependent on the approach employed for modelling light-saturated photosynthesis and demonstrated this conclusion using two temperature-dependent models and a new light-nutrient model. This later model produces ΣNPP estimates comparable with local-scale, daily ^{14}C-uptake results. Although the model was applied to global satellite C_{sat} data using a simple DIM formulation, its description at the level of chlorophyll synthesis and Calvin cycle reactions is ideally suited for application in a WRM. The primary advantage of using a spectrally resolved model will be the improved estimates of mixed-layer acclimation irradiances, which are a primary forcing factor for variability in P^b_{max}.

Over the 50 year history of productivity modelling, significant advances have been made toward assessing global-scale phytoplankton photosynthesis. Noteworthy accomplishments include the development of variable P^b_{max} models, characterisation of the underwater light field, description of vertical variability in phytoplankton biomass distributions, and the development of remote sensing capabilities that now provide global measurements of near-surface chlorophyll every two days. These achievements notwithstanding, we hope our discussion has impressed upon the reader the need for further improvements in primary productivity modelling. These improvements will require laboratory, field, computational and remote sensing components, all of which represent exciting challenges for future research programmes.

Acknowledgements

We thank Dorota Kolber (Rutgers University) for computational and programming assistance, Kendall Carder and Robert Chen (University of South Florida) for graciously supplying monthly nutrient depletion data, Charles McClain and Gene Feldman (NASA/GSFC) for assistance with SeaWiFS data, and Robert Frouin (Scripps Institute of Oceanography) for PAR data. This work was supported by the NASA Earth Science Enterprise through a grant to M.J. Behrenfeld (No. NAG5-9787) and through the EOS MODIS Instrument Team Member Investigation of W. Esaias. The SeaWiFS, MLD, and SST data were provided by the Goddard DAAC.

References

Antoine, D., André, J-M. & Morel, A. (1996) Oceanic primary production 2. Estimation at global scale from satellite (coastal zone color scanner) chlorophyll. *Global Biogeochemical Cycles* **10**, 57–69.

Balch, W.M. & Byrne, C.F. (1994) Factors affecting the estimate of primary production from space. *Journal of Geophysical Research* **99(C4)**, 7555–70.

Balch, W., Evans, R., Brown, J., Feldman, G., McClain, C. & Esaias, W. (1992) The remote sensing of ocean primary productivity: Use of new data compilation to test satellite models. *Journal of Geophysical Research* **97(C2)**, 2279–93.

Behrenfeld, M.J. & Falkowski, P.G. (1997a) A consumer's guide to phytoplankton primary productivity models. *Limnology and Oceanography* **42**, 1479–91.

Behrenfeld, M.J. & Falkowski, P.G. (1997b) Photosynthetic rates derived from satellite-based chlorophyll concentration. *Limnology and Oceanography* **42**, 1–20.

Behrenfeld, M.J. & Kolber, S. (1999) Widespread iron limitation of phytoplankton in the South Pacific Ocean. *Science* **283**, 840–43.

Behrenfeld, M.J., Marañón, E., Siegel, D.A. & Hooker, S.B. (2001b) A photoacclimation and nutrient based model of light-saturated photosynthesis for quantifying oceanic primary production. *Marine Ecology Progress Series* (in press).

Behrenfeld, M.J., Prasil, O., Kolber, Z.S., Babin, M. & Falkowski, P.G. (1998) Compensatory changes in photosystem II electron turnover rates protect photosynthesis from photo-inhibition. *Photosynthesis Research* **58**, 259–68.

Behrenfeld, M., Randerson, J., McClain, C., *et al.* (2001a) Biospheric primary production during an ENSO transition. *Science* **291**, 2594–7.

Carder, K.L., Chen, F.R., Lee, Z.P., Hawes, S. & Kamykowski, D. (1999) Semi-analytic MODIS algorithms for chlorophyll *a* and absorption with bio-optical domains based on nitrate-depletion temperatures. *Journal of Geophysical Research* **104(C3)**, 5403–21.

Ciais, P., Tans, P.P., Trolier, M., White, J.W.C. & Francey, R.J. (1995) A large Northern hemisphere terrestrial CO_2 sink indicated by the $^{13}C/^{12}C$ ratio of the atmosphere. *Science* **269**, 1098–102.

Claustre, H., Moline, M.A. & Prezelin, B.B. (1997) Sources of variability in the column photosynthetic cross-section for Antarctic coastal waters. *Journal of Geophysical Research* **102**, 25047–60.

Cullen, J.J., Lewis, M.R., Davis, C.O., & Barber, R.T. (1992) Photosynthetic characteristics and estimated growth rates indicate grazing is the proximate control of primary production in the equatorial Pacific. *Journal of Geophysical Research* **97**, 639–54.

Dierssen, H.M., Vernet, M. & Smith, R.C. (2000) Optimizing models for remotely estimating primary production in Antarctic coastal waters. *Antarctic Science* **12**, 20–32.

Doty, M.S. & Oguri, M. (1957) Evidence for a photosynthetic daily periodicity. *Limnology and Oceanography* **2**, 37–40.

Dubinsky, Z., Falkowski, P.G. & Wyman, K. (1986) Light harvesting and utilization by phytoplankton. *Plant Cell Physiology* **27**, 1335–49.

Durbin, E.G. (1974) Studies on the autecology of the marine diatom *Thalassiosira nordens-kiöldii* cleve: I. The influence of daylength, light intensity, and temperature on growth. *Journal of Phycology* **10**, 220–25.

Eppley, R.W. (1972) Temperature and phytoplankton growth in the sea. *Fisheries Bulletin* **70**, 1063–85.

Esaias W.E, Iverson, R.L. & Turpie, K. (1999) Ocean province classification from ocean color data: observing biological signatures of variations in physical dynamics. *Global Change Biology* **6**, 39–55.

Falkowski, P.G. (1981) Light–shade adaptation and assimilation numbers. *Journal of Plankton Research* **3**, 203–216.

Falkowski, P.G., Barber, R.T. & Smetacek, V. (1998) Biogeochemical controls and feedbacks on ocean primary production. *Science* **281**, 200–206.

Falkowski, P.G. & Raven, J. (1997) *Aquatic Photosynthesis*. Blackwell Science, Oxford.

Falkowski, P.G., Sukenik, A. & Herzig, R. (1989) Nitrogen limitation in *Isochrysis galbana* (Haptophyceae). II. Relative abundance of chloroplast proteins. *Journal of Phycology* **25**, 471–8.

Field, C.B, Behrenfeld, M.J., Randerson, J.T. & Falkowski, P.G. (1998) Primary production of the biosphere: Integrating terrestrial and oceanic components. *Science* **281**, 237–40.

Frouin, R. & Chertock, B. (1992) A technique for global monitoring of net solar irradiance at the ocean surface. Part 1: Model. *Journal of Applied Meteorology* **31**, 1056–66.

Gargas, E., Hare, I., Martens, P. & Edler, L. (1979) Diel changes in phytoplankton photosynthetic efficiency in brackish waters. *Marine Biology* **52**, 113–22.

Geider, R.J. (1987) Light and temperature dependence of the carbon to chlorophyll ratio in micgroalgae and cyanobacteria: Implications for physiology and growth of phytoplankton. *New Phytologist* **106**, 1–34.

Geider, R.J., Osborne, B.A. & Raven, J.A. (1985) Light dependence of growth and photosynthesis in *Phaeodactylum tricornutum* (Bacillariophyceae). *Journal of Phycology* **21**, 609–619.

Greene, R.M., Geider, R.J., Kolber, Z. & Falkowski, P.G. (1992) Iron-induced changes in light harvesting and photochemical energy conversion processes in eukaryotic marine algae. *Plant Physiology* **100**, 565–75.

Guikema, J.A. & Sherman, L.A. (1983) Organization and function of chlorophyll in membranes of cyanobacteria during iron starvation. *Plant Physiology* **73**, 250–56.

Harding, L.W., Fisher, T.R., Jr & Tyler, M.A. (1987) Adaptive responses of photosynthesis in phytoplankton: Specificity to time-scale of change in light. *Biology and Oceanography* **4**, 403–37.

Heber, U., Neimanis, S. & Dietz, K.-J. (1988) Fractional control of photosynthesis by the Q_B protein, the cytochrome f/b_6 complex and other components of the photosynthetic apparatus. *Planta* **173**, 267–74.

Kamykowski, D. & Zentara, S.-J. (1986) Predicting plant nutrient concentrations from temperature and sigma-t in the upper kilometer of the world ocean. *Deep-Sea Research* **33**, 89–105.

Keeling, R.F., Piper, S.C. & Heinmann, M. (1996) Global and hemispheric CO_2 sinks deduced from changes in atmospheric O_2 concentration. *Nature* **381**, 218–21.

Kirk, J.T.O. (1983) *Light and Photosynthesis in Aquatic Ecosystems*. Cambridge University Press, Cambridge.

Kok, B. (1956) On the inhibition of photosynthesis by intense light. *Biochimica et Biophysica Acta* **21**, 234–44.

Lapointe, B.E., Dawes, C.J. & Tenore, K.R. (1984) Interaction between light and temperature on the physiological ecology of *Gracilaria tikvahiae*. *Marine Biology* **80**, 171–8.

Leverenz, J.W., Falk, S., Pilström, C.-M., & Samuelsson, G. (1990) The effects of photo-inhibition on the photosynthetic light-response curve of green plant cells (*Chlamydomonas reinhardtii*). *Planta* **182**, 161–8.

Ley, A.C. & Mauzerall, D. (1982) Absolute absorption cross-sections for photosystem II and the minimum quantum requirement for photosynthesis in *Chlorella vulgaris*. *Biochimica et Biophysica Acta* **680**, 95–106.

Lindeman, R.L. (1942) The trophic-dynamic aspect of ecology. *Ecology* **23**, 399–418.

Longhurst, A. (1995) Seasonal cycles of pelagic production and consumption. *Progress in Oceanography* **36**, 77–167.

Longhurst, A., Sathyendranath, S., Platt, T. & Caverhill, C. (1995) An estimate of global primary production in the ocean from satellite radiometer data. *Journal of Plankton Research* **17**, 1245–71.

MacCaull, W.A. & Platt, T. (1977) Diel variations in the photosynthetic pararmeters of coastal marine phytoplankton. *Limnology and Oceanography* **22**, 723–31.

Malone, T.C., Garside, C. & Neale, P.J. (1980) Effects of silicate depletion on photosynthesis by diatoms in the plume of the Hudson river. *Marine Biology* **58**, 197–204.

Marañón, E. & Holligan, P.M. (1999) Photosynthetic parameters of phytoplankton from $50°$ N to $50°$ S in the Atlantic ocean. *Marine Ecology Progress Series* **176**, 191–203.

Marañón, E., Holligan, P.M., Varela, M., Mouriño, B. & Bale, A. (2000) Basin-scale variability of phytoplankton biomass, production, and growth in the Atlantic ocean. *Deep-Sea Research I* **47**, 825–57.

Megard, R.O. (1972) Phytoplankton, photosynthesis, and phosphorus in Lake Minnetonka, Minnesota. *Limnology and Oceanography* **17**, 68–87.

Michaels, A.F. & Knap, A.H. (1996) Overview of the U.S. JGOFS Bermuda Atlantic Time-series Study and the Hydrostation S program. *Deep-Sea Research II* **43**, 157–98.

Moline, M.A., Schofield, O. & Boucher, N.P. (1998) Photosynthetic parameters and empirical modelling of primary production: a case study on the Antarctic peninsula shelf. *Antarctic Science* **10**, 39–48.

Morel, A. (1991) Light and marine photosynthesis: A spectral model with geochemical and climatological implications. *Progress in Oceanography* **26**, 263–306.

Morel, A. & Berthon, J.-F. (1989) Surface pigments, algal biomass profiles, and potential production of the euphotic layer: Relationships reinvestigated in view of remote-sensing applications. *Limnology and Oceanography* **34**, 1545–62.

Morel, A. & Prieur, L. (1977) Analysis of variations in ocean color. *Limnology and Oceanography* **22**, 709–22.

Myneni, R.B., Tucker, C.J., Keeling, C.D. & Asrar, G. (1998) Interannual variations in satellite-sensed vegetation index data from 1981 to 1991. *Journal of Geophysical Research* **103**, 6145–60.

Platt, T., Caverhill, C. & Sathyendranath, S. (1991) Basin-scale estimates of oceanic primary production by remote sensing: The north Atlantic. *Journal of Geophysical Research* **96**, 15 147–59.

Platt, T. & Jassby, A.D. (1976) The relationship between photosynthesis and light for natural assemblages of coastal marine phytoplankton. *Journal Phycology* **12**, 421–30.

Platt, T. & Sathyendranath, S. (1988) Oceanic primary production: Estimation by remote sensing at local and regional scales. *Science* **241**, 1613–20.

Potter, C.S., Randerson, J.T., Field, C.B. *et al.* (1993) Terrestrial ecosystem production: a process model based on global satellite and surface data. *Global Biogeochemical Cycles* **7**, 811–41.

Reynolds, R.W. & Smith, T.M. (1994) Improved global sea surface temperature analyses using optimum interpolation. *Journal of Climate* **7**, 929–48.

Robins, D.B., Bale, A.J., Moore, G.F., *et al.* (1996) *AMT-1 Cruise report and preliminary results* (eds S.B. Hooker & E.R. Firestone), NASA Technical Memorandum 104566, Vol. 35, 87 pp. NASA Goddard Space Flight Center, Greenbelt, Maryland.

Rodhe, W. (1966) Standard correlations between pelagic photosynthesis and light. In: *Primary Productivity in Aquatic Environments* (ed. C.R. Goldman), pp. 367–81. University of California Press, Berkeley.

Rodhe, W., Vollenweider, R.A. & Nauwerk, A. (1958) The primary production and standing crop of phytoplankton. In: *Perspectives in Marine Biology* (ed. A.A. Buzzati-Traverso), pp. 299–322. University of California Press, Berkeley.

Ryther, J.H. (1956) Photosynthesis in the ocean as a function of light intensity. *Limnology and Oceanography* **1**, 61–70.

Ryther, J.H. & Yentsch, C.S. (1957) The estimation of phytoplankton production in the ocean from chlorophyll and light data. *Limnology and Oceanography* **2**, 281–6.

Sandmann, G. (1985) Consequences of iron deficiency on photosynthetic and respiratory electron transport in blue-green algae. *Photosynthesis Research* **6**, 261–71.

Sarmiento, J.L. & Wofsy, S.C. (1999) *A U.S. Global Carbon Cycle Plan*. US Global Change Research Program, Washington, DC.

Schopf, J.W. (1983) *Earth's Earliest Biosphere: its Origin and Evolution*. Princeton University, Princeton.

Siegel, D.A., Michaels, A.F., Sorensen, J., O'Brien, M.C. & Hammer, M.A. (1995a) Seasonal variability of light availability and its utilization in the Sargasso Sea. *Journal of Geophysical Research* **100**, 8695–713.

Siegel, D.A., O'Brien, M.C., Sorensen, J.C., Konnoff, D. & Fields, E. (1995b) *BBOP data processing and sampling procedures*. US JGOFS Planning Report Number 19, US JGOFS Planning and Coordination Office, 77 pp.

Siegel, D.A., Westberry, T.K., O'Brien, M.C., *et al.* (2001) Bio-optical modeling of primary production on regional scales: The Bermuda biooptics project. *Deep-Sea Research II* **48**, 1865–96.

Sournia, A. (1974) Circadian periodicities in natural populations of marine phytoplankton. *Advances in Marine Biology* **12**, 325–89.

Steemann Nielsen, E. (1952) The use of radio-active carbon (C^{14}) for measuring organic production in the sea. *Journal du Conseil pour la Conservation Internationale pour l'Exporation de la Mer* **18**, 117–40.

Steemann Nielsen, E. & Hansen, V.K. (1959) Light adaptation in marine phytoplankton populations and its interrelation with temperature. *Physiology of the Plant* **12**, 353–70.

Steemann Nielsen, E. & Jørgensen, E.G. (1968) The adaptation of plankton algae: I. General part. *Physiology of the Plant* **21**, 401–13.

Stitt, M. (1986) Limitation of photosynthesis by carbon metabolism, I. Evidence for excess electron transport capacity in leaves carrying out photosynthesis in saturating light and CO_2. *Plant Physiology* **81**, 1115–22.

Straus, N.A. (1994) Iron deprivation: physiology and gene regulation. In: *The Molecular Biology of Cyanobacteria* (ed. D.A. Bryant), pp. 731–50. Kluwer Academic, Dordrecht.

Sukenik, A., Bennett, J. & Falkowski, P.G. (1987) Light-saturated photosynthesis – limitation by electron transport or carbon fixation? *Biochimica et Biophysica Acta* **891**, 205–15.

Talling, J.F. (1957) The phytoplankton population as a compound photosynthetic system. *New Phytologist* **56**, 133–49.

Tans, P.P., Fung, I.Y. & Takahashi, T. (1990) Observational constraints on the global atmospheric CO_2 budget. *Science* **247**, 1431–8.

Tucker, C.J. & Nicholson (1999) Variations in the size of the Saharan desert from 1980 to 1997, *Ambio* **28**, 587–91.

Vassiliev, I.R., Kolber, Z., Wyman, K.D., Mauzerall, D., Shukla, V.K. & Falkowski, P.G. (1995) Effects of iron limitation on photosystem II composition and light utilization in *Dunaliella tertiolecta*. *Plant Physiology* **109**, 963–72.

Verity, P.G. (1981) Effects of temperature, irradiance, and daylength on the marine diatom *Leptocylindrus danicus* cleve: I. Photosynthesis and cellular composition. *Journal of Experimental Marine Biology and Ecology* **55**, 79–91.

Vollenweider, R.A. (1966) Calculation models of photosynthesis–depth curves and some implications regarding day rate estimates in primary production measurements. In: *Primary Productivity in Aquatic Environments* (ed. C.R. Goldman), pp. 426–57. University of California Press, Berkeley.

Weinbaum, S.A., Gressel, J., Reisfeld, A. & Edelman, M. (1979) Characterization of the 32,000 Dalton chloroplast membrane protein: probing its biological function in *Spirodela*. *Plant Physiology* **64**, 828–32.

Wright, J.C. (1959) Limnology of Canyon Ferry Reservoir: Phytoplankton standing crop and primary production. *Limnology and Oceanography* **4**, 235–45.

Yentsch, C.S. (1974) Some aspects of the environmental physiology of marine phytoplankton: a second look. *Oceanography and Marine Biology Annual Review* **12**, 41–75.

Yentsch, C.S. & Ryther, J.H. (1957) Short-term variations in phytoplankton chlorophyll and their significance. *Limnology and Oceanography* **2**, 140–42.

Yoder, J.A. (1979) Effect of temperature on light-limited growth and chemical composition of *Skeletonema costatum* (Bacillariophyceae). *Journal of Phycology* **15**, 362–70.

Chapter 8
On the Interannual Variability in Phytoplankton Production in Freshwaters

Colin S. Reynolds

8.1 Introduction: variability and constancy in phytoplankton seasonality

This chapter considers the phenomenon of year-to-year differences in the biomass and the species composition of the phytoplankton assembled in particular freshwater systems. Sometimes the interannual variation is relatively minor, with small differences in the maximum population of the same dominant species and small differences in the day of the year it is achieved. At other times, the interannual differences are startling: populations achieving bloom proportions in one year but with other species featuring in a more modest outburst the next. Inevitably, this variability can be traced back to interannual differences in the primary production achieved, in a way which is readily sensed using Steemann Nielsen's (1952) radio-labelling method of measurement. However, the observations are more deeply rooted, in differences in the starting populations, perhaps also in their provenance, in the environmental conditions obtaining and, frequently, in the dissipative loss processes to which they are subject. It is the aim of this chapter to convey something of the scope of interannual variability in the biomass and species composition of planktonic primary product assembled, by reference to some published examples. It will also seek generalised explanations for the extent of interannual variation observed and it attempts to discern patterns of community assembly with which the variability is integral. The consideration explores the theoretical links to the structure and organisation of communities in inherently variable environments.

By coincidence, this last objective provides an interesting point of entry to the problem. For the editors to determine that this topic merits a place in an appreciation of Steemann Nielsen's insights and approaches to phytoplankton physiology almost implies that some difficulty attaches to the idea that aquatic primary productivity should vary at all or by as much as it does. In truth, when the scales of phytoplankton production are matched to those of the key environmental constraints – temperature, insolation, vertical mixing – it would seem equally obvious that year-to-year variations in productivity relate to year-to-year differences in the sequence and intensity of fluctuations in the atmospheric weather. The surprise might then be that events in the production of phytoplankton showed any year-to-year reproducibility at all. The title of the essay could just as well have referred to the extent of interannual similarity as of variability.

I shall start by describing two contrasting but well-studied lakes – Lake Kinneret in Israel and Rostherne Mere in the English Midlands (see Tables 8.1 and 8.2). Kinneret provides one of the best-known and best-attested examples of year-to-year similarity in the abundance, distribution and composition of the phytoplankton. This warm monomictic lake is situated in the Rift Valley of the River Jordan, where it experiences a typically Mediterranean climate, with rainfall confined to the winter period (Serruya, 1978). The lake is isothermal between late December and February, with a minimum temperature of $\sim 13°C$; the lake is strongly stratified from April to December, with maximum epilimnetic temperatures of $\sim 30°C$. Its waters are alkaline and slightly saline; the maximum bioavailability of both nitrogen and phosphorus is modest ($\leq 200\,\mu g\ N\,l^{-1}$, $\leq 5\,\mu g\ P\,l^{-1}$), save in the late-summer hypolimnion. In each year of the 20-year period 1970–1989 inclusive, the phytoplankton developed in a characteristic way (Berman *et al.*, 1992). Starting with the autumnal breakdown of stratification, the phytoplankton generally comprised nanoplanktonic species (including the unicellular Cyanobacterium, *Chroococcus*, the cryptomonad *Plagioselmis* (formerly known as *Rhodomonas*) and the haptophyte, *Chrysochromulina*. With full turn-over and winter flooding, the plankton supported larger coenobial and filamentous forms (including the diatom *Aulacoseira granulata*). At this time, there would also be an excystment of the large dinoflagellate, now known

Table 8.1   Lake Kinneret.

Location	32° 50′ N, 35° 28′ E
Altitude	−209 m a.s.l.
Area	168 km^2
Volume	$4.3 \times 10^9\,m^3$
Mean depth	25.5 m
Maximum depth	43 m
Mean hydraulic retention time	5.2 years

Table 8.2   Rostherne Mere.

Location	53° 21′ N, 2° 23′ W
Altitude	32 m a.s.l.
Area	0.49 km^2
Volume	$6.5 \times 10^6\,m^3$
Mean depth	13.4 m
Maximum depth	30 m
Mean hydraulic retention time	2.2 years

as *Peridinium gatunense*, which then, under conditions of reduced vertical mixing, typically built to a maximum during March to May, when it constituted the greatest plankton biomass during the year. The termination of the bloom coincided with high surface temperatures (27 to 30°C) and severe nitrogen depletion, with many cells encysting and settling to the bottom. From June to September, phytoplankton biomass was generally rather low but dominated by picocyanobacteria (Malinsky-Rushansky *et al.*, 1995). A metalimnetic layer of green sulphur bacteria (*Chlorobium phaebacteroides*) would generally develop towards the summer end.

Seasonal averaging of the biomass (in Fig. 8.1(a) and (b)) and annual averaging of the composition of the biomass (Fig. 8.1(c)) confirmed the stability of the annual pattern, with no evident continuous trend. Berman *et al.* (1992) explored the interannual variability in relation to an increasing trend in phosphorus loading and to fluctuations in the biomass of herbivorous zooplankton but both seemed incidental to the central predominance of *Peridinium*. The success of this alga was attributed to its efficient perennation and to an ability to use its superior motility to 'scavenge' the resources in the stratified water column during the early part of the stagnation. Depletion of both nitrogen and phosphorus to the limits of detection emphasises the efficiency of uptake but also relates the crop size to the resources available. The phase of picoplankton dominance is similarly constrained by the resource paucity and is redolent of other highly oligotrophic pelagic systems where such organisms are the principal primary producers, within tightly coupled microbial food loops. Such interannual variations in the size of the *Peridinium* crop as were documented seemed to be inverse to the fluctuations in the *Aulacoseira* maxima: thus, years with larger *Aulacoseira* crops generally heralded smaller *Peridinium* maxima.

The fact that the stability of the pattern has been upset during the 1990s reinforces the impression of the erstwhile interannual regularity. Moreover, it provides some clues to the circumstances of recent increases in instability. Larger crops of *Peridinium* have been observed in some years and, in others, there has been relative failure (when green algae or *Microcystis* became briefly abundant). The appearance of *Aphanizomenon ovalisporum* in the summer of 1994 and in certain subsequent years also represents a departure from the stable pattern. The tempting explanation is that the ongoing increase in phosphorus loading has triggered these events, although analysis of the data shows other trends of decreasing winter concentrations and towards lengthening of the period of thermal stratification (Hambright *et al.*, 1994). Berman and Shteinman (1998) have commented on a trend of weakening turbulent diffusivity. The changes are not large but they may be considered sufficient to have affected the apparent selective exclusivity in favour of *Peridinium* at the critical point in its annual development.

The contrasting data set from Rostherne Mere (Fig. 8.2, from Reynolds and Bellinger, 1992) is put forward to illustrate the case of a lake supporting conspicuous year-to-year differences in productivity, biomass and composition of the phytoplankton. Properly classified as a warm monomictic lake (dimixis is rare in maritime

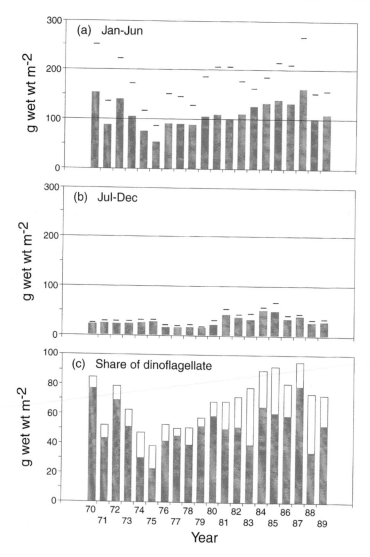

Fig. 8.1 Semi-annual average wet biomass of phytoplankton in Lake Kinneret, Israel (with standard deviations as horizontal lines) in successive years, for the period: (a) January–June; and (b) July–December. (c) Relative share of annual biomass in dinoflagellates (predominantly *Peridinium gatunense*). Redrawn from Berman *et al.* (1992).

climates), Rostherne Mere is rather smaller and shallower than Kinneret, and the annual rainfall falling on its hydrological catchment is spread through the year ($\sim 850\,\mathrm{mm}\ \mathrm{y}^{-1}$). Rostherne Mere is also much more eutrophic. The surface inflow into the lake is rather meagre, the catchment comprising kataglacial drift deposits and the hydrological balance in the lake is maintained by groundwater percolation. Besides damping the hydrological variability, this arrangement also maintains a substantial bicarbonate alkalinity ($\leq 2\,\mathrm{meq}\ \mathrm{l}^{-1}$) and a continuous supply of nutrients to the lake (some $2\,g\,P, 9\,g\,N\,m^{-2}\,y^{-1}$). The bioavailabilities of nitrogen ($\sim 1.5\,mg\,N$ l^{-1}) and, especially, phosphorus ($\leq 300\,\mu g\,P\,l^{-1}$) are such that they are unlikely to limit phytoplankton production, save perhaps when the water column is stratified in summer (when nitrogen is more vulnerable to exhaustion than is phosphorus). With typical winter temperatures close to $4°C$, stratification generally lasts from April to October and the mid-summer thermocline is centred at a depth of 4–6 m. However, Reynolds and Bellinger (1992) showed that there had been considerable interannual differences in the duration and stability of stratification, as well as in the frequency of downmixing events during the summer and in the intensity of restratification in their wake.

Over the period of years represented in Fig. 8.2, both the abundance of phytoplankton in Rostherne Mere and its species dominance showed conspicuous interannual variability. In five of the eighteen summers considered, a high plankton density ($>25\,mm^3\,l^{-1}$ by biovolume, $>100\,\mu g$ chlorophyll l^{-1}) was dominated (80–90% of the biomass) by the dinoflagellate, *Ceratium hirundinella.* In a further seven, the Cyanobacterium *Microcystis* dominated, albeit with a smaller biovolume. Any supposition of mutual exclusion is countered by the events in a thirteenth year (1989) when *Microcystis* followed *Ceratium* to dominance. In 1985, neither of these species were numerous but the Nostocalean *Aphanizomenon* maintained its prominence into a rather cool summer. In two of the years, diatoms (*Asterionella,* *Aulacoseira granulata* or *Stephanodiscus rotula*) dominated a small summer biomass. Throughout 1978, the plankton comprised little other than an enormous crop of the filamentous Cyanobacterium, *Planktothrix agardhii.* Oddest of all was the 1983 bloom of the chlorophyte, *Scenedesmus quadricauda,* which achieved a biomass of $44\,mm^3\,l^{-1}$.

Even the spring periods differed: in most years, there was a very modest growth ('bloom' would be a gross exaggeration), in which *Asterionella* and *Stephanodiscus rotula* featured, in variable proportions, with some *Cryptomonas.* However, larger crops of diatoms were eventually achieved in 1972 and in 1986.

All the species mentioned are firmly associated with eutrophic or very eutrophic hard-water lakes. All of them were present in the lake at least at some time in almost every year. With no obvious sign that nutrient availability should have influenced the initiation of any one species over the others, what factor(s) could possibly have determined this outcome? In fact, the analysis of Reynolds and Bellinger (1992) revealed that these interannual differences were not random events but were linked to initial or pervasive coherence with weather-generated external forcing. It is the severe light limitation in the mixed water column that restricts the development of large crops to the stratified periods. Weak early summer stratification prolongs the suspension opportunities for diatoms that may persist into the summer, or it facilitates their ready resurrection. More stable near-surface stratification selects against diatoms but promotes a succession that moves quickly from small flagellates and green algae to Nostocalean Cyanobacteria before passing to *Microcystis;* however, if the phase of its spring recruitment from the benthos is unsuccessful, the slower-

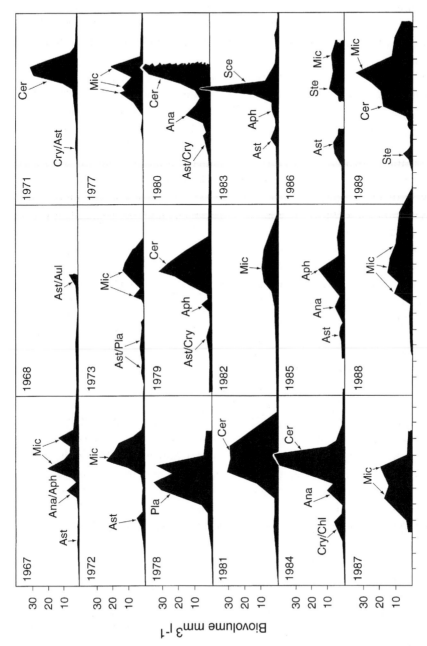

Fig. 8.2 Variability in the biovolume of phytoplankton in the upper 4.5 m of the water column of Rostherne Mere, UK, and the dominating species (Ana = *Anabaena circinalis*, Aph = *Aphanizomenon flos-aquae*, Ast = *Asterionella formosa*, Aul = *Aulacoseira granulata*, Cer = *Ceratium hirundinella*, Cry = *Cryptomonas* (mostly *C. ovata*), Mic = *Microcystis aeruginosa*, Pla = *Planktothrix agardhii*, Sce = *Scenedesmus quadricauda*, Ste = *Stephanodiscus rotula*). Redrawn from Reynolds and Bellinger (1992).

growing *Ceratium* comes through instead. When *Ceratium* recruitment or development is also limited by low temperatures (as happened in 1985), the early-summer Nostocalean populations (*Aphanizomenon* or *Anabaena* spp.) are left in the position of 'dominance'. Persistent episodes of downmixing apparently favoured the 1978 growth of a large *Planktothrix* inoculum. At the other extreme, the remarkable stable stratification of 1983 established pond-like conditions for a pond-like plankton.

Moreover, the patterns of dynamic responses agreed well with those diagnosed in another eutrophic lake in the region (Crose Mere: Reynolds and Reynolds, 1985). They have also been supported subsequently by observations on the sensitivity and selection of common species to contrived variations imposed by experimental mixing of limnetic mesocosms (see Reynolds *et al.*, 1983, 1984).

Between these two extremes of interannual variability and constancy lie numerous other behaviours that could be cited to support the impact upon the annual cycle of phytoplankton abundance and species composition. Fluctuations in fertility are best known in the context of eutrophication, or its reversal (oligotrophication) but there are conceivably cases where interannual fluctuations in external loading exact year-to-year differences in plankton populations. Leitão and Leglize (2000) have given an example of year-to-year differences in a small, relatively new, flow-balancing reservoir in an upper catchment in France. Here, the character of the phytoplankton assemblage fluctuates between oligotrophic and mesotrophic. Elsewhere, persistent nutrient poverty in clear, shallow oligotrophic lakes provides a ceiling on plankton-carrying capacity that vertical mixing and light limitation do not undercut. The deeper the lake, however, the higher the possibility that the carrying potential will react to fluctuations in physical mixing.

Reynolds (1997a) contrasts the minor interannual variability in a moderately shallow oligotrophic lake in the English Lake District (Derwent Water) with the scope for interannual variability in the oligotrophic great lakes, such as Baykal and Superior, based entirely upon the fragility of their stratification. Even in the case of regular events, such as the spring bloom of *Asterionella* in Windermere, the subject of Lund's (1949, 1950) classic studies, the inoculum, the onset and rate of growth and the eventual extent to which the silicon-regulated carrying capacity is attained, are all profoundly sensitive to the variations in temperature and energy income wrought by physical mixing (Maberly *et al.*, 1994). The more eutrophic is the lake and the greater the carrying capacity of its nutrients, then the greater is the concentration of plankton biomass and the greater is the vertical attenuation of light. In turn, episodes of stable stratification become increasingly important in generating populations that challenge the nutrient-determined carrying capacity. Stability, in the sense of interannual regularity in the opportunities for plankton growth, might be ideally sought in a shallow, low-latitude system, subjected to a constant inflow of nutrient-rich water. This ideal seems close to what might be contrived in a semi-continuous laboratory experiment. However, a near-natural equivalent, the remarkable, groundwater-fed Montezuma Well in Arizona, does come close to simulating the ideal conditions (see Boucher *et al.*, 1984).

Generally, it is possible to uphold the supposition of Reynolds and Bellinger (1992) that the physical environment acts like a filter in allowing particularly advantageous adaptations and traits and processing specialisms among certain groups of algae to be selected, through the resource- and energy- harvesting opportunities that variability provides. Indeed, it should be possible to match the biotic responses, from photosynthetic regulation through to the assembly of phytoplankton communities, to the proximal drivers influenced by the variability. It is this hypothesis which is further explored below.

8.2   Variability in environmental forcing

The first priority is to remind ourselves what is variable, with what predictability and with what effects and at what scales those responses might be perceived. Listed in Table 8.3 are some properties and forcing phenomena characterising open-water systems which are relevant to the behaviours of phytoplankton. The main sections are arranged across the page in order of ascending stochasticity and, within the vertical blocks, factors are listed in ascending order of the scale of variation. Part of the relevant variability involves factors that are constant or so nearly so on the scales considered that they have not been listed. An extreme example might be the wide variation in the radiant intensity of sunlight reaching the water surface but which may always be related back to a maximal flux density of electromagnetic energy emanating from the sun and the angle of its incidence to the Earth's outer atmosphere. Other such fixed 'invariables' – substantial and colligative properties of water and its solutes, including those influenced by its molecular polymerisation, its specific heat, the latent heat of its state changes, the coefficient of thermal expansion, the solubility products of gases and many salts of biological importance, and their interactions with density, boiling and freezing points – are also not listed. Gravitational acceleration, molecular diffusion, the stoichiometry of carbohydrate and protein synthesis and the energetic yields of their molecular oxidation are also assumed to be 'constants' in the present consideration. Moreover, in the context of interannual variability within a single lake system, year-to-year differences in basin shape, or in its relationship to its hydrological catchments, would usually be disregarded.

The least predictable events mostly occur on very short time-scales. Averaging over longer time-scales tends to a greater level of estimable probability: thus, the precise rainfall to fall on any given future day is unpredictable, except within the probability limits set by past experiences (hence, expressions referring to such as a 'one-in-a-hundred-years storm'). However, the aggregate rainfall over the year may vary around a long-term annual mean by about $\pm 30\%$. Care is needed in the ascription 'long-term', which is measured on purely anthropocentric scales: in

Table 8.3 Drivers of environmental variability in aquatic systems.

Time-scale	Robustly predictable variables	Variables of estimable probability	Strongly stochastic variables
< Second	—	—	Wave breaking Cell-specific PAR flux
Second–minute	—	—	Wind speed Wind direction Cloud cover Net heat flux Depth-specific PAR flux
Minute–hour	—	Water temperature Surface mixed-layer depth	Turbulent intensity
Hour–day	Day length	Period of thermal stratification	Daily wind run $I**$ Direct rainfall Hydraulic load
Day–week	—	Depth of epilimnion *chl* (I) Carbon flux Nutrient flux	—
Week–year	—	*chl* (K) Seasonal duration of stratification	—
Year–decade	—	Nutrient load Hydrological balance GAtC effects	—
> Decade	—	Cyclical climate change Milankovitch oscillation	Anthropogenic climate change

Notes: chl(K) = the phytoplankton carrying capacity provided by nutrients; *chl*(I) = phytoplankton carrying capacity provided by PAR income; GAtC = global atmospheric circulation, including those of the El Niño and North Atlantic Oscillations; $I**$ is a measure of light availability in the water (see text); PAR = photosynthetically-available radiation.

reality, annual rainfall has fluctuated very much more widely during the great climatic shifts of the ten postglacial millennia.

This nesting of effects nevertheless helps us to appreciate the scale linkages of forcing functions. Thus, the principal factors regulating the onset and maintenance of thermal stratification (solar heating, net of long-wave and evaporative heat losses, generating buoyancy; the work of wind and convection, in its mechanical dissipation) can each vary on scales of seconds to minutes. This means that any tendency for the surface waters to stratify or to propagate turbulent mixing may change from minute to minute and, certainly, from hour to hour, during the course of a single day. Moreover, because the combination is unique, the relative magnitudes of the components of forcing may be quite different the next day, with a quite different outcome in terms of the resultant structure. The longer is the period of integration (several days, a week or two), the more the within-year variability is expected to even out the supposedly self-compensating extremes, towards some interannual norm. Yet there are important residuals that make up significant year-to-year differences in spring water temperatures, the speeds and direction of the wind, the date of the eventual stagnation and the depth of the seasonal thermocline. Also, the warmer the hypolimnion at the time of its isolation then, not surprisingly, the less its resistance to late-summer convectional downmixing and the earlier the date of the autumnal 'overturn' (holomixis) is likely to be.

These interannual differences in aggregate forcing, distinguishing (say) mild and cold winters, or cool, wet springs from sunny warm ones, are now generally recognised to be driven by quasi-cyclical hemisphere-scale variability in the fluid motion of the oceans and of the atmosphere. None of these variations seems likely to be wholly stochastic but the precise causes of the oscillations and their frequencies are imperfectly understood. The El Niño cycle that pervades the pattern of rainfall and storm intensity in the Pacific Ocean and adjacent land masses is mediated by changes in the energy and moisture released to the atmospheric circulation in the region of the intertropical convergence zone. They follow a cyclical net increase in sea temperature in the equatorial Pacific until it exceeds that of the air above. The balance is rapidly redressed through storm activity that peaks periodically, about every four to seven years. The El Niño Southern Oscillation underpins much of the interannual variability in rainfall, runoff and storm frequency in the land masses on both the western and eastern sides of the Pacific (Strub *et al.*, 1985; Harris *et al.*, 1988; Anderson *et al.*, 1996).

The courses of the main ocean currents and their role in regulating temperature and precipitation in the adjacent continental seaboards are well-understood. Yet even quite small fluctuations in the original trajectory can be shown to result in wide disparities of impact at the other side of the ocean. George and Taylor (1995) have related differences in the stability of lake stratification in maritime Europe to interannual oscillations in the angle at which the warm North Atlantic Drift current (or Gulf Stream) is deflected at Cape Hatteras; variations of a degree or two are magnified into a north–south variation in its eastern landfall of several hundreds of kilometres. The further south its trajectory, then the less western Europe benefits (in terms of landfall) from the warming effects and the less stable is the vernal temperature stratification of the maritime lakes.

In the atmosphere, superficially small season-to-season differences in the upper-atmosphere path of the circumpolar jetstream (geostrophic wind) and in the gradients between accumulated latitudinal differences in atmospheric pressure propagate subsidiary effects to the level of local weather patterns. Local air temperatures, cloud and precipitation have long been known to reflect the provenance and moisture content of the predominant air-masses to which the locality is exposed (McIntosh and Thom, 1969). Thus, year-to-year weather differences depend most on the relative penetration and persistence of the most influential air masses. There is considerable current interest in the influence of the North Atlantic Oscillation (NAO) on weather variability in northern and central Europe. The 'oscillation' refers to the variable gradient in atmospheric pressure between the subtropics (the 'Azores High') and the subarctic (the 'Iceland Low'). A high positive index corresponds to a steep gradient, with greater penetration of mild, moist oceanic air. The lowered incidence and truncated duration of winter ice cover on Continental lakes has been one of the obvious effects (Hurrell, 1995). Supposing an extended open-water phase and a warmer average winter air temperature to be beneficial to plankton production, analysis of relevant data sets is expected to show more subtle biological responses to the oscillation. Certainly, some striking variations in the vernal biomass of *Daphnia* in Bodensee (or Lake of Constance; 47° N; over 1000 km from the Atlantic seaboard) have been shown to be correlated to the NAO index in the preceding midwinter (Straile and Geller, 1998). Straile (2000) has since demonstrated a cascading negative impact on the survival of phytoplankton with extension of the 'clear-water phase'.

Accepting the sources of large-scale pelagic environmental variability, the response scales of phytoplankton population dynamics are mainly contained within the middle rows of Table 8.3 (days to weeks). It is on these differential outcomes that the focus of this chapter is now mainly directed.

8.3   Patterns of community responses to variable pelagic forcing

The challenge is to explain the means whereby interannual variability is transmitted to year-to-year differences in the community structures actually supported. This transmission is a two-part process. There is a need to discern the selective patterns that associate particular species assemblages with particular sets of dominant environmental variables: what properties of a few species were ultimately favoured by the average conditions obtaining? Then there is a need to clarify the resonances of the biological responses to the multiplicity of small-scale variations itemised in Table 8.3 that are accommodated at the ecological scales of population growth and community assembly. The second part is integral to the first but, as an exercise in reconstruction, it is easier to deal with the large-scale pattern first.

Current theories about how planktonic communities are assembled revolve around two main philosophies. One supposes that competition among species for the resources to build new cells and to sustain population increase determines that the best-adapted species is most likely to rise to dominance. Planktonic species show demonstrably differential affinities for inorganic nutrients, including carbon, that frequently occur at low concentrations (Tilman *et al.*, 1982). There are also significant interspecific differences in the carbon processing efficiencies of algae, reflected in their abilities to grow at continuous low light intensities or in deep-mixed water columns that afford low integral doses (Reynolds, 1987; Huisman *et al.*, 1999a, b). Thus, communities are seen to comprise species, each superior to the others in gathering one of its resources but inferior to at least one of them in gathering its other requirements. The composition of the community is determined by the relative concentrations in which resources are supplied. The more resources that are determined to be limiting, the more species that can coexist. This, the essence of Tilman's (1992) resource-ratio hypothesis, can be expanded to include the effects of light limitation (Huisman and Weissing, 1994). The hypothesis has been backed by persuasive evidence from laboratory experiments on algae grown in semi-continuous cultures (Tilman *et al.*, 1982). However, outcomes are less well clear in simulations using several species competing for more than two limiting resources (Huisman and Weissing, 1999).

The alternative philosophy differs in supposing that there are occasions when resources are relatively abundant and interspecific competition for their seques-tration is relatively weak, so that the growth of each of several species is simulta-neously maintained, independently of each other and independently of the ratios in which resources are supplied. This view does not obscure the possibility that one or other of these resources is exhaustible or, indeed, that the first to be exhausted sets an upper limit on the biomass-carrying capacity. Neither does it preclude the like-lihood that, as the critical resource availability is approached, distinct competitive advantage falls to the species (strictly of those present) with the highest affinity for the limiting resource. Prior to that time, the species most advantaged are the more exploitative of those present, usually having dispersive life histories and potentially very fast rates of growth. This non-equilibrium view of planktonic communities has been advocated by Harris (1983, 1986) and it is also backed by strong experimental evidence from field-scale manipulations using mesocosms (Reynolds, 1986, 1988a).

The trouble with both theories is their failure to explain the numbers of phyto-plankton species that make up the global pool (some 4000–5000; Reynolds, 1997a), or indeed, the species richness to be found in a single lake during the year (probably well over 100: Padisák, 1992), or even the numbers of those comprising the assemblage at any one point in time (anywhere between four and twenty species may be needed to contribute the first 95% of algal biomass obtaining: Reynolds, 1997a). If all the others are inferior, why has competition not eliminated them aeons ago?

This is one of the most perplexing questions in phytoplankton ecology, mainly because it confounds intuition, Darwinian logic and the 'principle of competitive exclusion' (Hardin, 1960). G.E. Hutchinson's (1961) eloquent diagnosis of the plankton paradox has tantalised plankton ecologists ever since. The most widely accepted explanatory theory is the one that Hutchinson himself suggested: whereas complete competitive exclusion takes a long time to achieve (equivalent to some 17–20 consecutive algal generations), the plankton environment changes over generally shorter periods (see section 8.2). If the change is of sufficient magnitude and of the appropriate temporal scale (seconds to weeks), population growth (and, so, community development) may well be slowed, arrested or redirected, to an extent which *disturbs* the progress towards the outcome of competitive exclusion. If such disturbances are rare, local competitive exclusion is more nearly approached and species diversity is depressed. On the other hand, if such disturbances are frequent, only the most tolerant or most resilient species will survive. At intermediate frequencies, however, many more species survive but few are ever excluded – high species diversity is favoured.

This principle has also been cast formally (the intermediate disturbance hypothesis, or IDH: Connell, 1978) and has been supported by carefully analysed field observations (notably those of Haffner *et al.*, 1980; Trimbee and Harris, 1980; Padisák, 1993), laboratory (Sommer, 1985, 1995) and field experiments (Reynolds, 1988a). If the pelagic environment is modified sufficiently to raise the growth rate and, eventually, the standing biomass of a new contender over the present dominant, that is to force a response at the level of species composition, it may take four or more generations of the contender to bring it to dominance. By the same logic, a relaxation in the forcing may switch the dynamic advantage again. Thus, a disturbance frequency of two to three generations seems likely to yield the highest levels of diversity (Reynolds *et al.*, 1993).

This leaves open important questions about the intensity and frequency of forcing necessary to invoke the community response. We will return to those later. For the moment, we may accept that linkages exist between environmental variability and the potential to invoke variations in species composition. However, it has proved singularly difficult, even using sophisticated model approaches, to simulate the conditions resulting in the dominance of particular systems. Detailed analyses of data sets (using statistical software such as CANOCO, DECORANA or TWIN-SPAN) rarely manage to relate species dominance to environmental descriptions in a way which conveys the preferred conditions of a given species or the species likely to dominate the simulated conditions (Rojo *et al.*, 2000). Models applying resource-ratio hypothesis to the simulation of community assembly quickly reveal random, non-equilibrial dynamics when more than two species are faced with more than two dimensions of variability (Huisman and Weissing, 1999). Community models that seem to work do so either because the system modelled is regulated overwhelmingly by just one factor, as in the case of the relation of phytoplankton growth rates and the integral mixed-layer light dose in Windermere (Reynolds, 1990) and of *Planktothrix* blooming in a monomictic shallow lake (Jiménez Montealegre *et al.*, 1995), or because the community is reduced to a small selection of phylogenetic representation (Steel, 1995; Steel and Duncan, 1999), or because the community is reduced to functional respondents (Elliott *et al.*, 2000).

The last-named approaches carry the important message that community assembly is not chaotic or random but, rather, that the patterns in which several species are active simultaneously may be attributable to some general feature(s) of the contemporaneous environment which are either tolerated by the active species or are obligately required. This is emphatically quite different from the supposition that individual species are each selected by particular environmental qualities.

Following this line of reasoning, we may suppose that the assembly process should also be viewed in this alternative perspective: thus, the species composition of the emerging assemblage may be fortuitous, save that they are all tolerant of the set of environmental conditions obtaining (Reynolds, 2000). Then, the next logical deduction is that the species composition and abundance in assembling communities are subject to a hierarchy of general governing restrictions. They range from the non-random, at the level of the pool of available species, to the increasingly stochastic determination of the local species composition (Belyea and Lancaster, 1999). Although recalling concepts rooted in the origins of ecology (Pianka, 1999), current interest in formulating the rules governing community assembly (Keddy, 1992; Weiher and Keddy, 1995) is generating reasonable and promising conceptual models of the way in which variation in production is linked to the extent of environmental variability. Simply, if the co-occurrence and abundance of two species is not random but is attributable to their simultaneous satisfaction of an assembly rule which excludes a third (and, doubtless, many others), then those mechanisms should be 'explicit and quantitative'.

Such explicit attributes of the freshwater phytoplankton *have* been recognised and quantified in recent years. Using a variety of quantitative and statistical techniques to analyse co-respondent species in changing lake environments (Reynolds, 1980; Seip and Reynolds, 1995), blocks of species (I called them 'associations') have been delimited and given provisional identities (so far, only by alphanumerics: Reynolds, 1997a; see also Table 8.4). Moreover, common attributes of the species thus aggregated have been discerned and they have been shown to have consistent affinities for certain kinds of limnetic environment, distinguished by seasonal or hydrographic constraints, as well as by their ionic composition and trophic state (Reynolds, 1988b, 1995; Seip and Reynolds, 1995).

Table 8.4 Trait-based associations of fresh water phytoplankton[1].

		Variable[2]							
Assn	Key members[3]	h_m <3	I^* <1.5	θ <8	[P] <10^{-7}	[N] <10^{-6}	[Si] <10^{-5}	[CO_2] <10^{-5}	f <0.4
A	*Urosolenia* *Cycl. comensis*	–	?	+	+	+	+	–	–
B	*Asterionella* *Cycl. menghiniana* *Aulac. italica*	–	+	+	+	–	–	–	–
C	*Asterionella* *Steph. rotula* *Aulac. ambigua*	–	+	+	–	–	–	?	–
D	*Nitzschia* *Steph. hantzschii* *Synedra acus*	+	+	+	–	–	–	+	–

Continued

Table 8.4 *Continued.*

Assn	Key members[3]	Variable[2]							
		h_m < 3	I^* < 1.5	θ < 8	[P] < 10^{-7}	[N] < 10^{-6}	[Si] < 10^{-5}	[CO_2] < 10^{-5}	f < 0.4
N	*Tabellaria* *Cosmarium* *Staurodesmus*	–	–	–	+	–	+/–	–	?
P	*Fragilaria* *Aulac. granulata* *Staurastrum pingue* *Closterium*	–	–	–	–	–	+/–	+	+
E	*Dinobryon* *Chrysophaerella* *Glenodinium*	+	+	+	+	–	+	–	–
F	*Sphaerocystis* *Gemellicystis* *Botryococcus*	+	–	+	+	–	+	–	–
G	*Eudorina, Volvox*	+	–	+	–	–	+	+	+
H	*Aphanizomenon* *Anabaena flo-aq* *Gloeotrichia*	+	–	–	–	+	+	+	+
J	*Pediastrum* *Coelastrum* *Scenedesmus*	+	?	+	–	–	+	?	–
K	*Aphanothece* *Aphanocapsa* *Synechocystis*	+	?	–	–	–	+	+	?
L_M	*Ceratium hirundinella* *Microcystis*	+	–	–	–	–	+	+	+
L_O	*Woronichinia* *Peridinium inconspic.* *Ceratium* *Merismopedia*	+	–	–	+	–	+	–	+
M	Perennial *Microcystis*	+	–	–	–	–	+	+	+
U	*Uroglena*	+	–	?	+	–	+	–	+
T	*Geminella* *Binuclearia* *Tribonema*	–	?	–	+/–	–	+	?	+
S_1	*Limnothrix* *Pseudanabaena* *Planktothrix agardhii*	+	+	+	–	–	+	+	+
S_2	*Spirulina, Arthrospira*	+	+	–	–	–	+	+	+
S_N	*Cylindrospermopsis* *Anabaena minutissima*	+	+	–	–	+	+	+	+
R	*Planktothrix rubescens* *Planktothrix mougeotii* *Lyngbya limnetica*	+	+	–	–	–	+	?	+
V	*Chromatium* *Chlorobium*	+	+	–	–	–	+	–	–

Continued

Table 8.4 *Continued.*

Assn	Key members[3]	Variable[2]							
		h_m <3	I^* <1.5	θ <8	[P] <10^{-7}	[N] <10^{-6}	[Si] <10^{-5}	[CO_2] <10^{-5}	f <0.4
W	*Euglena* *Gonium* *Synura*	+	+	+	–	–	+	?	–
Y	*Cryptomonas*	+	+	+	–	–	+	?	–
X₁	*Chlorella, Ankyra* *Monoraphidium* Eutrophic picoplankton	+	–	+	–	–	+	+	–
X₂	*Rhodomonas* *Chrysochromulina*	+	–	+	?	–	+	?	–
X₃	*Koliella* *Chrysococcus*	+	–	+	+	–	+	–	–
Z	*Synechococcus* Eukaryotic picoplankton	+	–	+	+	+	+	?	–

Notes:

(1) As proposed by Reynolds (1997a) with modifications of Padisák and Reynolds (1998) and Huszar *et al.* (2000). Entries in table are to denote tolerance (+) or no positive benefit (–) of the environmental condition set; (+/–) is used to denote that some species in the association are tolerance (?) denotes that tolerance is suspected but not proven.

(2) Variables signified are: depth of surface mixed layer (h_m, in metres from surface); mean daily irradiance levels experienced (I^*, in mol photons m^{-2} d^{-1}); water temperature (θ, in °C); the concentration of soluble reactive phosphorus ([P]), in mol l^{-1}); the concentration of dissolved inorganic nitrogen ([N]), in mol l^{-1}); the concentration of soluble reactive silicon ([Si], in mol l^{-1}); the concentration of dissolved carbon dioxide ([CO_2], in mol l^{-1}); and the proportion of the water processed each day by rotiferan and crustacean zooplankton (*f*).

(3) Some representative genera or species only are listed.

In Fig. 8.3, these threads are drawn together to demonstrate the match between the metabolic classification of limnetic environments, the trait boundaries of a small selection of representative planktonic photoautotrophs and the generalised distribution of some of the alphanumeric associations, or trait-based categories. It is emphasised again that while such broad distinctions as the ability to fix nitrogen or the obligate requirement for silicon set very powerful restrictions to community participation, and while the size and shape of organisms present restrictions with respect to light harvesting, growth rate and survival strategies, the approach *does not* predict or explain the dominance of a particular species in an emerging community. This is not a failure of the approach but a recognition that the abundance of particular species in particular communities is ultimately unpredictable, save on the grounds of probability. Instead, it works at a higher level of certainty, anticipating the attributes of species which will not be restricted from participating and, hence, the trait-based category that the dominant species represent.

To return to the examples introduced in section 8.1, the plankton sequence in Kinneret might be summarised, in terms of the alphanumeric classification (Table 8.4), as X2→P→L→Z. That in Rostherne Mere normally conforms to

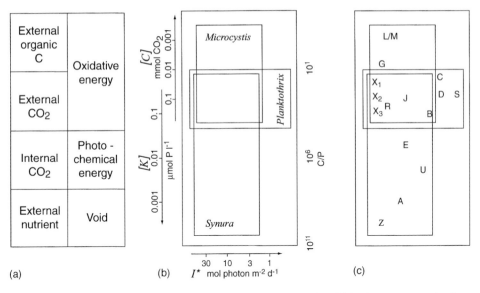

Fig. 8.3. (a) Limnetic habitat matrix, representing domains in which the ecosystem is characteristically constrained (either by processing energy, carbon or nutrient resources) by the factors named, as proposed in Reynolds (1999b): the vertical axis corresponds to carbon synthesis; at the lower left-hand corner, poverty of nutrients is restrictive; half-way up, nutrients match carbon fixation potential; above that carbon supplies become critical or supplemented by external organic carbon. Moving rightwards, processing constraints are progressively imposed: less light energy for fixing carbon or less oxidative potential. (b) Provisional habitat template, using analogous axes but quantified in terms of nutrient concentration ([K], actually P concentration), carbon concentration ([C]) and a daily integral photon flux ($I*$); the boxes show the ecological ranges of three specialist phytoplankters (*Microcystis aeruginosa* and *Planktothrix agardhii*, drawn from data collected in Reynolds, 1997a; and *Synura petersenii*, from Saxby-Rouen *et al.*, 1997), which nevertheless overlap in nutrient-replete, carbon-replete, well-insolated waters. (c) The same template but now populated with some trait-based associations of phytoplankton (categories as in Table 8.4).

$C \rightarrow H \rightarrow L$, though it managed only $C \rightarrow H$ in 1985; the outright oddities were the $C \rightarrow S$ sequence in 1978 and $C \rightarrow H \rightarrow J$ in 1983. The phytoplankton in the Vieux-Pré Reservoir changed progressively with diminishing trophy from $C/X1 \rightarrow E \rightarrow L$ to $A \rightarrow F \rightarrow U$ (Leitão and Leglize, 2000), while Montezuma's Well is kept at X1 for substantial periods (Boucher *et al.*, 1984).

8.4   Response thresholds of phytoplankton to variable pelagic forcing

Diagnosing response thresholds

If the overriding role of externally generated atmospheric events in setting the selective restrictions is accepted and the intermediacy of disturbance events is acknowledged to be a principal agent in maintaining diversity, then the next priority is to clarify the mechanisms of the community response, with particular reference to the intensity and the frequency of forcing. There is a related problem, exposed in several studies of the impact of physical variability on phytoplankton composition, which is that structural responses are not always precipitated by a given level of physical forcing, any more than every abrupt change in composition is necessarily brought about through the intervention of external force (see, for instance, Eloranta, 1993; Jacobsen and Simonsen, 1993; Barbiero *et al.*, 1999). Even when the external causes (the forcing functions) are adequately distinguished from the internal reactions (the responses: Juhász-Nagy, 1993), there remain difficulties in sorting out the variability of impacts owing to variations in the extent, intensity, duration and return frequency of the forcing itself (Romme *et al.*, 1998). The relative importance of infrequent large events (Turner *et al.*, 1998), especially the ways in which responses might be influenced by persistent, small-scale responses (Paine *et al.*, 1998), taxes the terrestrial theoreticians.

However, I believe that plankton biology has already evolved, at least in a simplified form, the practical means to overcome the quantitative inadequacies of the intermediate disturbance hypothesis and to define the scale linkages between forcing and response (Reynolds, 1997b). Its essential component is the direct comparability of the mechanical energy of the forcing with the capacity of an assembled structure to absorb it (and so, to survive intact or 'undisturbed') or, failing that, to recover structure once again (i.e. to be resilient to forcing events). The quantification of the relevant components provides the relevant time and space scales.

The approach to the quantification emanates from the structural dynamic models of ecosystems (e.g. Jørgensen, 1992) and Nielsen's (1992) clever thermodynamic working of the energy exchanges of ecosystems. This recognises that useful, short-wave energy comes from the sun and some of this is intercepted by photoautotrophs to generate and transfer the reductive power to synthesised carbohydrates and proteins (see Box 8.1). The incipient ecosystem does not acquire energy so much as it intervenes in the abiotic dissipation of radiant solar energy. Like a water-wheel in a stream, the biotic component diverts part of the general (entropic) flux into useful work. The capacity to build and organise ecosystems comes via the energy that is harvested in primary production. So long as the biotic energy harvesting rate exceeds its biotic dissipation (in respiration, maintenance and transfer), there is a net assembly of biomass. In other words, while the balance of its exchanges is positive, the ecosystem holds the capacity to invest in the increase in its mass and complexity.

In Reynolds' (1997b) application, the energy of external forcing required to slow, arrest or overcome a net positive balance was calculated, using the same units of energy flux ($W\ m^{-2}$ or $MJ\ m^{-2}\ d^{-1}$). The mechanical energy required to deepen a mixed layer and, more importantly, the light energy lost to absorption in the greater thickness of water, were set against the optimal energy-harvesting and photosynthate-allocation properties of the alga *Chlorella*. The relevant findings are that, at low biomass, the overall energy interception is modest (and so, difficult to

Box 8.1   Exergy.

The exergy concept provides a simple and convenient framework for judging ecosystem behaviour, although its derivation and quantification come from thermodynamics. Exergy is a dynamic measure of the deviation of a system from its tendency towards thermodynamical equilibrium. Whereas all the incoming solar radiation is reflected, reradiated and otherwise dissipated irreversibly by an abiotic world, biological systems function by extracting some of the incoming energy flux into reducing power and carbon bonds. Subsequent controlled oxidation of carbohydrates and proteins, either by the system's autotrophs (expended in growth and maintenance) or by its heterotrophic consumers, releases the energy (mostly as long-wave heat). Thus, the thermodynamic laws are upheld: ecosystems dissipate heat as do abiotic systems. However, the flow of entropy is impeded. So long as the ecosystem is able to intercept more energy than its consumptive maintenance costs, it is able to assemble biomass and grow. Thus, it is the nature of an ecosystem to move away from thermodynamic equilibrium, to an extent that is equivalent to negative entropy. This is sometimes called the negentropy or, better, the exergy of the system (Jørgensen, 1992; Nielsen, 1992). The exergy flux is the difference between the potential harvest of incoming solar radiation by the ecosystem's autotrophs and the maintenance costs of the existing biomass. The relationship is depicted in Fig. 8.4(a), evaluated on the basis of the photosynthetic and respiration rates of *Chlorella* (Reynolds, 1997b).

    However, the concept applies to individual plants and animals as well as in their emergence in whole ecosystems. Extension of the essential logic takes the energy-allocation model towards a quantitative summary of ecosystem integrity and resilience. A net positive exergy flux is indicative of a building ecosystem, with increasing biomass, complexity and information. Failure of the energy income, or of the ability of the system to harvest sufficient of it to be able to meet its accumulated running costs, results in an unsustainable imbalance, necessitating redress through rapid restructuring and loss of redundant biomass. This is manifest as disturbance.

overcome) but it is large compared with the burden of respiration and maintenance. As biomass increases, however, self-shading affects the biomass-specific harvesting rate but the biomass-specific respiration rate is effectively constant: the greater is the biomass, the more of its total energy harvest is allocated to maintenance. As the energetic climax is approached, when allocation and maximal harvest are balanced ($\sim 12.6\,MJ\ m^{-2}\ d^{-1}$, which will sustain a biomass equivalent to $\sim 10.4\,mol$ cell $C\ m^{-2}$; Reynolds, 1997b) so the more the system is sensitive to forced fluctuations in the income of harvestable energy and the enforced inability to meet its own metabolic costs (Fig. 8.4(a)). Of course, when consumers of producer biomass are written into the sensitivity of the harvesting capacity, the relationships become more complex but the simplified ecosystem of energy-income/plant biomass serves to illustrate the principle that the excess of energy harvesting rate over its rate of dissipation as a cost of biomass is decisive in whether the system builds or is energetically unsustainable and must shed biomass.

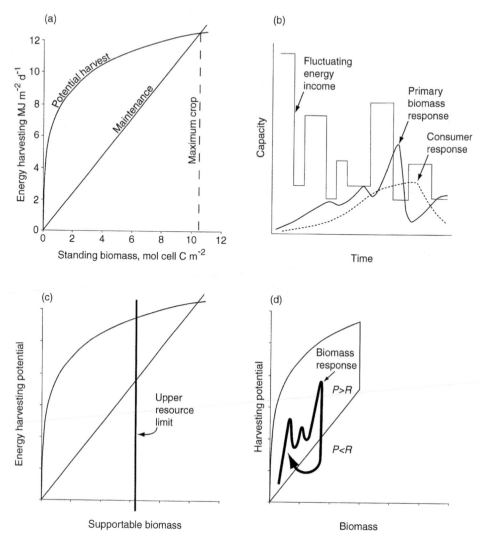

Fig. 8.4 (a) The potential energy harvested by active producer biomass in comparison with fixed, mass-dependent maintenance; the excess of energy income over maintenance expenditure is the *exergy flux*. (b) Plot showing biomass responses to variable energy income: provided harvesting capacity exceeds the maintenance requirement, an increase in primary-producer biomass is supported; when exergy is exceeded, biomass has to readjust to sustainable levels; the response of herbivore biomass is shown, on the same principle of comparing energy income and expenditure. (c) The imposition of an upper attainable biomass by resource availability. (d) Synthesis, showing a hypothetical time track of community assembly (bold, arrowed line) through fluctuating energy and resource availability: the fall to unsustainability, when harvesting capacity is exceeded by maintenance costs ($P<R$), and the leftward lurch to recover energetic balance (where $P> R$) constitutes a disturbance reaction.

What emerges is a simple model of the acquired resistance and metabolic resilience of an ecosystem to external forcing and the criteria for its structural disturbance. Inevitably, the concept introduces its own terminology, which is nevertheless helpful to its assimilation and application. It is more generally helpful, perhaps, to regard the envelope between the maximum harvesting and minimal maintenance fluxes as an 'exergy cushion' (Reynolds, 1997b), which effectively damps out the relatively high-frequency fluctuations in energy income and harvesting capacity (Fig. 8.4(b)), so providing the resilience of the system to external forcing. If the allocation costs consistently exceed the harvestable income, the system loses its cushion of resilience and it becomes energetically unsustainable. It must readjust its mass in order to bring it back into balance, at a lower and probably simpler biomass.

Exceedence of thresholds

The outstanding task is to relate exceedence of the exergy flux, when forcing is sufficient to evoke a disturbance response, to the mechanics of its manifestation. To do this, it is necessary first to recognise where tangible exergy fluxes are located, how they are exceeded and how the response is transferred to higher levels of expression. When examined from the point of view of their capacities rather than their products, almost all anabolic functions in the cells of phytoplankton are demonstrably capable of saturating the maximal growth requirement. Thus, taking something as tangible (and, incidentally, as highly reproducible) as the growth rate of a laboratory-cultured plankton alga, it is possible to determine the energy and resource fluxes required to fuel the alga-specific maximal performance. The requirement is always within the maximal capacity of the resource-gathering function (Reynolds, 1997a). Relevant data for *Chlorella*, set out in Box 8.2, illustrate the point well.

The corollary has also been made (Reynolds, 1997a) that it is not reasonable to refer to a rate of growth being resource- or process-limited before the supply of any of the resources or the rates of their processing falls below the demand of the maximal attainable growth rate. For instance, for reasons of poor light or low carbon availability, the *Chlorella* cell in Box 8.2 takes more than 9 h to fix the carbon needed to double the cell complement. Similarly, it may well be that, owing to extreme external deficiencies, the actual rate of phosphorus uptake fails to balance the consumption during the division cycle. Under these circumstances, growth slows to the rate that is permitted by the weakest supply – the rate is limited by the supply of the least available resource. Moreover, the rates of nutrient uptake (and, if less familiarly so, of carbon fixation) remain sensitive to internal demand as well as to external availability, in conformity with Droop-type dynamics (Droop, 1974). The essential point is that while adequate external supplies remain, the cell is able to damp out fluctuations in resource intake and consumption in anabolic construction and still meet the requirement posed by the maximum sustainable growth rate.

Box 8.2   Resource gathering and growth in phytoplankton.

When examined from the point of view of their capacities rather than their products, almost all anabolic functions in the cells of phytoplankton are capable of saturating the requirement of maximal cell growth. The exponential rates of biomass increase of laboratory-cultured plankton algae, given conditions of constant temperature and continuously saturating light and nutrient supplies, are readily measured and, for any given species, highly reproducible. In several previous papers (see Reynolds, 1997a, for a synthesis), I have been able to show that this is not dependent upon the maximum capacity of any of the cell's resource-gathering functions. The point may be illustrated by data for *Chlorella*, whose maximum exponential growth rate at 20°C, 1.84 d^{-1} or 21.3×10^{-6} s^{-1}, is sufficient to effect a doubling of biomass (i.e. 1 mol carbon assimilated per 1 mol cell carbon) in 32 550 s, i.e. a little over 9 h. With its supposedly typical chlorophyll complement of 1 g per 50 g cell C (or ~ 0.24 g chlorophyll (mol cell C)$^{-1}$), the photosynthesis will deliver a maximum of 2.1 mg reduced C (g chlorophyll)$^{-1}$ s^{-1}, which is theoretically sufficient to sustain a doubling of cell carbon in 16 500 s, or 4.6 h. If the area projected by the chlorophyll coincides with that of the cell, the photon flux needed to saturate this rate of photosynthesis is 42 µmol photons m^{-2} s^{-1}. Meanwhile, assuming carbon dioxide is present at the air–water equilibrium at 20°C (11 µmol CO_2 l^{-1}), the requisite new carbon can be absorbed in 2300 s (38 min). In order to maintain its internal (Redfield) stoichiometry (106C: 16N: 1P), the growing *Chlorella* requires 0.151 mol N and 0.0094 mol P (mol cell C)$^{-1}$. At the maximum measured uptake rates, the cell can gain the nitrogen requirement (as nitrate) in 44 min and those for phosphorus in just 440 s (7.3 min)!

Subject to the same condition of external adequacy, the cell retains a biochemical buffering capacity which ensures that responses to high-frequency forcing (entrained traverses of the natural light gradient, encounters with nutrient micropatches) are retained at the biochemical level and a higher-order impact on growth rate is avoided.

The molecular-level exergy residing in cellular resource-gathering and resource-processing is also exhaustible. Persistent deficiencies in the supply of any one of them, be it a nutrient or carbon or light, will bring growth rate into resource dependency. Such inadequacies are, of course, ultimately responsible for the non-linearity of phytoplankton growth in nature, although the behaviour may be attributable to the control set by the one resource whose relative unavailability fails the growth requirement by the widest margin. Thus, we need to be aware of the relative modesty of the resource levels that are able to supply the harvesting abilities of most phytoplankton. As a general guide, the maximum temperature-specific self-replication rates of most species continue to be saturated at daytime light levels of $\geqslant 100$ µmol photons m^{-2} s^{-1} and resource concentrations of $\geqslant 20$ µmol l^{-1} dissolved inorganic carbon, $\geqslant 7$ mmol l^{-1} dissolved inorganic nitrogen, and $\geqslant 0.5$ µmol l^{-1} dissolved phosphorus (Reynolds, 1997a). Skeletal silicon ($\geqslant 10$ µmol l^{-1}) and bioavailable iron ($\geqslant 0.01$ µmol l^{-1}) may be added to that list, which is, in any case, probably rather conservative (see Reynolds, 1999a). Specialist phylogenetic traits (such as the ability to fix atmospheric nitrogen), predispositions to functional efficiency (as influenced by cell-size surface-volume constraints) and high-affinity adaptations of uptake systems (especially with respect to phosphorus and carbon) provide some species with the ability to function far below these general levels. Such specialists, of course, then constitute the 'diminishing pool' of available species that is able to exploit environments chronically deficient in the relative resource.

It may be further noted that many limnetic environments are supposed to be chronically deficient in phosphorus, the supply of which, though it may well be internally recyclable, is ultimately dependent upon processes in the watershed. A characteristically resource-deficient biomass may be only weakly susceptible to variability in energy income, as the maintenance requirement of the resource-limited biomass is correspondingly modest. On the other hand, the restriction on biomass assembly in nutrient-rich environments is more consistently related to light availability. Moreover, it is in the variable, fluid environment of the surface layers, where the poverty of a steady state and the highest frequencies of physical forcing are most apparent, that the mechanisms of resilience are most severely tested. Here, variable forcing particularly affects the vertical distance that entrained algae are transported and, hence, the proportion of the daylight period that is spent in saturating or in subsaturating light. At the scale of hours to days, the change in frequency varies with the wind speed; over a few days to weeks it might experience a longer cycle of variation with changing turbidity. At any time, the column may stagnate, parking cells in darkness or in high light.

Physiological responses to threshold exceedence

At all times, the response of the entrained cell will be towards restoring or maintaining the energy supply to provide fixed carbon to maintenance and growth or, alternatively, to protect the photosynthetic apparatus from damage by high light or by oxidative poisoning. Thus, the anabolic process is provided with damping mechanisms and alternative pathways (Table 8.5). Starting at the level of millisecond bursts of wave-flash focused, high-intensity irradiation of a handful of light-harvesting complexes in a single algal cell, and proceeding to the level of the few moments during which the passage of a small cloud obscures the sun, the photosynthetic apparatus is exposed to strikingly high-frequency forced variability in the photon flux. This is modulated at the biochemical level within the individual chlorophyll-protein light-harvesting complexes (LHCs) of Photosystem II (PSII): a finite capacity of the plastoquinone pool to accept the electrons stripped in the light reaction works like a surge tank by controlling the return of oxidant to reopen the photon receptors. A burst of electron flow to the plastoquinone pool prevents excess electron generation; the overflow energy flux is filtered off and reradiated (as fluorescence; see Falkowski, 1992). With diminishing photon flux, the electron

Table 8.5 Adaptive responses of phytoplankton to variable light levels.

Time-scale	Driving variables	Respondent system	Nature of response
< Seconds	Enhanced photon flux	Light-harvesting centres	Variable fluorescence
Minutes	Variable photon flux	Chromophore	Contraction/ enlargement
Minutes–hours	Change in mean light intensity Mehler reaction	PSI carboxylation	Extracellular production Photorespiration
		Cell movement	Direction of swimming Buoyancy change Altered sinking rate
Hours–days	Change in mixing intensity	Pigmentation	Chlorophyll synthesis Accessory pigmentation Chlorophyll destruction Xanthophyll reactions
> Days	Variation in average light climate	Cell replication	Growth rate changes Photooxidative death Respirational death
2–4 algal generations	Major shift in light climate	Population recruitment	Altered assemblage composition
Weeks–months	Seasonal change	Periodicity of stratification	Seasonal succession
> Years	Annual cycles	Selection of assemblages	Interannual variability

transport subdues, the plastoquinone pool drains of electrons, leaving more receptor reaction centres open, poised to accept the next burst of photons.

With more sustained bursts of high light (of the order of minutes to hours), the cell institutes physiological and behavioural responses. Although planktonic cells are considered to be too small for plastid relocation to have any significance, contraction of the chromophores housing the LHCs follows several minutes of exposure to high photon fluxes (Neale, 1987). Many flagellates move quickly away from a source of high light, while what the normal function of buoyancy-regulatory mechanism of planktonic cyanobacteria provides is to achieve a downward migration of colonies, away from the water surface and the proximal source of high light. The sinking rates of diatoms are sharply enhanced by exposure to high light intensities, supposedly as an avoidance reaction, though the mechanism is still unexplained (Reynolds and Wiseman, 1982; Neale *et al.*, 1991).

Meanwhile, maximal photosynthesis results in the accumulation of carbohydrate at a rate which may exceed its assimilation by a factor of up to 2, or, if the cells are simultaneously resource-limited, very much more. Some intracellular storage, as starch or glycogen or other condensed carbohydrate, is possible (indeed, it is essential in the context of maintaining growth in intermittent light) but plankton cells have no recognised means of accumulating photosynthate, although this could be one of a number of possible functions of external mucilage. In the short term, there is little else to do but to vent the photosynthate: most algae can produce glycollic acid by substituting an oxidative reaction for photosynthetic carboxylation. In green algae, as in higher plants, the latter may be further oxidised to produce carbon dioxide at an accelerated rate (photorespiration). Release of photosynthetic intermediates into the water is one of the most common means by which cells (Sharp, 1977; Azam *et al.*, 1995) regulate the delivery of fixed carbon to supply the growth requirement. The primary product is lost to the alga but not necessarily to the pelagic system, as this becomes the primary source of organic carbon to the bacterioplankton.

Continued or frequent exposure to high light intensities (hours to days) brings structural responses in the photosynthetic apparatus. Conversion of xanthophylls (violoxanthin to zeaxanthin in green algae, diadinoxanthin to diatoxanthin in diatoms) protects PSII from excessive light, siphoning of energy as heat, and allows net production to be maintained. (Demmig-Adams and Adams, 1992).

Population responses to threshold exceedence

At the scale of generation times, the amount of chlorophyll may be reduced so that cell-specific photosynthetic carbon fixation is more closely attuned to carbon-specific growth. Despite the sophistication of the various responses, prolonged exposure of phytoplankton cells, especially those previously attuned to weak or intermittent light doses, frequently results in photooxidative damage to the photosynthetic apparatus, impairment of cell function, perhaps even cell lysis and death. By this stage, growth impairment and cell mortalities will feed through to the composition of the entire plankton assemblage.

An analogous suite of responses is associated with a diminution in the photon flux, although differentiation is made between the effects of sustained low intensities and those of increasingly short or infrequent periods of exposure to high light as the depth or the turbidity of a mixed layer is increased. In both cases, the generation-scale response is to raise light-harvesting capacity but, in the former case, it is optimised by widening the spectrum of photosynthetically-active wavelengths through the deployment of accessory pigments. The enhanced phycobilin content in deep-stratified Cyanobacteria is one of the most familiar cases of chromatic photoadaptation (Tandeau de Marsac, 1977; Post *et al.*, 1985). In entrained populations exposed intermittently to high light, the most advantageous attribute is to fix as much carbon as possible during the brief windows of opportunity. More, or larger, photosynthetic units, involving more chlorophyll deployed in more photoreceptor units, make for higher cell-specific photosynthetic rates and the harvest of sufficient photosynthate to supply growth throughout the dark period. The dark interludes allow the receptors to be fully opened and for the electromotive pathways to be cleared for maximum activity when the next opportunity arises. Indeed, it is the rapid saturation of samples of deep-mixed populations that are most likely to show 'photoinhibition' when they are enclosed in small bottles and captured for several hours under near-surface light intensities (Harris, 1983; Long *et al.*, 1994). Expressed in terms of carbon fixed per photon received while dispersed, the photosynthetic efficiency of deep-mixed populations of photoadapted diatoms and *Planktothrix* is among the highest recorded (Tilzer, 1984). Moreover, it is among such deep-mixed populations that the greatest proportion of carbon fixation is routed into growth (Talling, 1957; Reynolds, 1990).

Assemblage responses to threshold exceedence

These various considerations confirm that the extent of the phytoplankton response depends just as much on the periodicity of the dominant forcing function as upon its amplitude or even its frequency. It has been noted that persistent nutrient scarcity rather precludes fluctuations in biomass outside the carrying capacity of the resources. Yet deep mixing in an oligotrophic lake may well impose an even lower, energy-determined capacity for substantial periods that may be mistaken for nutrient regulation. The capacities are not difficult to calculate and generally establish the current sensitivity (Reynolds, 1997a). So long as the restriction resides with a single production constraint (be it light or phosphorus or other resource or even hydraulic flushing), variation in other factors becomes increasingly less likely to evoke an alternative response. Moreover, while the more severe the selective limitation becomes, the narrower is the range of tolerant species and the more predictable is the community outcome: a low-diversity population dominated by a specialist survivor. This may be an S-type *Planktothrix* in permanently light-deficient environments, a Z-type picoplankton in the face of chronic nutrient deficiency, or an L-type *Ceratium* or *Peridinium* in a strongly segregated environment. Given a steadily flushed, nutrient system, even an X-type nanoplankton can persist more or less indefinitely. Examples of each outcome are noted in Section 8.1. Each is somewhere close to ecological stability, and each is the product of at least the last 12 to 18 cell doublings of the dominant species (usually requiring some 35 to 60 days) required to effect competitive exclusion.

  The principle that this process is liable to be interrupted, reversed or diverted by forcing – manifesting as a disturbance – is established. The decisive criterion of exergy exceedence has also been promoted. Nevertheless, the period of forcing is often crucial to the exceedence at the higher levels of species-specific growth, dynamic intervention and structural subversion of the existing assemblage, for the large-scale response is the failure of the small-scale buffers. This can be readily understood by following the progress of a major forcing through a scale-sequence, as proposed in Table 8.5. The events associated with the onset of thermal stratification of a water column provide an appropriate example, when, with a strengthening heat flux and a weakening wind, buoyancy forces dominate the upper water column and the surface mixed layer shrinks rapidly. Everywhere beyond the truncated surface circulation, the existing population is rapidly disentrained. Non-motile forms, especially diatoms, inevitably begin to sink, into lower light intensities and, for many of them, to depths where saturation of photosynthesis is possible, to depths where photosynthesis fails to support rapid growth and, eventually, to depths where photosynthesis fails to compensate respiration. At sinking rates of 0.2–0.4 m d^{-1} (Reynolds and Wiseman, 1982) a trophogenic zone of 5–10 m depth will lose growing diatoms at a rate of up to 8% per day. The cells retained in the surface circulation, however, are coping with different problems, namely those of dealing with super-saturating light fluxes. Near the surface, at least, regulation of carbon flux through fluorescence and chromophore contraction will not, over scales of more than a few minutes, have prevented a flood of unassimilable carbon and compensatory glycolate production. The cells will have also accelerated their sinking rates, to up to 1 m d^{-1} (Reynolds and Wiseman, 1982), leading to a massive leakage of cells from suspension and then rapid elimination from the stagnating layers beneath. While all this is happening, disentrained but motile nanoplankton will be optimising their position in the light gradient and adjusting their biochemical filters so that growth can be fuelled throughout what amounts to an extended daytime period. Within 24 h, the cells will have been able to accelerate to the maximum sustainable under the temperature-, photoperiod- and the resource-conditions obtaining. These are quite likely to be in the range of a doubling per day.

Such tangible adjustments in species-specific growth- and loss-rates are crucial to the assemblage response but the latter still has to clear the inertia of the existing biomass. If, on the next day, the wind strengthens and the sun is occluded by cloud, and kinetic forces once again dominate the trophogenic layer, the nanoplankton are once again entrained and their growth potential slowed while the diatoms are re-entrained and may begin to repair their photosynthetic efficiencies. The massive inoculum effect ensures that the diatoms do not surrender dominance to nano-plankton and the assemblage structure will be seen to have 'survived the dis-turbance'. In effect, the exergy margin of the existing structure has not been overcome and the accumulated resilience to external forcing has held the commu-nity structure, without serious disturbance. If, on the other hand, the second day is similar to the first, as, too, is the third, the population responses of the first are followed through, towards the potential replacement of the existing dominant by the newly-selected upstart. With the former in rapid decline and the latter in the ascendant, a change in community structure is inevitable. The deciding component in the change (as opposed to a mere biomass reduction of the former dominant) is the size of the inoculum of the growing species: the larger the inoculum, the sooner the change in dominance. On the basis of field experiments (Reynolds *et al.*, 1983, 1984; Reynolds, 1988a) and the synthesis of many analyses of 'succession rate' (see

Jassby and Goldman, 1974), it requires some two to four generations of the ascendant species to alter the community structure. Given the rate at which algae can double their numbers, the translation to conventional time-scale is around 5 to 15 days (Reynolds *et al.*, 1993).

8.5   Understanding interannual variability in phytoplankton

It is now possible to bring the complex array of interacting assembly mechanisms that lead to the formation of planktonic communities, from molecular synthesis to the recruitment to specific populations, to the question of interannual differences in outcome. These can be relatively minor (as was the case in 20 consecutive years in Lake Kinneret) or apparently very large (as observed over 18 years in Rostherne Mere). Both examples nevertheless comply consistently with emergent principles governing the production and synthesis of phytoplankton assemblages. The differences between them are therefore ones of degree. On the one hand, no one year results in a carbon-copy of another, simply because the environmental circumstances are scarcely identical from year to year. A wetter winter here, a colder spring there or an unusually windy episode one summer inevitably modifies the nutrient resource base and the opportunities for plankton to process them. It is inevitable that small interannual differences in supportive capacity and the rate of its attainment result are unpredictable (but, it is proposed, explicable in retrospect). In general, however, the main sequence of key driving events in all years (lengthening days, higher irradiance inputs, altered hydraulic balance and hydrographic structure) is determined by latitude and local climate. The assembly response to weather fluctuations takes place around a behaviour that it is conditioned to major cyclical intra-annual environmental changes. The assemblage is a series of dynamic outcomes, attributable to fluctuations in inoculum size, growth rate and maxima attained, superimposed on the inertia of the same (more or less) species doing similar things each year. Interannual variability in the community responses of the phytoplankton normally tends to a stable cycle: it conforms to the notion of the Lorenzian attractor of chaos theory (see, for instance, Gleick, 1987).

On this basis, the interannual differences in weather variability might seem to be of insufficient magnitude to bring about the wholesale interannual differences in the dominant phytoplankton observed in Rostherne Mere. A stable basic attractor cycle, set by latitude and climate, determines an annual cycle of thermal stratification, while the lake has a well-established, perennating plankton flora that should be capable of supplying adequate inocula of key species from year to year. Reasonably, a stronger element of randomness is that the assembly response must be suspected. However, a closer investigation (Reynolds and Bellinger, 1992, and above) reveals an overriding conformity to a basic pattern that is describable in the notation of functionally grouped species associations (as C → H → L; see Table 8.4). Species-specific differences in the reinfection strategies of *Ceratium* and *Microcystis* proved decisive in which of the two predominated in any one year but that either did so at all is related to their common (L–) adaptive traits of self-regulation, biomass conservation and investment in perennation (the same ones, incidentally, that favour the dominance of *Peridinium* in Kinneret). Nevertheless, the sensitivities of both species to low temperatures and poor light leave their growth dynamics more vulnerable to vernal mixing events. However, it is the two- to four-generation 'lagging' of the community response that most influences the anabolic setback: the more frequent are the episodes which take the growth conditions outside the boundaries tolerated by L–species, the greater is the cumulative impact upon their growth and the less is their overall abundance. With the more frequent passage of westerly storms and frontal systems which, together, characterise the weather in north-west Europe when a high, positive NAO index obtains, stratification is sufficiently delayed or weakened to impair the recruitment of *Ceratium* and *Microcystis*. In contrast, the diatoms and *Planktothrix agardhii* are more tolerant of the mixed, turbid conditions and recruit relatively more consistently in these cooler, windier years. In the opposite direction, persistent near-surface stratification of this nutrient-rich lake in a year of low NAO indices favoured the pond-like J-flora dominated by *Scenedesmus*.

All that really separates the assembly responses of Rostherne Mere from those of Lake Kinneret is a greater sensitivity to variability in processing opportunities and a lesser sensitivity to the supply of resources. There may be a greater annual oscillation in insolation at Rostherne (owing to latitude) and a greater frequency of critical forcing events (owing to its maritime location) but the essential criteria at both locations remain the frequency and persistence with which forcing exceeds the cushion of a positive exergy flux. Transgression of the trait boundaries of the species of the existing assemblage set off revised patterns of reassembly. Beyond that, these various examples reveal the extent of interannual variability in community structure that may arise through the assembly responses of phytoplankton to interannually variable forcing functions. This view accommodates the following principles.

(1) At the level of species specificity, neither individual populations nor their contribution to the assemblage as a whole are predictable, save in the very short term (two days or fewer ahead) and based upon a full knowledge of today's assemblage. At a higher level of specificity (trait-based associations), the compositional response to defined environmental conditions is increasingly predictable (Table 8.4).

(2) Change in community composition is the aggregate of the dynamic responses of each species present to altered environmental conditions; the significance and extent of the changes are determined by the intensity and persistence of the exceedence of the trait boundaries of the various species concerned.

(3) The annual pattern of changing species composition is a sequence of assembly responses to cyclical shifts in the environmental conditions with respect to the trait boundaries of the species present.

(4) Year-to-year variability in assemblage composition is related to the efficacy of superimposed forcing of sufficient intensity to exceed the trait boundaries of the species present and of sufficient duration to impact continuously on two or more successive generations.

(5) Interannual variability is conditioned by the timing and frequency of boundary exceedences.

In the context of the general perceptions of interannual variability, addressed at the outset of this chapter, assembly responses underpin the anticipation that there will be broad year-to-year reproducibility in the seasonal pattern of planktonic primary productivity and in the functional composition of its contributors. It is no less evident that the same set of assembly responses is also implicated in the year-to-year variations imposed upon these cyclical patterns. The apparent perversity of these conclusions is overcome by the recognition that the assembly responses are stimulated by externally generated atmospheric events but which are accommodated and manifested at differing ecological scales (see Siegel, 1998). By distinguishing among the suite of biotic responses, from the molecular level of biochemistry feeding through to the growth and reproduction at the physiological scale of cell growth and to the community level of population replacement, it is possible to account for the periodicity in abundance and composition of phytoplankton assemblages, at least at the level of functional groups.

References

Anderson, W.L., Robertson, D.M. & Magnusson, J.J. (1996) Evidence of recent warming and El Niño-related variations in the ice breakup dates of Wisconsin lakes. *Limnology and Oceanography* **41**, 815–21.

Azam, F., Smith, D.C., Long, R.A. & Steward, G.F. (1995) Bacteria in oceanic carbon cycling as a molecular problem. In: *Molecular Ecology of Aquatic Microbes* (ed. I. Joint), pp. 39–54. Springer, Berlin.

Barbiero, R.P., James, W.F. & Barko, J.W. (1999) The effects of disturbance events on phytoplankton community structure in a small temperate reservoir. *Freshwater Biology* **42**, 503–12.

Belyea, L.R. & Lancaster, J. (1999) Assembly rules within a contingent ecology. *Oikos* **86**, 402–16.

Berman, T. & Shteinman, B. (1998) Phytoplankton development and turbulent mixing in Lake Kinneret (1992–1996). *Journal of Plankton Research* **20**, 709–26.

Berman, T., Yacobi, Y.Z. & Pollingher, U. (1992) Lake Kinneret phytoplankton: stability and variability during twenty years (1970–1989). *Aquatic Sciences* **54**, 104–27.

Boucher, P., Blinn, D.W. & Johnson, D.B. (1984) Phytoplankton ecology in an unusually stable environment (Montezuma Well, Arizona, USA). *Hydrobiologia* **119**, 149–60.

Connell, J.H. (1978) Diversity in tropical rain forests and coral reefs. *Science* **199**, 1302–10.

Demmig-Adams, B. & Adams, W.W. (1992) Photoprotection and other responses of plants to high light stress. *Annual Reviews of Plant Physiology and Plant Molecular Biology* **43**, 599–626.

Droop, M.R (1974) The nutrient status of algal cells in continuous culture. *Journal of the Marine Biological Association (UK)* **54**, 825–55.

Elliott, J.A., Irish, A.E., Reynolds, C.S. & Tett, P. (2000) Modelling freshwater phytoplankton communities: an exercise in validation. *Ecological Modelling* **128**, 19–26.

Eloranta, P. (1993) Diversity and succession of the phytoplankton of a small lake over a two-year priod. *Hydrobiologia* **249**, 25–32.

Falkowski, P.G. (1992) Molecular ecology of phytoplankton photosynthesis. In: *Primary Productivity and Biogeochemical Cycles in the Sea*, pp. 47–67. Plenum Press, New York.

George, D.G. & Taylor, A.H. (1995) UK lake plankton and the Gulf Stream. *Nature* **378**, 139.

Gleick, J. (1987) *Chaos*. Heiemann, London.

Haffner, G.D., Harris, G.P. & Jarai, M.K. (1980) Physical variability and phytoplankton communities. III. Vertical structure in phytoplankton populations. *Archiv für Hydrobiologie* **89**, 363–81.

Hambright, K.D., Gophen, M. & Serruya, S. (1994) Influence of long-term climatic changes on the stratification of a subtropical, warm monomictic lake. *Limnology and Oceanography* **39**, 1233–42.

Hardin, G. (1960) The competitive exclusion hypothesis. *Science* **131**, 1292–7.

Harris, G.P. (1983) Mixed-layer physics and phytoplankton populations. Studies in equilibrium and non-equilibrium ecology. In: *Progress in Phycological Research, Vol. II* (eds F.E. Round & D.J. Chapman), pp. 1–52. Elsevier, Amsterdam.

Harris, G.P. (1986) *Phytoplankton Ecology: Structure, Function and Fluctuation*. Chapman and Hall, London.

Harris, G.P., Davies, P., Nunez, M. & Meyers, G. (1988) Interannual variability in climate and fisheries in Tasmania. *Nature* **333**, 754–7.

Huisman, J., Jonker, R.R., Zonneveld, C. & Weissing, F.J. (1999a) Competition for light between phytoplankton species: experimental tests of mechanistic theory. *Ecology* **80**, 211–22.

Huisman, J., van Oostveen, P. & Weissing, F.J. (1999b) Species dynamics in phytoplankton blooms: incomplete mixing and competition for light. *American Naturalist* **154**, 46–68.

Huisman, J. & Weissing, F.J. (1994) Light-limited growth and competition for light in well-mixed aquatic environments: an elementary model. *Ecology* **75**, 507–20.

Huisman, J. & Weissing, F.J. (1999) Biodiversity of plankton by species oscillation and chaos. *Nature* **402**, 407–10.

Hurrell, J.W. (1995) Decadal trends in the North Atlantic Oscillation: regional temperatures and precipitation. *Science* **269**, 676–9.

Huszar, V.L.M., Silva, H.L.S., Marinho, M., Domingos, P. & Sant'Anna, C.L. (2000) Cyanoprokaryote assemblages in eight productive tropical Brazilian waters. *Hydrobiologia* **424**, 67–77.

Hutchinson, G.E. (1961) The paradox of the plankton. *American Naturalist* **95**, 137–47.

Jacobsen, B.A. & Simonsen, P. (1993) Disturbance events affecting phytoplankton biomass, composition and species diversity in a shallow eutrophic temperate lake. *Hydrobiologia* **249**, 9–14.

Jassby, A.D. & Goldman, C.R. (1974) A quantitative measure of succession rate and its application to the phytoplankton of lakes. *American Naturalist* **108**, 688–93.

Jiménez Montealegre, R., Verreth, J., Steenbergen, K. & Machiels, M. (1995) A dynamic simulation model for the blooming of *Oscillatoria agardhii* in a monomictic lake. *Ecological Modelling* **78**, 17–24.

Jørgensen, S.-E. (1992) *Integration of Ecosystem Theory: A Pattern.* Kluwer, Dordrecht.

Juhász-Nagy, P. (1993) Notes on compositional diversity. *Hydrobiologia* **249**, 173–82.

Keddy, P.A. (1992) Assembly and response rules: two goals for predictive community ecology. *Journal of Vegetation Science* **3**, 157–64.

Leitão, M. & Leglize, L. (2000) Long-term variations of epilimnetic phytoplankton in an artificial reservoir during a 10-year survey. *Hydrobiologia* **424**, 39–49.

Long, S.P., Humphries, S. & Falkowski, P.G. (1994) Photoinhibition of photosynthesis in nature. *Annual Review of Plant Physiology and Plant Molecular Biology* **45**, 633–62.

Lund, J.W.G. (1949) Studies on *Asterionella*. I. The origin and nature of the cells producing seasonal maxima. *Journal of Ecology* **37**, 389–419.

Lund, J.W.G. (1950) Studies on *Asterionella formosa* Hass. II. Nutrient depletion and the spring maximum. *Journal of Ecology* **38**, 1–35.

Maberly, S.C., Hurley, M.A., Butterwick, C., *et al.* (1994) The rise and fall of *asterionella formosa* in the South Basin of Windermere: analysis of a 45-year series of data. *Freshwater Biology* **31**, 19–34.

McIntosh, D.H. & Thom, A.S. (1969) *Essentials of Meteorology.* Wykeham, London.

Malinsky-Rushansky, N., Berman, T. & Dubinsky, Z. (1995) Seasonal dynamics of pico-phytoplankton in Lake Kinneret, Israel. *Freshwater Biology* **34**, 241–54.

Neale, P.J. (1987) Algal photoinhibition and photosynthesis in the aquatic environment. In: *Photoinhibition*, pp. 39–65. Elsevier, Amsterdam.

Neale, P.J., Heaney, S.I. & Jaworski, G.H.M (1991) Responses to high irradiance contribute to the decline of the spring diatom maximum. *Limnology and Oceanography* **36**, 761–8.

Nielsen, S.N. (1992) *Application of maximum energy in structural dynamic models.* PhD thesis, ISBN 87 7772 056 3, National Research Institute of Denmark, Miljøministeriet, København. (Copies available from Department of Policy Analysis, Frederiksborgvej 399, DK-4000 ROSKILDE, Denmark.)

Padisák, J. (1992) Seasonal succession of phytoplankton in a large, shallow lake (Balaton, Hungary) – a dynamic approach to ecological memory, its possible role and mechanisms. *Journal of Ecology* **80**, 217–30.

Padisák, J. (1993) The influence of different disturbance frequencies on the species richness, diversity and equitability of phytoplankton in shallow lakes. *Hydrobiologia* **249**, 135–56.

Padisák, J. & Reynolds, C.S. (1998) Selection of phytoplankton associations in Lake Balaton, Hungary, in response to eutrophication and restoration measures, with special references to the cyanoprokaryotes. *Hydrobiologia* **384**, 41–53.

Paine, R.T., Tegner, M.J. & Johnson, E.A. (1998) Compounded perturbations yield ecological surprises. *Ecosystems* **1**, 535–45.

Pianka, E.R. (1999) Putting communities together. *Trends in Ecology and Evolution* **14**, 501–502.

Post, A.F., de Wit, R. & Mur, L.R. (1985) Interactions between temperature and light intensity on growth and photosynthesis of the cyanobacterium, *Oscillatoria agardhii*. *Journal of Plankton Research* **7**, 487–95.

Reynolds, C.S. (1980) Phytoplankton assemblages and their periodicity in stratifying lake systems. *Holarctic Ecology* **3**, 141–59.

Reynolds, C.S. (1986) Experimental manipulations of the phytoplankton periodicity in large limnetic enclosures. *Hydrobiologia* **138**, 43–64.

Reynolds, C.S. (1987) The responses of phytoplankton communities to changing lake environments. *Schweizerische Zeischrift für Hydrologie* **49**, 220–36.

Reynolds, C.S. (1988a) The concept of ecological successsion applied to the seasonal periodicity of phytoplankton. *Verhandlungen der internationale Vereinigung für theoretische und angewandte Limnologie* **23**, 683–91.

Reynolds, C.S. (1988b) Functional morphology and adaptive strategies of freshwater phytoplankton. In: *Growth and Reproductive Strategies of Phytoplankton* (ed. C.D. Sandgren), pp. 388–433. Cambridge University Press, New York.

Reynolds, C.S. (1990) Temporal scales of variability in pelagic environments and the responses of phytoplankton. *Freshwater Biology* **23**, 25–53.

Reynolds, C.S. (1995) Successional change in planktonic vegetation: species, structures, scales. In: *The Molecular Ecology of Aquatic Microbes* (ed. I. Joint), pp. 115–32. Springer-Verlag, Berlin.

Reynolds, C.S. (1997a) *Vegetation Processes in the Pelagic*. ECI, Oldendorf.

Reynolds, C.S. (1997b) Successional development, energetics and diversity in planktonic communities. In: *Ecological Perspectives of Biodiversity* (eds T. Abe, S.R. Levin & M. Higashi), pp. 167–202. Springer, New York.

Reynolds, C.S. (1999a) With or against the grain: responses of phytoplankton to pelagic variability. In: *Aquatic Life-cycle Strategies* (eds M. Whitfield, J. Matthews & C. Reynolds), pp. 15–43. Marine Biological Association, Plymouth.

Reynolds, C.S. (1999b) Metabolic sensitivities of lacustrine ecosystems to anthropogenic forcing. *Aquatic Sciences* **61**, 183–205.

Reynolds, C.S. (2000) Plankton designer – or how to predict compositional responses to trophic change. *Hydrobiologia* **424**, 123–32.

Reynolds, C.S. & Bellinger, E.G. (1992) Patterns of abundance and dominance of the phytoplankton of Rostherne Mere, England: evidence from an 18-year data set. *Aquatic Sciences* **54**, 10–36.

Reynolds, C.S., Padisák, J. & Sommer, U. (1993) Intermediate disturbance in the ecology of phytoplankton and the maintenance of species diversity. *Hydrobiologia* **249**, 183–8.

Reynolds, C.S. & Reynolds, J.B. (1985) The atypical seasonality of phytoplankton in Crose Mere, 1972: an independent test of the hypothesis that variability in the physical environment regulates community dynamics and structure. *British Phycological Journal* **20**, 227–42.

Reynolds, C.S. & Wiseman, S.W. (1982) Sinking losses of phytoplankton in closed limnetic systems. *Journal of Plankton Research* **4**, 489–522.

Reynolds, C.S., Wiseman, S.W. & Clarke, M.J.O. (1984) Growth- and loss-rate responses of phytoplankton to intermittent artificial mixing and their potential application to the control of planktonic biomass. *Journal of Applied Ecology* **21**, 11–39.

Reynolds, C.S., Wiseman, S.W., Godfrey, B.M. & Butterwick, C. (1983). Some effects of artificial mixing on the dynamics of phytoplankton populations in large limnetic enclosures. *Journal of Plankton Research* **5**, 203–234.

Rojo, C., Ortega-Mayagoitia, E. & Alvarez Cobelas, M. (2000) Lack of pattern among phytoplankton assemblages. Or, what does the exception to the rule mean? *Hydrobiologia* **424**, 133–9.

Romme, W.H., Everham, E.H., Frelich, L.E., Moritz, M.A. & Sparks, R.E. (1998) Are large, infrequent disturbances qualitatively different from small, frequent disturbances? *Ecosystems* **1**, 524–34.

Saxby-Rouen, K.J., Leadbeater, B.S.C. & Reynolds, C.S. (1997) The growth response of *Synura petersenii* (Synurophyceae) to photon-flux density, temperature and pH. *Phycologia* **36**, 233–43.

Seip, K.L. & Reynolds, C.S. (1995). Phytoplankton functional attributes along trophic gradient and season. *Limnology and Oceanography* **40**, 589–97.

Serruya, C. (1978) *Lake Kinneret.* (Monographiae Biologiae Series No. 32.) W. Junk, Den Haag.

Sharp, J.H. (1977) Excretion of organic matter by marine phytoplankton: do healthy cells do it? *Limnology and Oceanography* **22**, 381–99.

Siegel, D.A. (1998) Resource competition in a discrete environment: why are plankton distributions paradoxical? *Limnology and Oceanography* **43**, 1133–46.

Sommer, U. (1985) Comparisons between steady-state and non-steady-state competition; experiments with natural phytoplankton. *Limnology and Oceanography* **30**, 335–46.

Sommer, U. (1995) An experimental test of the intermediate disturbance hypothesis using cultures of marine phytoplankton. *Limnology and Oceanography* **40**, 1271–7.

Steel, J.A. (1995). Modelling adaptive phytoplankton in a variable environment. *Ecological Modelling* **78**, 117–27.

Steel, J.A. & Duncan, A. (1999) Modelling the ecological aspects of bankside reservoirs and implications for management. *Hydrobiologia* **395/396**, 133–47.

Steemann Nielsen, E. (1952) The use of radioactive carbon (^{14}C) for measuring organic production in the sea. *Journal du Conseil International pour l'Exploration de la Mer* **18**, 117–40.

Straile, D. (2000) Meteorological forcing of plankton dynamics in a large and deep continental European lake. *Oecologia* **122**, 44–50.

Straile, D. & Geller, W. (1998) The response of *Daphnia* to changes in trophic status and weather patterns: a case study from Lake Constance. *ICES Journal of Marine Science* **55**, 775–82.

Strub, P.T., Powell, T. & Goldman, C.R. (1985) Climatic forcing: effects of El Niño on a small, temperate lake. *Science* **227**, 55–7.

Talling, J.F. (1957) The phytoplankton population as a compound photosynthetic system. *New Phytologist* **56**, 133–49.

Tandeau de Marsac, N. (1977) Occurrence and nature of chromatic adaptation in cyanobacteria. *Journal of Bacteriology* **130**, 82–91.

Tilman, D. (1992) *Resource Competition and Community Structure.* Princeton University Press, Princeton.

Tilman, D., Kilham, S.S. & Kilham, P. (1982) Phytoplankton community ecology: the role of limiting nutrients. *Annual Reviews in Ecology and Systematics* **13**, 349–72.

Tilzer, M.M. (1984) The quantum yield as a fundamental parameter controlling vertical photosynthetic profiles of phytoplankton in Lake Constance. *Archiv für Hydrobiologie (Supplementband)* **69**, 169–98.

Trimbee, A.M. & Harris, G.P. (1980) Use of time-series analysis to demonstrate advection rates of different variables in a small lake. *Journal of Plankton Research* **5**, 819–33.

Turner, M.G., Baker, W.L., Peterson, C.J. & Peet, R.K. (1998) Factors influencing succession: lessons from large, infrequent natural disturbances. *Ecosystems* **1**, 511–23.

Weiher, E. & Keddy, P.A. (1995) Assembly rules, null models and trait dispersion: new questions from old patterns. *Oikos* **74**, 159–64.

Chapter 9

Sustained and Aperiodic Variability in Organic Matter Production and Phototrophic Microbial Community Structure in the North Pacific Subtropical Gyre

David M. Karl, Robert R. Bidigare and Ricardo M. Letelier

9.1 Introduction

On 27 February 1940, a research team from University of California at Berkeley led by Martin D. Kamen discovered ^{14}C, a radioactive isotope of the element carbon (Ruben and Kamen, 1940a). Prior to the availability of this relatively long-lived radioisotope (half-life = 5730 yrs), $^{11}C-CO_2$ (half-life = 21.5 ± 0.5 min) produced in the Berkeley cyclotron by bombardment of amorphous boron with 8 Mev deuterons had been used as a tracer in the study of photosynthesis (Ruben *et al.*, 1939). In a series of papers entitled *Photosynthesis with Radioactive Carbon*, Kamen and his colleagues staggered through the maze of experimental limitations on the use of ^{11}C, eventually producing a 'tentative theory of photosynthesis' (Ruben and Kamen, 1940b). They concluded their paper by noting 'the production of a long-lived radioisotope of carbon will make feasible a more detailed and extensive investigation'.

The general availability of ^{14}C, beginning in 1945 after the termination of World War II, equipped scientists with an invaluable research tool (see Chapter 2). It quickly led to an explicit understanding of the carbon pathways in photoautotrophic, chemoautotrophic and heterotrophic micro-organisms. It also contributed enormously to a growing understanding of carbon cycle processes in the global ocean, including rates of organic matter production, especially following the development of a novel ^{14}C-based method for the measurement of photosynthesis in marine plankton (Steemann Nielsen, 1951; 1952a, b; Steemann Nielsen and Aabye Jensen, 1957). The improved sensitivity of the ^{14}C method, relative to the previously employed light–dark bottle oxygen technique, provided the opportunity to survey photosynthesis in all regions of the world ocean.

On 29 March 1952, during a New Zealand to San Francisco ocean transect, the

Danish research vessel *Galathea* occupied station No. 698 located at 23°00′ N, 155°25′ W approximately 200 nautical miles NE of Station ALOHA, the current open ocean time-series site for the Hawaii Ocean Time-series (HOT) programme. At this location, for the first time in the North Pacific Subtropical Gyre (NPSG), the new ^{14}C method was used to measure rates of primary production (Steemann Nielsen and Aabye Jensen, 1957). Great care was taken during both sampling and incubation in order to achieve reliable results (see Box 9.1).

Box 9.1   Comparison of the ^{14}C method initially described by Steemann Nielsen[1] with the HOT programme protocols[2].

Objective	Steemann Nielsen *circa* 1952	HOT programme *circa* 1990
1. Collection of uncontaminated water sample	Non-toxic, Pyrex® glass sampler in teak frame	Non-toxic, Teflon®-coated polyvinyl chloride sampler fitted with silicone or Viton® O-rings
2. Sample incubation	*In situ* (if possible) or shipboard with simulated *in situ* light/temperature	*In situ*
3. Sample filtration	20 cm² filtration area, 0.5 µm porosity collodion filter facilitated by suction	2.5 cm² filtration area, 0.7 µm nominal porosity glass fibre filter facilitated by suction
4. Dissolved inorganic carbon estimation	Calculation from measurements of pH and alkalinity	Direct measurement by coulometry

---

(1)   Steemann Nielsen (1952a, b); Aabye Jensen and Steemann Nielsen (1952).
(2)   Karl *et al.* (1996); Letelier *et al.* (1996); Karl *et al.* (1998).

So began the 'modern' era of biogeochemical studies in the NPSG. In the USA, Maxwell S. Doty of the University of Hawaii was among the first to routinely employ Steemann Nielsen's new ^{14}C method. With decade-long funding (1953–1964) from the US Atomic Energy Commission (AEC), Doty and his colleagues initiated a systematic, long-term study entitled 'Algal productivity of the tropical Pacific as determined by isotope tracer technique' (AEC contract No. AT-04-3-15), including:

(1)   shipboard methodological improvements,
(2)   study of island mass effects on regional patterns of ocean productivity, and
(3)   pan-Pacific observations of primary productivity.

This last task (item (3)) was accomplished largely via collaboration with the Bureau of Commercial Fisheries. Doty was also responsible for the establishment of the first open ocean time-series programme that included routine determinations of both ^{14}C-based primary production and photosynthetic pigments. He also described a novel non-toxic (lucite and stainless steel) 'high-speed' water sampler that could be used to collect surface sea water from commercial vessels travelling at speeds up to 23 knots (Doty and Oguri, 1958). This sampler was required to enlist merchant ships of opportunity in an attempt to enhance the spatial coverage of ^{14}C-based primary production estimates throughout the Pacific basin. This coherent research plan of a combined time-series measurement program at fixed station locations with periodic global ocean surveys was quite visionary for that time period, and even today has never been fully implemented.

In 1961, at a University of Hawaii-sponsored symposium of the 10th Pacific Science Congress, the subject of 'Marine productivity in the Pacific' was exhaustively debated (Doty, 1961). At that time, the rates of primary production in the NPSG and their spatial and temporal variations were not well-constrained. A compilation of the relatively small extant field data base concluded that the NPSG had a relatively low sustained rate of primary production of $< 100 \, \text{mg C m}^{-2} \, \text{d}^{-1}$. Both seasonal and interannual variability were thought to be negligible; mean annual production for the gyre was estimated to be approximately $24 \, \text{g C m}^{-2}$ compared to approximately $240 \, \text{g C m}^{-2}$ for coastal neritic waters.

For the next two decades our view of the NPSG remained essentially unchanged. However, today we recognise this habitat as a fundamentally different ecosystem. Primary production appears to be much higher than previously thought possible, and there is both regular (seasonal) and stochastic variability in plankton rates and processes. This chapter will examine primary production at Station ALOHA (22°45′ N, 158° W) in the NPSG over time-scales ranging from sub-seasonal (days to weeks) to interannual–decadal. Following a brief introduction to the NPSG habitat and its planktonic inhabitants, we will explore the degree to which regional and larger-scale physical forces are responsible for observed changes in phytoplankton standing stocks and rates of organic matter production in the gyre by investigating various time–space scales. We will end our presentation with a discussion of several remaining, first-order ecological questions, and with a brief research prospectus for the future.

Terms and definitions

Many terms in contemporary studies of the ocean's carbon cycle are ambiguous, for example, one of the most commonly employed words – 'production' – is equivocal. For this chapter we adopt the explicit definitions presented by Strickland (1960), Williams (1993) and Sakshaug *et al.* (1997), including the following:

(1) *Gross primary carbon production (GPCP).* This is the quantity of organic carbon, produced by the reduction of carbon dioxide as a consequence of the photosynthetic process, per unit volume or surface area (integrated to a specified reference depth) per unit time. In the ^{14}C method, GPCP includes the sum of dissolved and particulate organic matter (which can both be measured) plus ^{14}C-labelled organic carbon that is respired during the incubation procedure (which cannot be measured). Consequently, GPCP cannot be measured directly using the ^{14}C method.

(2) *Net primary carbon production (NPCP).* This is GPCP minus organic carbon losses due to photoautotrophic respiration per unit volume or surface area (integrated to a specified reference depth) per unit time. Because it is impossible to isolate photoautotrophs from other micro-organisms prior to field incubations, NPCP cannot be measured directly using the ^{14}C method.

(3) *Net community carbon production (NCCP).* This is GPCP minus all autotrophic and heterotrophic organic carbon losses due to respiration per unit volume or surface area (integrated to a specified reference depth) per unit time. In theory, NCCP can be measured by the ^{14}C method only when all carbon pools are in $^{14}C/^{12}C$ equilibrium; this rarely occurs during the incubation time periods employed and is not easily assessed in field samples.

(4) *Phototrophic carbon production (PCP).* This is the production of cell carbon using energy derived from light. This production of organic tissue need not involve the reduction of carbon dioxide; for example, photoheterotrophic assimilation of preformed dissolved organic compounds and their subsequent incorporation into macromolecules might also be involved.

Gordon Riley, in response to a sharp criticism of the oxygen method made by Steemann Nielsen that Riley considered to be both 'empty and violent', did praise the ^{14}C method for superior sensitivity and accuracy but he also concluded that:

'... the results of the ^{14}C method fall somewhere between the net phytoplankton production [i.e. NPCP, item (2) above] and total photosynthesis [i.e. GPCP, item (1) above], but exact evaluation of the meaning of the experiments will require an extensive experimental programme.'

Unfortunately, this uncertainty dogs us today. If both particulate and dissolved organic carbon pools are measured, we believe that the ^{14}C method as employed in our field programmes approximates, but does not equate with, NPCP.

North Pacific Subtropical Gyre: a climax community?

Within the marine environment, oceanic provinces far removed from land account for a majority of total global ocean primary production. The most extensive open-ocean regions, termed subtropical gyres, collectively occupy approximately 40% of the surface of the Earth; the largest is the NPSG, which at the ocean's surface extends from approximately 15° N to 35° N latitude and 135° E to 135° W longitude (Fig. 9.1). With a surface area of approximately $2 \times 10^7 \, km^2$, the NPSG is the largest circulation feature on our planet and the Earth's largest contiguous biome.

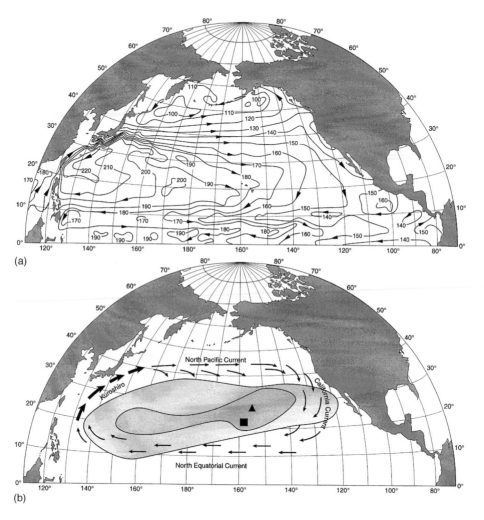

(a)

(b)

Fig. 9.1 Map of the North Pacific Ocean showing several important features of the NPSG. (a) Dynamic topography of the sea surface in dyn-cm relative to 1000 dbar based on historical hydrographic observations. Arrows show the direction of geostrophic flow. Redrawn from Wyrtki (1975). (b) Major circulation features in the North Pacific that collectively define the approximate boundaries of the NPSG. Also shown is the estimated range of central Pacific Ocean mesozoo-plankton based on an analysis of the fidelity of mesozooplankton species. The dark area is the core of the Central Pacific fauna (100% fidelity) and the lighter shade defines the 60% fidelity boundary. Redrawn from McGowan (1974). Note that Stations ALOHA (■) and Climax (▲), two ocean time-series locations discussed in this chapter, are both located in the core region.

The coherent and relatively homogeneous waters that occupy the NPSG are separated from neighbouring habitats by the large anticyclonic circulation features that characterise this and other subtropical ocean gyres (Fig. 9.1). The relatively deep, permanent pycnocline insulates the deep nutrient-rich waters from the euphotic zone. The anticyclonic circulation also ensures downwelling at the gyre boundaries, further isolating the light-sufficient but nutrient-starved plankton assemblages in the surface ocean from the large supply of inorganic nutrients below.

The NPSG ecosystem is very old; present boundaries have persisted since the Pliocene Period (approximately 10^7 years before present), or earlier. This great age and relative isolation were the primary factors leading to the development of a hypothesised 'climax-type' community and the motivation behind the establishment in 1965 of a multi-year observational program centred near 28° N, 155° W in an area dubbed the 'Climax region' (Venrick, 1995). A climax community refers to a more or less permanent and final stage of succession that is nearly in equilibrium with its environment and is inseparably connected with both habitat characteristics and climate (Clements, 1916). Succession is usually driven by changes in the physical environment, including climate, and generally culminates in a stable, terminal ecosystem – the 'climax community.' This assemblage of organisms achieves a maximum utilisation of resources (see Chapter 11). A key aspect of climax community theory is that, under present ruling climate, the community does not change. As a corollary, changes in the physical habitat, sustained by local or regional climate variations, will drive the ecosystem away from its steady state. In subsequent sections of this chapter we will present evidence for climate-driven changes in NPSG community structure and photosynthesis. Climate affects the seascape and, hence, impacts both biodiversity and biodynamics.

The Climax region ^{14}C-based primary production data sets, collected between 1968 and 1980, generally confirmed the anticipated low rates of primary production (Hayward *et al.*, 1983). These field results reinforced the existing perception that the low nutrient, euphotic zones of anticyclonic ocean gyres were aquatic analogues of terrestrial deserts. Unfortunately, the coarse sampling frequency was unable to resolve seasonal or interannual variability, if any existed at all.

These historical results contrast sharply with more recent observations from the NPSG documenting a higher than expected sustained rate of primary production in addition to a vigorous field of mesoscale variability, including discrete eddies, near-inertial motions and internal tides. For example, wind-forced currents and meso-scale eddies can interact to produce vigorous vertical motions that deliver short-lived pulses of nutrients to habitats that are otherwise nutrient-starved (Letelier *et al.*, 2000; Seki *et al.*, 2001a). These physical processes can abruptly alter the ecological status quo, resulting in pulses of primary production, particle export and, hence, brief opportunities for carbon sequestration. Because of the stochastic nature of these events, the tradition of ship-based oceanographic expeditions to the NPSG, usually during summer, has failed to observe these non-steady-state processes. Furthermore, there is now ample recent evidence to suggest that major changes in the structure of the NPSG habitat can occur over interannual to decadal time periods. Because of the role of the ocean as a potential sink for the increasing burden of atmospheric carbon dioxide, it is imperative that ecological processes in open ocean ecosystems such as the NPSG are understood.

As early as 1953, Riley (1953) had reported large variations in his oxygen method-based production estimates for both subtropical and tropical oceanic waters. He concluded that only through careful time-series investigations will we achieve an accurate estimate of annual production. This 'call to arms' for systematic and repeated measurements of the subtropical gyres was not taken seriously for several decades. In 1988, a deep ocean station was established north of Hawaii within the NPSG to conduct a high-frequency time-series study of the oceanic carbon cycle, including core measurements of microbial community structure, primary production and export. By the end of 2000, the HOT program had completed 121 approximately monthly research cruises to Station ALOHA with comprehensive seasonal coverage. This ongoing serial measurement program, with a focus on physical–biogeochemical coupling and microbial processes, has yielded unexpected results that challenge past views of NPSG biogeochemical cycles.

Below we present selected summaries of the extensive HOT program core measurements, including a presentation of the ^{14}C primary production and pigment biomarker data sets. We will not debate the validity of the various, and generally untested, assumptions of the ^{14}C technique or its accuracy. These important matters are covered elsewhere in this volume (see Chapter 4). We will argue, however, that gross primary production in the NPSG is likely to be at least twice as high as the ^{14}C data would indicate. We follow these data presentations with the discussion of three case studies of aperiodic oceanic variability.

9.2   NPSG primary production and microbial community structure climatology and variability: the HOT programme data set

In his original application of the ^{14}C method, Steemann Nielsen described elaborate procedures that were necessary at that time to prepare the stock ^{14}C solutions by ^{14}C-carbon dioxide distillation, from the barium ^{14}C-carbonate solid reagent obtained from Oak Ridge. Fortunately, today we simply order ready-to-use solutions of chemically-pure sodium ^{14}C-carbonate solutions of certified specific radioactivity. One must still be cognisant of potential contamination of the ^{14}C stocks, either by ^{14}C-DOC (Williams *et al.*, 1972) or toxic trace metals (Fitzwater *et al.*, 1982). We also use modern liquid scintillation counters instead of the relatively inefficient Geiger-Müller tubes that were used at the time the ^{14}C method was first developed.

Beyond these technical aspects of the ^{14}C method, there is a remarkable similarity between the protocols used by Steemann Nielsen 50 years ago and those employed in the HOT program today (Box 9.1; Karl *et al.*, 1996; Letelier *et al.*, 1996). The most critical features include:

(1) collection of uncontaminated water samples from pre-selected reference depths,
(2) *in situ* incubation of ^{14}C-spiked water samples,
(3) post-incubation processing of samples, using filters of appropriate porosity, and
(4) careful determinations of the total dissolved inorganic carbon concentration for accurate calculation of specific radioactivity and, hence, rate of *in situ* photosynthesis (Box 9.1).

Chlorophyll determinations

Chlorophyll *a* (Chl *a*) functions as the primary light-harvesting pigment for all prokaryotic and eukaryotic marine oxygenic photoautotrophs. Even though the Chl *a*:carbon ratio of photoautotrophic cells varies considerably as a function of environmental conditions and growth rate (Laws *et al.*, 1983), measurements of Chl *a* have been used extensively to estimate the biomass of photoautotrophic microorganisms in the sea. Typically, cells are concentrated by filtration, extracted into an organic solvent (usually acetone or methanol) and the pigments detected by fluorescence, sometimes after chromatographic separation (Bidigare and Trees, 2000).

At Station ALOHA, total Chl *a* (sum of monovinyl Chl *a* and divinyl Chl *a*) concentration is low in near-surface waters (< 0.1 mg Chl *a* m^{-3}), with slightly elevated concentrations observed during winter months (Plate 9.1 and Figs 9.2 and 9.3). These enhanced wintertime Chl *a* concentrations are a result of photoadaptation – specifically enhanced pigmentation at low light levels (Letelier *et al.*, 1993; Winn *et al.*, 1995). Between approximately 40 m and 110 m there is a region defined by a steep positive gradient in Chl *a* concentration with increasing water depth to maximum concentrations ≤ 0.2 mg Chl *a* m^{-3} (Plate 9.1). The Chl *a* concentration in this deep euphotic zone Chl *a* maximum layer (DCML) varies both seasonally with a maximum in summer, and interannually. The DCML is not a region of enhanced photoautotroph biomass, but rather increased pigmentation per unit cell carbon. During the decade-long surveillance of plankton processes at Station ALOHA, 1989 and 1999 stand out as years when the Chl *a* concentration in these deep layers is enhanced relative to the climatological mean conditions (Plate 9.1c). The DCML at Station ALOHA is well below the approximately 1 optical depth (~ 25 m) detection by satellite remote sensing, so these aperiodic dynamics can only be observed by ship or ocean mooring based measurement programmes.

The position of the DCML closely tracks isolume contours; the DCML shoals in winter and deepens in summer (Fig. 9.2 and Plate 9.1). The water sampling protocols employed in the HOT programme target the DCML on every cruise regardless of

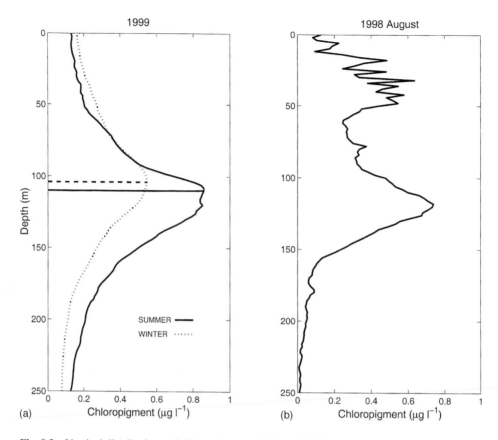

Fig. 9.2 Vertical distributions of chloropigments (chlorophyll plus pheophytin) determined from *in vivo* fluorescence measurements and bottle calibrations. (a) Average distributions at Station ALOHA for summer (June–August) versus winter (December–February) for 1999, showing changes in both total concentration of chloropigments and in the depth of the DCML. (b) Anomalous distribution of chloropigments characteristic of aperiodic blooms of diatoms that are typically encountered in near-surface waters during late summer. See text for additional details.

depth of occurrence. On much shorter time-scales, the depth of the DCML varies considerably but irregularly due to inertial period (~ 31 h) and semi-diurnal tidal oscillations (Fig. 9.4). For example, during HOT-90 (February 1998) the position of the peak of the DCML varied systematically with depth over the 31 h inertial period; however, when plotted in density space the DCML was always at or near the 24.5 potential density surface. The HOT program 'burst sampling' is designed to record energetic vertical motions so that a reliable mean cruise density versus depth profile can be estimated; this prevents aliasing in the long-term data set (Karl and Lukas, 1996). Both the amplitude of motion and its timing relative to local apparent noon varies for each cruise. These vertical displacements of phototrophic organisms within a depth-fixed exponentially changing light field can have a significant impact on daily integrated primary production, as discussed in Case Study No. 1 (see

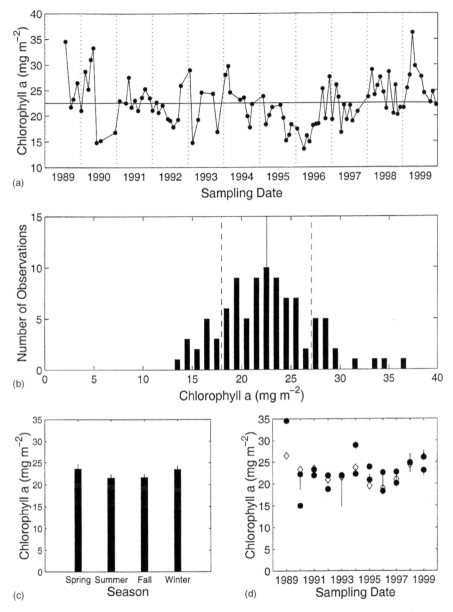

Fig. 9.3 Euphotic zone (0–200 m) depth-integrated concentrations of Chl *a* (mg m^{-2}) measured at Station ALOHA during the first 11-year HOT program observation period. (a) Each datum in the upper graph is the cruise integral based on a profile collected from approximately 12 individual reference depths. (b) These same data presented as a frequency histogram with mean (solid vertical line) ± 1 standard deviation (SD) (dashed vertical lines), (c) as seasonal means (winter = December–February, spring = March–May, summer = June–August, autumn = September–November), and (d) as annual/seasonal means. In (d), the open diamond represents the mean value for the year indicated. The solid circle with the + 1 SD shown is the mean summer (June–August) value for that year and the solid circle with the − 1 SD shown is the mean winter (December–February) for that year.

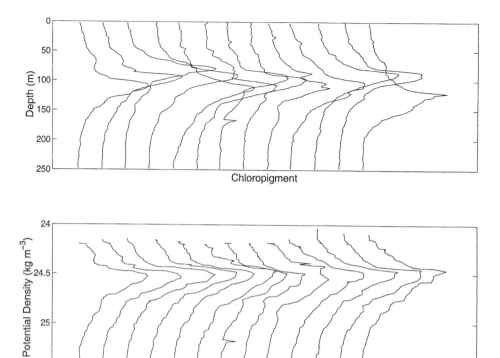

Fig. 9.4 Continuous fluorescence (arbitrary units) versus water column depth profiles at Station ALOHA during HOT-90 (February 1998). Each trace is from a separate *in vitro* fluorescence profile obtained using a submersible fluorometer calibrated to extracted chloropigments using bottle measurements. These consecutive CTD casts cover an approximately 36 h period from 20.9 hours on 18 February to 9.0 hours on 20 February 1998. Data presented as fluorescence versus water depth and fluorescence versus potential density. The individual traces have been offset for presentation clarity. Note the systematic vertical oscillation of the DCML in 'depth space' compared to the relative constancy in 'density space'.

Section 9.3); these vertical motions are not reproduced during our *in situ* primary production experiments.

Less well understood are the interannual dynamics, especially the Chl *a* concentrations in the DCML during 1989 and 1999 (Plate 9.1 and Fig. 9.3). It is very unlikely that this is a result of undersampling (see sample density in Plate 9.1). Higher than average Chl *a* concentrations between years could be related to total solar insolation (i.e. a 'cloudy' year versus a 'sunny' year) or to temporal changes in the phytoplankton community structure or biomass. None of these processes seems likely to occur at the magnitudes observed for Station ALOHA. Despite the temporal dynamics of the DCML, this region has little impact on total euphotic zone depth-integrated rates of ^{14}C-based primary production because the majority of the primary production occurs in the upper 100 m of the water column (Letelier *et al.*, 1996; Ondrusek *et al.*, 2001). However, because the position of the DCML nearly coincides with the depth of increasing inorganic nutrients, habitat changes in this region could fuel aperiodic growth and export of organic matter. Below the DCML, there is an equally steep negative gradient in Chl *a* concentration with increasing water depth. In this region of the water column, inorganic nutrients are present but light is insufficient for sustained rates of photosynthesis.

Over the 11-year observation period at Station ALOHA, the total euphotic zone (0–200 m) depth-integrated Chl *a* concentrations varied more than two-fold, from minima of < 15 mg Chl *a* m^{-2} in June 1990, February 1993 and March–May 1996 to > 30 mg Chl *a* m^{-2} in July 1989, April–May 1990 and April 1999; the overall mean and standard deviation for euphotic zone depth-integrated Chl *a* was 22.5 ± 4.5 mg m^{-2} (range 13.6 to 36.2; $n = 94$), (Fig. 9.3). The time-series data set reveals several interesting and as yet unexplained phenomena including: (1) rapid sub-seasonal changes in Chl *a* inventories (e.g. May 1990 to June 1990 and January 1993 to February 1993) and (2) multi-season, even multi-year, periods of systematic Chl *a* increases (e.g. 1996–1999) or Chl *a* decreases (e.g. 1994–1996). Overall 1989 and 1999 stand out as relatively high Chl *a* years, and 1995 and 1996 stand out as relatively low years. These low-frequency dynamics in euphotic zone Chl *a* inventories are probably related to changes in habitat which may impact microbial community structure, but at the present time we lack a complete understanding of possible cause-and-effect relationships.

Primary production

Estimation of primary production at Station ALOHA, using the *in situ* ^{14}C technique, revealed several unexpected results including (Plate 9.2 and Fig. 9.5):

(1) A relatively high and sustained rate of organic carbon production of 484 ± 129 mg C m^{-2} d^{-1} ($n = 94$) for the 0–200 m depth region, compared to historical estimates from the period 1965–1980 that were two- to threefold lower.

(2) A predictable seasonal variation in production with maxima in summer months (especially May–August) and minima in winter months (especially December–January).

(3) Large aperiodic variations from the long-term mean production rate of 484 mg C m^{-2} d^{-1} with values ranging from < 200 mg C m^{-2} d^{-1} (e.g. December 1998) to > 900 mg C m^{-2} d^{-1} (e.g. April 1995).

(4) Significant interannual variations in organic matter production (e.g. 1990 versus 1995).

The relatively high rates of organic matter production that we routinely estimate by the 'particulate' ^{14}C method employed in the HOT program (see Box 9.1) are similar to those predicted using an independently calibrated optical model (see Box 9.2 and Fig. 9.6(a) and (b)). The mean observed euphotic zone production rate of 484 ± 129 mg C m^{-2} d^{-1} is statistically indistinguishable from the model estimate of

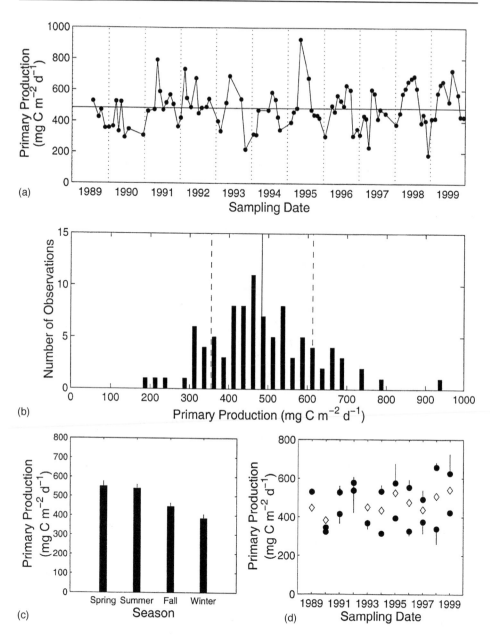

Fig. 9.5 Euphotic zone (0–200 m) depth-integrated primary production (mg C m^{-2} d^{-1}) measured at Station ALOHA during the first 11-year HOT program observation period. Data presentation as in Fig. 9.3.

Box 9.2   Modelling primary production rates

(1)   *Optical modelling of primary production rates*

A versatile model (Bidigare *et al.*, 1987) was used for the estimation of primary production rates (PP_{model}, mg C m^{-3} t^{-1}):

$$PP_{model} = \Phi_C \times E_{abs} \times 12\,000 \tag{9.I}$$

where, Φ_C is the *in situ* quantum efficiency of carbon fixation (mol C mol quanta^{-1}), E_{abs} (mol quanta m^{-3} t^{-1}) is the absorbed quanta by phytoplankton per unit time, and $12\,000$ has units of mg C mol C^{-1}. Parameterisation of E_{abs} was accomplished by splitting the solar irradiance into spectral components via the 5S radiative transfer code (Tanré *et al.*, 1990) and propagating the spectral light field through the ocean using very simple two-stream approximations (Ondrusek *et al.*, 2001). Absorption and scattering coefficients were obtained from the published literature (Petzold, 1972; Smith and Baker, 1981; Gordon and Morel, 1983; Pope, 1993; Bricaud *et al.*, 1995). E_{abs} values were calculated from measured sea-surface E_{PAR} and measured sub-surface Chl *a* distribution. E_{abs} was then related to the total quanta absorbed at a given depth and the fraction of the total absorption coefficient accounted for by phytoplankton absorption. Absorption calculations were carried out spectrally and integrated over the visible region of the electromagnetic spectrum (400–700 nm) and with respect to time and depth to obtain areal production rates during the photoperiod (mg C m^{-2}). *In situ* quantum efficiency was calculated from maximum quantum yield for C-fixation, $(\Phi_{max})_C$, a light-dependent term. The light dependency is taken from Bidigare *et al.* (1992) with a modification to parameterise the term E_{PAR}/E_k from Waters *et al.* (1994). A $(\Phi_{max})_C$ value of 0.026 mol C mol quanta^{-1} was used for the calculation of primary production rates (Ondrusek *et al.*, 2001). The latter was determined from particulate ^{14}C-based P versus E measurements and has an uncertainty of $\pm 35\%$ (Ondrusek and Bidigare, 1997).

(2)   *Estimation of Prochlorococcus primary production*

Daily primary production contributions by *Prochlorococcus* (PP_{pro}, mg C m^{-2} d^{-1}) in the mixed layer (0–25 m) can be estimated as:

$$PP_{pro} = \mu C_o(\mu - g)^{-1} (e^{(\mu - g)t} - 1) \tag{9.II}$$

where μ and g are the growth and grazing rates for *Prochlorococcus* (d^{-1}), and C_o is *Prochlorococcus* carbon biomass (mg C m^{-2}) in the upper 25 m. This relationship was derived from equation (1) in Laws (1984) and does not include fixed carbon excreted as DOC. As the $(\mu - g)$ terms in this equation approach zero (i.e. *Prochlorococcus* growth and grazing rates are in proximate balance over the daily time-scale), PP_{pro} becomes equal to the product of μ and C_o. Previous studies performed in the equatorial Pacific (Vaulot *et al.*, 1995) and at Station ALOHA (Liu *et al.*, 1995) have documented that μ is of the order of g in the mixed layer. For the purposes of this study, we have assumed that μ is of the order of g and μ in the mixed layer ranged from 0.5 to 0.7 d^{-1} (Liu *et al.*, 1995; Vaulot *et al.*, 1995). C_o was estimated using flow cytometry data collected during HOT-22 through HOT-110 and a carbon conversion ratio of 30 fg C cell^{-1} (Björkman *et al.*, 2000).

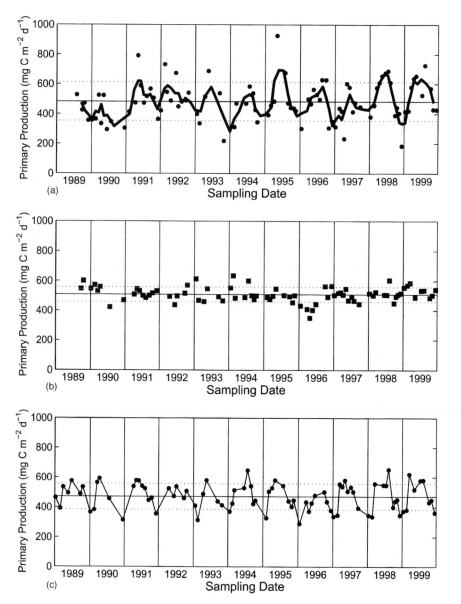

Fig. 9.6 Comparison of measured (^{14}C method) primary production at Station ALOHA with output from bio-optical model estimations (see Box 9.2) for the same time period. (a) Each datum in the ^{14}C graph is the cruise integral determined from *in situ* measurements at eight separate reference depths. The solid curve is the three-point running mean for the ^{14}C data set. The solid and dashed horizontal lines are the measured mean $\pm$ 1 SD (484 ± 129 mg C m^{-2} d^{-1}) for the ^{14}C data set. (b) Primary production estimated using a bio-optical model. Model structure, input and assumptions are described in Ondrusek *et al.* (2001) and in Box 9.2. The solid and dashed horizontal lines are the modelled mean $\pm$ 1 SD (509 ± 50 mg C m^{-2} d^{-1}). (c) Primary production (closed circles, solid line) estimated using a bio-optical model and seasonally adjusted ϕ_{max} values (see text for details). The solid and dashed horizontal lines are the modelled mean $\pm$ 1 SD (471 ± 86 mg C m^{-2} d^{-1}).

509 ± 50 mg C m^{-2} d^{-1}; the mean difference (measured/modelled) is -33 mg C m^{-2} d^{-1} (SD $= 136$ mg C m^{-2} d^{-1}). However, the optical model does not capture the seasonal trends apparent in the measured rates, and at times, modelled rates differ by as much as 50% from the measured rates (Fig. 9.6(b)). While the optical model accounts for temporal variations in Chl *a* and Photosynthetically Available Radiation (PAR), it assumes a constant maximum quantum yield ($\phi_{max} = 0.026$ mol C mol quanta^{-1}) for the calculation of primary production (see Box 9.2). Unfortunately, the P versus E measurements used to estimate ϕ_{max} were only made on select cruises at a restricted number of depths, and therefore did not allow Ondrusek *et al.* (2001) to parameterise the temporal variability of this key model parameter.

In order to account for intra-annual variations in ϕ_{max}, we binned the HOT cruises by season and derived ϕ_{max} values from the seasonal differences between modelled and measured rates of primary production. This exercise yielded ϕ_{max} values of 0.017, 0.028, 0.028 and 0.023 mol C mol quanta^{-1} for cruises conducted during winter (December–February), spring (March–May), summer (June–August) and autumn (September–November) periods, respectively. When these latter values were used in place of the default value of 0.026 mol C mol quanta^{-1}, the measured and modelled production rates display similar temporal patterns (Fig. 9.6(c)). Future HOT cruises will include vertical profiles of fast repetition rate (FRR) fluorometer-determined photosynthetic parameters and seasonally resolved P versus E measurements, which will allow seasonal variations in ϕ_{max} to be incorporated in the optical models.

Another limitation of the present model is the calibration using ^{14}C values derived from particulate matter following time incubations. The particulate ^{14}C method can underestimate GPCP in the NPSG by at least a factor of 2 (Karl *et al.*, 1998). Laws *et al.* (2000) reported similar findings for the Arabian Sea and determined that ^{14}C uptake accounted for only 45 ± 10% of the gross photosynthesis. Probable reasons for this disparity are the combined effects of the Mehler reaction, photorespiration, dark respiration, DOC excretion and grazing. Direct measurements of ^{14}C-DOC production at Station ALOHA confirm exudation as a key ecophysiological parameter (Karl *et al.*, 1998) affecting the interpretation of particulate ^{14}C data. Regardless of the mechanism, the production of large amounts of ^{14}C-DOC during dawn-to-dusk *in situ* experiments implies that primary production in this and probably other low-nutrient habitats has been significantly underestimated – we believe, by at least a factor of 2. Sustained primary production rates of ≤ 1 g C m^{-2} d^{-1} are unprecedented for the NPSG 'desert' and this highlights our limited understanding of the processes controlling productivity and carbon fluxes in this and, probably, other oligotrophic ocean habitats.

With regard to deviations from the long-term mean rate of organic matter production, there are both predictable seasonal changes and aperiodic variations. However, the aperiodic maxima ($\geq +1$ SD from the mean value) are generally in summer (12 of the 14 'high production' observations occurred between May and August; Fig. 9.5), and the minima (≤ -1 SD from the mean) are generally in winter

(13 of the 15 'low production' observations occurred between October and April; Fig. 9.5). This supports the hypothesis that light intensity and day length, even in this low-latitude subtropical habitat, are probably the most critical ecological determinants for organic matter production. It is also interesting to note that the HOT program ^{14}C primary production data set appears to have a 'basal level' of approximately 200 mg C m^{-2} d^{-1}. This is approximately equal to the sustained production estimates reported for the NPSG from the period 1952–1980 (Karl *et al.*, 2001a). We hypothesise that there has been a 'blooming' of the NPSG over the past two decades as a consequence of variations in climate forcing. This topic is discussed further in Case Study No. 3 (see Section 9.3).

Photosynthetic assimilation numbers

The light-saturated rates of carbon fixation normalised to total Chl *a* (i.e. referred to as the assimilation number, productivity index, P^B_{max}, or photoperiod assimilation ratio, and usually reported in units of grams of C fixed per gram Chl *a* h^{-1}) is a measure of the physiological state of the photoautotrophic microbial assemblage under *in situ* conditions. This parameter reflects both the maximum photosynthetic efficiency and the maximum specific growth rate (Falkowski, 1981). Because the seasonal patterns in rates of ^{14}C incorporation at Station ALOHA are directly out of phase with the concentrations of Chl *a* (i.e. high summertime rates of ^{14}C incorporation during conditions supporting low ambient Chl *a* concentrations, for reasons discussed above), the assimilation number for surface assemblages has an amplified seasonality compared to either of the two measured parameters (Fig. 9.7). The mean and standard deviation of the light-saturated assimilation number at Station ALOHA is 6.7 ± 2.8 (range = 2.2 to 15.5; $n = 89$) with seasonal maxima in spring and summer periods. However, both the mean value and the seasonal range of values within a given year are variable. Furthermore, for nearly a four-year period from 1993–1996, the mean annual assimilation number at Station ALOHA systematically increased, both in winter and in summer, abruptly returning to a lower and less variable value for the following three-year period (Fig. 9.7). However, these dramatic changes in the light-saturated, Chl-normalised rates of ^{14}C primary production had little apparent impact on total water column particulate primary production (see Plate 9.2 and Fig. 9.5).

Based on theoretical considerations including the maximum turnover time for a photosynthetic unit (PSU) of 1 ms, Falkowski (1981) suggested that 1 g of Chl *a* could potentially fix 25 g C h^{-1} at light saturation, or a theoretical maximum assimilation number of about 25. Assimilation numbers measured for natural populations of light-saturated phytoplankton are typically 10–50% of this theoretical value, suggesting slower PSU turnover times because of limitations of light, nutrients or temperature. The photoautotrophic plankton communities in the warm (annual temperature fluctuation at Station ALOHA is 23–27°C), upper water column (0–25 m) of the NPSG are not likely to be temperature- or light-limited

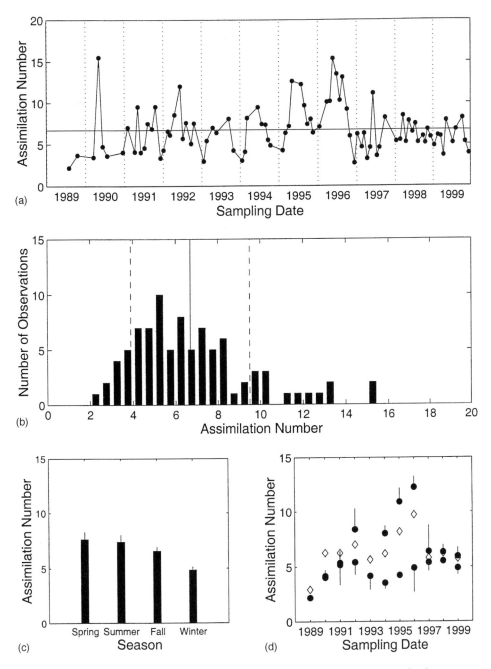

Fig. 9.7 Light-saturated (5 m reference depth) assimilation number (mg C mg chl^{-1} h^{-1}) measured at Station ALOHA during the first 11-year HOT programme observation period. Data presentation as in Fig. 9.3.

(Letelier *et al.*, 1996). The low assimilation numbers previously reported from oligotrophic regimes (summarised in Karl *et al.*, 1998) have been interpreted as evidence for nutrient (generally N) limitation. Because Chl *a* contains N and is also generally bound *in vivo* to protein which also contains N (Falkowski, 1981), inorganic N limitation is predicted to lead to low assimilation numbers and low rates of total primary production. However, Cullen (1995) argued that it is not appropriate to use Chl-normalised photosynthesis as a diagnostic indicator of nutrient limitation. Because this parameter is the product of specific growth rate (μ, d^{-1}) and the C:Chl ratio, and nutrient addition leads to increases in μ and decreases in C:Chl (Laws and Bannister, 1980), the assimilation number may be insensitive to changes in phytoplankton nutrient status. Consequently, the low frequency variations in assimilation number at Station ALOHA are not well understood and may be related to photoadaptation in the upper euphotic zone as well as changes in phototrophic microbial community structure. Furthermore, if there is a significant accumulation of ^{14}C-DOC during the incubation period then the assimilation numbers we present are underestimates of the actual production maxima. Karl *et al.* (1998) have concluded that ^{14}C-DOC can equal the ^{14}C-POC in surface waters at Station ALOHA. Under these conditions, the measured assimilation numbers may exceed 18 mg C mg Chl *a* h^{-1}, a value that is $>70\%$ of the theoretical maximum value. In support of this suggestion, it may be noted that the *in vitro* and *in situ* oxygen-flux based observations made during the 1985 PRPOOS study (Williams and Purdie, 1991) gave assimilation numbers commonly in the range 1–2 mmol O_2 Chl *a* h^{-1} (i.e. $\sim$ 10–20 mg C mg Chl *a* h^{-1}). The oxygen technique does not suffer the complication of the separation between particulate and dissolved production.

Phototrophic microbial community structure

The microscopic, photosynthetic organisms in the ocean have traditionally been referred to as phytoplankton ('plant' plankton), a term initially reserved for eukaryotic algae (Strickland, 1965). This terminology is dated, but is still used extensively in the scientific literature. Biologists now recognise three major lines of evolution, e.g. the domains *Bacteria*, *Archaea* and *Eukarya*. Although the terms 'bacteria' and 'algae' have been used to refer to the 'heterotrophic bacteria' (bacterioplankton) and the 'photosynthetic eukaryotic algae' (phytoplankton), it is now well-known that most of the algae in the NPSG are actually *Bacteria* (i.e. cyanobacteria, formerly known as blue-green algae).

Before 1978, photoautotrophic plankton populations in the NPSG were thought to be dominated by eukaryotic micro-organisms, especially monads, flagellates and to a lesser extent diatoms (Beers *et al.*, 1982). During the Climax program era, Venrick (1993) conducted comprehensive analyses of diatom species distributions and diversity, but no assessment was made of the relative importance of this subcomponent of the total phytoplankton assemblage to total primary production. This traditional view of eukaryotic dominance began to change in 1979 with the discovery of picophytoplankton (Johnson and Sieburth, 1979; Waterbury *et al.*, 1979), and even today our views of the NPSG microbial food web are in transition.

Several attempts have been made to quantify the various photoautotrophic components of marine plankton based on analyses of class-specific chlorophyll and carotenoid pigments. Letelier *et al.* (1993) used an empirical approach to estimate the Chl *a* contributions from major phytoplankton groups within the DCML at Station ALOHA. Their results, based on a three-year (1989–1991) data set, confirmed the prokaryote dominance with a mean % Chl *a* composition of: *Prochlorococcus* spp. (39%), other cyanobacteria (24%), prymnesiophytes (22%) and chrysophytes (13%). Diatoms and dinoflagellates comprised < 3% of the mean Chl *a* inventory. Based primarily on pigment and electron microscopic observations of water samples collected from Station ALOHA during a single cruise (April 1993), Andersen *et al.* (1996) concluded that *Prochlorococcus*-like micro-organisms constituted approximately 50% of total euphotic zone Chl *a*. *Synechococcus*, and two picoeukaryotic algal groups, prymnesiophytes and pelagophytes (*sensu* Andersen *et al.* (1993) referred to as chrysophytes in Letelier *et al.* (1993)) together accounted for most of the remainder of the standing stock of phototrophic cells (Andersen *et al.*, 1996). Cells larger than 8 μm (e.g. diatoms and dinoflagellates) were rare.

Recently, a more extensive analysis has been conducted using the CHEMTAX program developed at the CSIRO Marine Laboratory in Australia (Mackey *et al.*, 1996) to the HPLC-analysed pigment data sets from Station ALOHA in an effort to identify and to quantify key taxonomic groups based on specific biomarkers. A few preliminary results are already available. First, the photoautotrophic assemblages in the upper euphotic zone at Station ALOHA are dominated by *Prochlorococcus*-like micro-organisms with divinyl Chl *a*, Chl *b* and zeaxanthin as key pigments. This is the only taxonomic group known to contain divinyl Chl *a*. Other algal groups (e.g. chlorophytes and prasinophytes) contain Chl *b* and all cyanobacteria contain zeaxanthin, so these two biomarkers are not unique to *Prochlorococcus*. Zeaxanthin content per unit cell volume in *Prochlorococcus* and *Synechococcus* does not vary as a function of growth irradiance (Moore *et al.*, 1995). Thus, this biomarker may be useful as a proxy for cyanobacteria biovolume at Station ALOHA.

Prymnesiophytes contain significant amounts of 19'-hexanoyloxyfucoxanthin (19'-hex), and lower concentrations of 19'-butanoyloxyfucoxanthin (19'-but) and fucoxanthin. Fucoxanthin is also found in diatoms where it is the major accessory pigment. Pelagophytes contain 19'-but as their primary diagnostic pigment. Theoretically, these distinct but sometimes co-occurring groups of photosynthetic prokaryotic and eukaryotic micro-organisms can be resolved into specific groups using multiple equations based on educated values of accessory pigment-to-Chl *a* ratios for each taxon. The application of the CHEMTAX algorithm to the decade-long Station ALOHA data set indicates a dominance by *Prochlorococcus*, other cyanobacteria (including *Synechococcus* and *Trichodesmium*), prymnesiophytes and pelagophytes. The relative abundances of diatoms and dinoflagellates, once thought to dominate the NPSG phytoplankton assemblage, are negligible. All of the diagnostic biomarker pigments, and therefore the abundance of all of the key taxa, displayed sub-seasonal, seasonal and interannual variability (e.g. Fig. 9.8 and 9.9(a)). These changes in pigment concentrations imply changes in the community structure of the phototrophic microbial assemblage. Relative to the individual 11-year mean inventories, there are multi-year periods of specific biomarker dominance (e.g. a doubling of 19′-hex for the period 1996–1999); selected biomarkers co-vary in their relative abundances (e.g. 19′-but and fucoxanthin), suggesting group selection. Other pigment biomarkers vary independently, indicating changes in community structure. As discussed later in this review chapter, the relatively low standing stocks of diatoms, as indicated by the biomarker fucoxanthin, are aperiodically enhanced many-fold leading to significant open ocean blooms, especially of diatom species containing endosymbiotic N_2-fixing cyanobacteria. Because most of the major plankton groups have unique physiological characteristics, growth requirements and trophic interactions, these changes in functional groups are likely to have major ecological implications. These and other sustained and aperiodic changes in phytoplankton standing stocks portend an ecosystem that is poised for habitat variability. In many ways, this is the antithesis of a climax community conceptualisation – an anti-climax as it were.

Campbell and colleagues (Campbell and Vaulot, 1993; Campbell *et al.*, 1994), first enumerated *Prochlorococcus* and *Synechococcus* cells throughout the water column at Station ALOHA using laser-based flow cytometry. Since their pioneering efforts, these measurements have been retained as core parameters in our field program. Flow cytometry provides quantitative information on numbers only; extrapolation to biomass demands additional, largely unavailable information on precise cell dimensions to estimate total biovolume, and on biovolume-to-biomass and biomass-to-carbon extrapolation factors. Rough estimates of biomass-C have been obtained using single conversion factors of $53 \, \text{fg C cell}^{-1}$ and $250 \, \text{fg C cell}^{-1}$ for *Prochlorococcus* and *Synechococcus*, respectively (Campbell *et al.*, 1994; Liu *et al.*, 1997), but such an approach ignores spatial and temporal variations in cell size and carbon content (which could vary seasonally and with depth as a function of growth rate or light and nutrient stress), and has unknown error. Björkman *et al.* (2000) recently used $30 \, \text{fg C cell}^{-1}$ and $100 \, \text{fg C cell}^{-1}$ as the biomass-C conversion factors for these two key groups of photoautotrophs, revised downward based largely on field data obtained by Chavez *et al.* (1996) and Zubkov *et al.* (1998). Björkman *et al.* (2000) also pointed out that when the Campbell *et al.* (1994) conversion factors were applied to the field data from Station ALOHA the estimated *Prochlorococcus* biomass-C exceeded the measured total particulate carbon (PC) at this location. Clearly PC is an upper constraint on the living carbon pool, which *Prochlorococcus* is one component thereof. Our new estimates of the contribution of *Prochlorococcus* biomass to total PC (Fig. 9.9(b)) indicate a range < 10% to > 35%, with an approximate doubling in biomass between the period 1992–1995, remaining high thereafter. We believe that these low-frequency, sub-decade scale changes in phototrophic microbial community structure are a result of large-scale climate

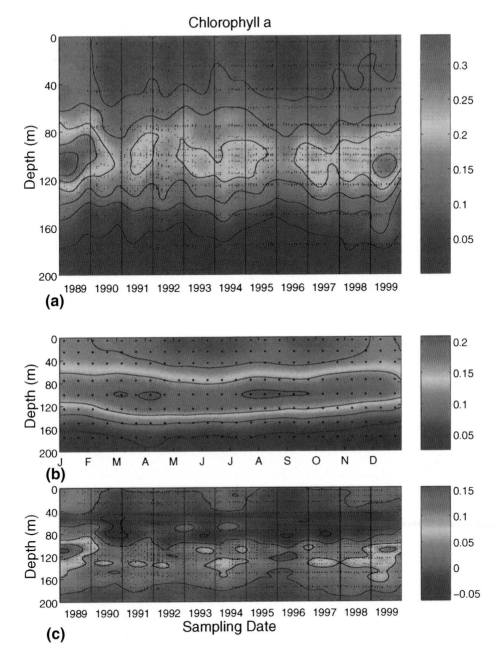

Chlorophyll a

(a)

(b)

(c)

Sampling Date

Plate 9.1. Contour plots of (a) the 11-year time-series record of approximately monthly Chl *a* concentrations (mg Chl *a* m^{-3}) versus depth at Station ALOHA, (b) the annual Chl *a* climatology binned by two-week intervals, and (c) the signed differences from the long-term 11-year average conditions presented as observed minus climatology.

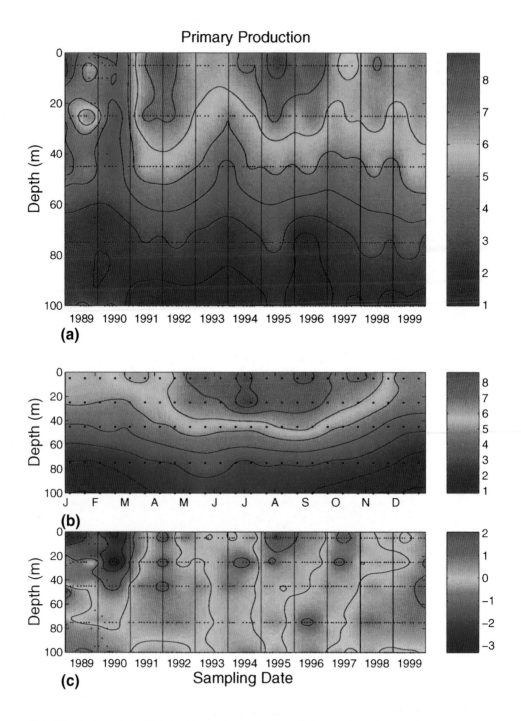

Plate 9.2. Contour plots of (a) the 11-year time-series record of approximately monthly primary production (mg C m^{-3} d^{-1}) versus depth at Station ALOHA based on the ^{14}C method, (b) the annual primary production climatology binned by two-week intervals, and (c) the signed differences from the long-term 11-year average conditions presented as observed minus climatology.

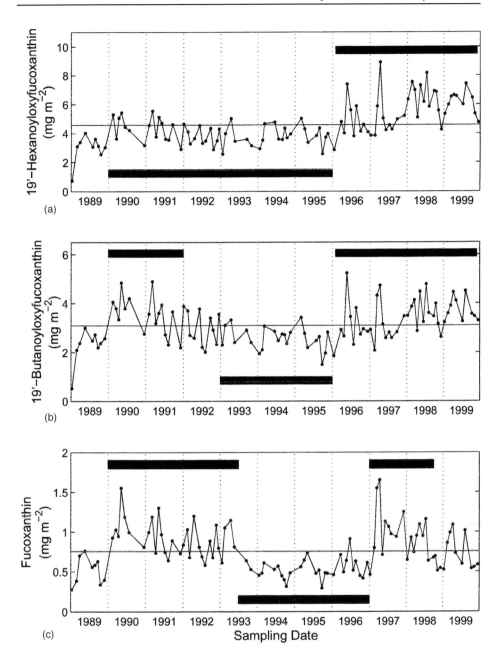

Fig. 9.8 Temporal changes in concentrations of diagnostic pigment biomarkers at Station ALOHA. Each datum represents the 50–200 m depth-integrated inventory for a given cruise. The solid horizontal line is the mean value for each biomarker data set. The dark bars highlight extended periods of deviation from the mean values.

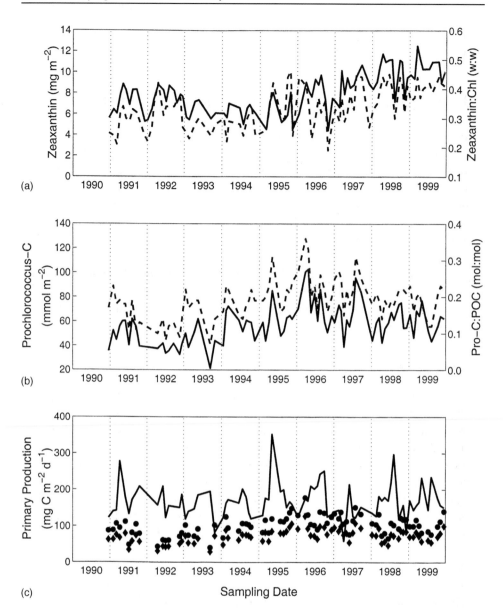

Fig. 9.9 Time-series of *Prochlorococcus* biomass and productivity parameters at Station ALOHA: (a) Zeaxanthin (mg m^{-2}; solid line) and the mean Zeax:Chl ratio (w:w; dashed line) for the upper 175 m, (b) *Prochlorococcus* carbon biomass (mmol m^{-2}; solid line) and the mean Pro-C:POC ratio (mol:mol; dashed line) for the upper 175 m, and (c) range of modelled *Prochlorococcus* production rates (mg C m^{-2} d^{-1}, ♦ = 0.5 d^{-1} and ● = 0.7 d^{-1}; see Box 9.2) for the upper 25 m of the water column. The ^{14}C-determined primary productivity for the same region is shown by the solid curve.

variations in the NPSG (Karl, 1999; Karl *et al.*, 2001a; also see Case Study No. 3 in section 9.3).

9.3   Varying time-scales of change: three case studies

Scales of variability affecting photosynthetic activity in open waters range from less than 1 s to greater than a decade (see Chapter 8). Understanding the role of forcing at different scales in shaping the community structure is at the core of ecology. However, this is a difficult task given the nested nature of environmental perturbations and the relevant scale of response of distinct populations to transient changes in the environment. In the following sections we discuss three scales of aperiodic ocean variability:

(1)   high-frequency (days to weeks) variability in physical forcing of primary production,
(2)   sub-seasonal, stochastic variability in time and in space, and
(3)   low-frequency, decade-scale processes.

Case Study No. 1: Day-to-day variations in primary production

Sharp vertical gradients in nutrient, light and temperature, represent a clear frame of reference for organisms in the euphotic zone. In this context, cloud cover, internal waves and near-inertial oscillations are all processes that may have a strong influence on PAR and, hence, on phototrophic micro-organisms. These processes will have their strongest effect on populations located at the base of the euphotic zone (Banse, 1987; Letelier *et al.*, 1996). The reason for this localised effect is that daily-integrated photosynthetic rates in the upper euphotic zone of the NPSG appear to be light-saturated (Letelier *et al.*, 1996). Furthermore, isopycnal displacements affect principally the community located in the upper pycnocline (see Fig. 9.4).

Banse (1987) noted that the effect of cloud cover could enhance the flux of nutrients into the lower euphotic zone but also emphasised that this effect would not reach the region above the DCML. For this reason, and because >50% of the primary production takes place in the upper euphotic zone, supported mainly by nutrient regeneration, he concluded that the variability in light availability would not have a significant effect on the depth-integrated productivity of offshore tropical and subtropical oceans, even though it could have an effect on the short-term uncoupling between nutrient uptake by photoautotrophs and nutrient fluxes (new and regenerated) into the lower euphotic zone.

Letelier *et al.* (1996) calculated the effect of near-inertial oscillations in the light field of the deep euphotic zone chlorophyll maximum layer at Station ALOHA (Fig. 9.10). In this region, short-term vertical displacements of the DCML can reach

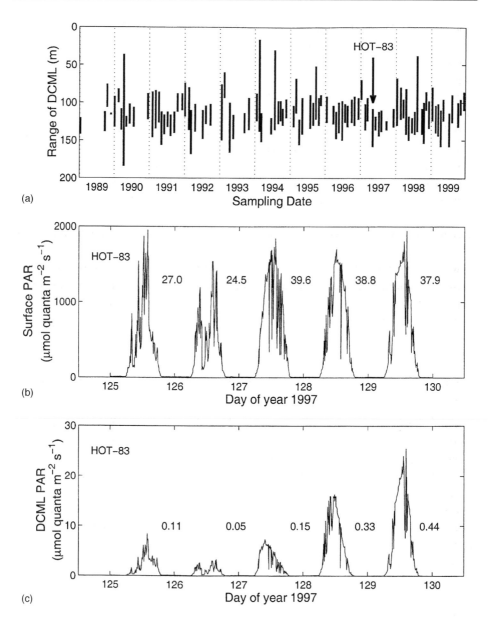

Fig. 9.10 Effect of isopycnal vertical displacements in accounting for day-to-day variability of photosynthetically available radiation (PAR) at the DCML at Station ALOHA. (a) Observed minimum and maximum depth range distribution of the DCML for each HOT cruise based on continuous fluorescence trace profiles obtained from 12 CTD casts deployed over a 36 h sampling period. (b) Surface PAR measured at the HALE ALOHA mooring location during HOT-83 (5–9 May 1997). (c) Estimated PAR at the DCML based on the vertical displacement of the DCML, surface PAR, and assuming $k_{PAR} = 0.04 \, m^{-1}$. Daily integrated PAR values (in mol quanta $m^{-2} \, d^{-1}$) are displayed next to each light cycle in (b) and (c). These day-to-day variations in light caused by inertial period oscillations of the DCML and variations in surface PAR caused by clouds are certain to have significant effects on rates of *in situ* photosynthesis.

> 60 m (range 14–69 m; mean $= 33$ m (SD $= 10.6$ m); median $= 30$ m, $n = 107$) within an inertial period (approximately 31 h at Station ALOHA). When considering the additional variability of cloud cover, daily irradiance at this reference level can vary eightfold on consecutive days as a result of the difference in the period of isopycnal displacement and solar irradiance (Fig. 9.10). These observations suggest that the combined effect of clouds and isopycnal displacements may result in day-to-day variability in PAR of nearly an order of magnitude for phototrophic microbial populations located in the upper layers of the pycnocline. However, these effects are not necessarily additive and in some instances may cancel out.

How important are these short-term perturbations in shaping the community structure at the base of the euphotic zone? Venrick's studies of phytoplankton diversity in the NPSG (Venrick, 1988, 1990, 1999) suggest that the highest effect may be found in a transition region located in the upper part of the DCML. This is also consistent with the role of the DCML as a trap to the nutrients that are diffusing upward from the nutricline (Banse, 1987). If this region represents the transition between communities physiologically adapted to grow on recycled and new nutrients, then the day-to-day variability in available PAR may result in a similar variability on the rate of supply of new nutrients into these layers, allowing the coexistence of both communities in this region of the water column.

Case Study No. 2: Cyclonic mesoscale eddies and meandering frontal boundaries

Despite the fairly 'predictable' plankton climatology in the NPSG, repeated shipboard measurements as well as both satellite remote sensing and continuous in-ocean mooring-based surveillance of this habitat have detected aperiodic blooms of phytoplankton that can usually be traced to changes in physical forcing often leading to an increased supply of inorganic nutrients from below. However, it is important to emphasise that not all increases in photosynthetic cell biomass and rates of primary production are controlled by enhanced forcing. For example, the well-documented near-surface accumulation of the N_2-fixing filamentous cyanobacterium *Trichodesmium* is a result of decreased, not increased, turbulent mixing (Karl *et al.*, 1992; Cullen *et al.*, 2001). In August 1989, a massive bloom of the filamentous cyanobacterium *Trichodesmium* was detected near Station ALOHA (Karl *et al.*, 1992). Based on ancillary mooring-based measurements, the authors concluded that the bloom was a manifestation of stratified water column conditions and an extended period of slack winds in the region. These habitat conditions are known to be conducive for accumulation and growth of *Trichodesmium*, and these environmental conditions are common in summer months at Station ALOHA and throughout the NPSG.

On the other hand, enhanced upwelling events, enhanced wind mixing and the subsequent entrainment of nutrient-enriched subeuphotic zone water can also lead to biomass accumulation and increased rates of primary production (Letelier *et al.*,

2000; Leonard *et al.*, 2001). In March–April 1997, upwelling of inorganic nutrients resulted in a mesoscale bloom of diatoms with attendant changes in primary and export production. The upwelling was traced to the combined local effects of strong wind divergence and passage of a cyclonic eddy through the HOT programme study area (Letelier *et al.*, 2000; Fig. 9.11), and these environmental conditions are common in winter months at Station ALOHA. Consequently, sporadic blooms can appear in both summer and in winter, but for fundamentally different reasons.

The accumulation of N_2-fixing micro-organisms in the upper water column during summer months is a phenomenon of particular interest from both an ecological and a biogeochemical perspective. Historically, transient increases in the carrying capacity of a pelagic oligotrophic marine ecosystem have been interpreted as the result of the injection of inorganic nutrients into the euphotic zone caused by physical forcing. Under habitat conditions that are favourable for growth and reproduction of the N_2-fixers, the 'normal' nutrient recharge processes that supply dissolved inorganic carbon, nitrate, phosphate, silicate and trace elements in proper stoichiometric proportions to balance long-term export, are disrupted in favour of a local, near-surface supply of new nitrogen only during the fixation of N_2 into organic matter. This leads to a decoupling of nutrient stoichiometry, and a pathway for net sequestration, by particulate matter export, of carbon, phosphorus and silica. It has also been hypothesised that atmospheric deposition of bioavailable iron may help to trigger these blooms of N_2-fixing micro-organisms (Wu *et al.*, 2000). Iron is an essential co-factor for nitrogenase, the key enzyme in the N_2 reduction pathway. Because the summertime seas are stable and well-stratified, any deposited iron dissolves into a fairly shallow mixed layer, leading to a significant enrichment in concentration relative to the mean state (Bruland *et al.*, 1994). The result is a surface enrichment in cells, most often diatoms containing endosymbiotic N_2-fixing cyanobacteria. During bloom-forming periods, the vertical distribution of Chl *a* is very different from the climatological profile, especially with regard to surface enhancements of Chl *a* and the general presence of dual Chl *a* peaks – one near the surface (associated with the base of the mixed layer) and the other, as expected, at depth (associated with the top of the nutricline; see Fig. 9.2). Furthermore, the enhanced 'noise' in the *in situ* fluorescence trace is caused by the presence of the large chain-forming cells and diatom mats that create small-scale spatial structure that is readily observed within these continuous profiles. This leads to the possibility that these blooms may be detected and mapped by remote sensing (e.g. moorings, autonomous underwater vehicles, or satellites).

The continued resupply of phosphorus (P) and, perhaps, iron remains a matter of debate. Because of their capacity to regulate buoyancy, and the high critical turgor pressure of gas vacuoles (Walsby and Bleything, 1988), *Trichodesmium* may store energy in the upper euphotic zone in the form of carbohydrate and migrate downward into the upper nutricline where limiting nutrients are available. Under this scenario there would be a temporal and spatial uncoupling between autotrophic carbon fixation and phosphorus uptake, in addition to the disruption of N supply

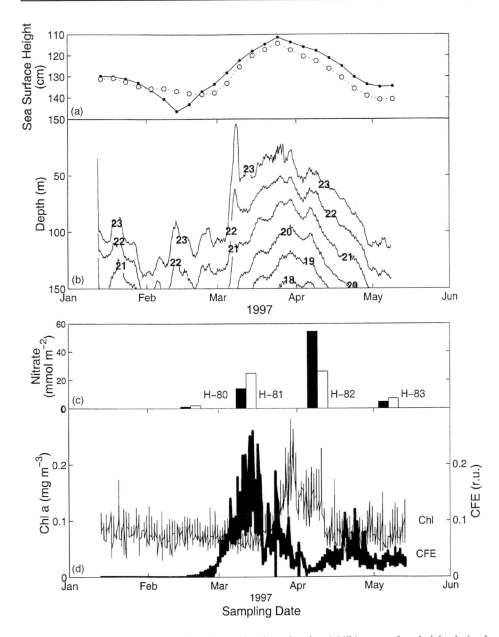

Fig. 9.11 HALE ALOHA mooring time-series data showing (a)/(b) sea surface height derived from TOPEX/ERS 2 and AVISO (solid) and TOPEX/ERS 2 and Pathfinder (open dashed) along with the 0–150 m depth distribution of temperature recorded by moored thermistors deployed at 2, 38, 50, 60, 80, 100, 120, 130 and 150 m; (c)/(d) temporal variations in the Chl concentration in mg m^{-3} and Chl fluorescence efficiency (CFE) in relative units, based on moored spectrophotometric observations at 25 m depth, plotted along with upper water column nitrate inventories for 0–125 m depth-integrated (solid bars) or surface to 1% light penetration depth-integrated (open bars) regions of the water column.

mentioned above. The migration behaviour required for this nutrient shuttle hypothesis, proposed by Karl *et al.* (1992), is similar to the behaviour found for some freshwater filamentous cyanobacteria. However, while in lakes these migrations rarely exceed 30 m, the depth of the nutricline in the NPSG is >100 m. Model simulations of *Trichodesmium* migrations (Kromkamp and Walsby, 1990), as well as field observations (Letelier and Karl, 1998) provide evidence regarding the feasibility of *Trichodesmium* migrations into the upper nutricline. Still, energy balances and migration frequency need to be taken into account when testing this shuttle hypothesis.

The development of monospecific diazotrophic blooms in the NPSG represents an important transient shift in the pelagic community structure. Not only do these blooms contribute to the uncoupling among C-N-P cycles, but they also induce an uncoupling between photosynthesis and community respiration, as well as a change in the size distribution of photoautotrophs. For these reasons, aperiodic blooms can have significant consequences for energy flux pathways and for the fates of the newly produced organic matter. Furthermore, *Trichodesmium* blooms can trigger ecological successions dominated by diatoms, the so-called 'echo' blooms (Devassy *et al.*, 1979). In the NPSG, increases in diatom abundance and diatom blooms are also observed during summer months. However, the ecological role of *Trichodesmium* in triggering these diatom blooms is unresolved.

Alternatively, diatom assemblages found in the euphotic zone of the NPSG during summer months may be dominated by species that are known to vertically migrate (Villareal *et al.*, 1999) or to develop associations with the endosymbiont N_2-fixing cyanobacterium *Richelia intracellularis* (Heinbokel, 1986; Scharek *et al.*, 1999a). Hence, it is not clear if the increase in diatom abundance results from having characteristics similar to *Trichodesmium* (nitrogen fixation and vertical migrations) or as an echo bloom phenomenon. These two pathways are not mutually exclusive, and the presence of multiple dominant species during these open ocean blooms (Scharek *et al.*, 1999a) suggests multiple selection mechanisms. Whatever the causes for the aperiodic increases in diatom abundance during summertime, this increase appears to be tightly coupled to an enhanced flux of organic matter out of the euphotic zone. The analyses of sediment trap material at 4000 m depth at Station ALOHA for 1992 and 1994 indicate that the flux of cytoplasm-containing diatom cells increases by >500-fold during late July (Scharek *et al.*, 1999b). If ultimately supported by 'new' nitrogen in the form of N_2 gas, these export events may be very important in the net sequestration of carbon into the deep ocean environment.

The sporadic injection of nutrients into the euphotic zone resulting from enhanced physical forcing has long been recognised as an important source of variability in the productivity of oligotrophic pelagic environments (e.g. Hayward, 1987). The analysis of the temporal distribution of nitrate in the upper 100 m of the water column at Station ALOHA displays the highest concentration and highest variability during winter months (Karl *et al.*, 2001b). While the increase in winter concentrations may be explained by a reduction in the depth penetration of light, the variability can only be attributed to short-lived nutrient injections into the euphotic zone as a result of upwelling and mixing events. These short-lived nutrient increases (<1 month) are captured sporadically during HOT cruises. However, there is no clear effect of these events in the temporal trend of euphotic zone integrated primary productivity in the HOT program data. One possible explanation for this apparent uncoupling is the observed depth distribution of nitrate during these events. While >50% of the integrated primary production takes place in the upper 45 m of the water column, mesoscale events such as cyclonic eddies will tend to raise the nutrient concentration at the base of the euphotic zone. Furthermore, it should be emphasised that if export (new) production is <10% of total production, then even a doubling in nutrient supply will only increase total production by about 10%, at most. It is doubtful whether field measurements using the standard HOT program ^{14}C technique would even reliably detect these productivity enhancements.

Even though an increase in productivity during winter months resulting from nutrient injections has not clearly been detected using the ship-based sampling protocols, there appears to be ample evidence suggesting that these sporadic events may cause a transient but significant shift in the community structure and the carrying capacity of the system. In 1997, a physical/biochemical/optical mooring was deployed in the vicinity of Station ALOHA. The data recorded by the moored instruments allowed the characterisation of the upper water column (0–25 m) phytoplankton assemblage response to the combined effect of the passage of a strong cyclonic eddy and a wind-driven upwelling event during March and April (Letelier *et al.*, 2000; Fig. 9.11). Chlorophyll concentration in the upper 25 m increased from approximately 0.08 mg m^{-3} to 0.23 mg m^{-3} between 21 March and 1 April 1997. Phytoplankton samples collected for pigment analyses towards the end of this event indicate that the relative abundance of diatoms doubled. Hence these results suggest that the absolute diatom biomass in the upper 25 m of the water column increased sixfold. Because diatoms contribute significantly to the downward flux of organic matter (Legendre and Le Fèvre, 1989), this transient event may have significantly altered elemental fluxes in the euphotic zone. It may also help to explain the observed wintertime peak in sediment-trap-collected particle export (Karl *et al.*, 1996).

The most intense eddies in the NPSG occur in the lee of the main Hawaiian Islands, and have diameters of 50–100 km and surface currents up to 100 cm s^{-1}. Interactions of the easterly trade-winds and the North Equatorial Current with the island of Hawaii result in the formation of westward-propagating cyclonic and anticyclonic eddies (Patzert, 1969; Lumpkin, 1998). These eddies are typically generated on time-scales of 50–70 days. Of particular interest are the cyclonic eddies which vertically displace the underlying nutricline into the overlying, nutrient-depleted euphotic zone. In the region just west of the major islands, the eddies dominate the mean annual currents and this region as a whole dominates the velocity variance signal of the entire NPSG. These near-shore-formed eddies generate upwelling with vertical velocities up to 0.01 cm s^{-1} in the centres where surface temperatures can be up to 2°C cooler than outside the eddies. These eddies are most readily detected with satellite ocean thermal imagery. During the investigation of two cyclonic eddies formed in the lee of Hawaii in November 1999, up to twofold enhancements in pigment biomass and primary productivity were observed at the stations sampled within the eddies (Seki *et al.*, 2001a).

Increases in nutrient availability caused by wind-driven upwelling and mixing would also be expected to alter the composition of primary producers since different species have different micro- and macro-nutrient requirements. Seki *et al.* (2001a) used HPLC pigment analysis to detect changes in phytoplankton community structure within two cyclonic eddies sampled in the lee of Hawaii during November 1999. Depth-integrated concentrations of divinyl Chl *a* and zeaxanthin at eddy stations were similar to control stations, suggesting that the growth of cyanobacteria in the NPSG is not strongly limited by nutrient availability. Nevertheless, greater than twofold increases in the concentrations of 19'-hex (prymnesiophytes), 19'-but (pelagophytes), fucoxanthin (diatoms) and peridinin (dinoflagellates) were observed within eddy stations. The several-fold increases in pigment biomarkers associated with larger phytoplankton taxa (i.e. diatoms and dinoflagellates) set the stage for enhanced particle export in these mesoscale features (Legendre and LeFèvre, 1989). These results are also consistent with the observations made at Station ALOHA (Letelier *et al.*, 2000).

During January–May, multiple individual planetary-scale fronts accentuate the central North Pacific Subtropical Frontal system. Data obtained during a recent series of cruises and by satellite remote sensing elucidate a coupling between the physics and biology associated with these fronts, and provide new insights into the seasonal variability of phytoplankton dynamics in the NPSG (Leonard *et al.*, 2001; Seki *et al.*, 2001b). The most prominent of these fronts is located at 32–34° N (subtropical front, or STF) and 28–30° N (south subtropical front, or SSTF), although considerable interannual variability in position and intensity is observed. The STF marks the transition from low Chl *a*, nutrient-depleted surface waters to the south to a more productive regime to the north. A sharp increase in depth-integrated Chl *a* is also observed at the SSTF and is caused by an increase in the magnitude of the DCML stimulated by the shoaling of the nutricline and thermocline structure into the euphotic zone (Seki *et al.*, 2001b).

Springtime composites of Chl *a* provided by SeaWiFS show a threefold increase in Chl *a* (0.1 to 0.3 mg m^{-3}) in the 30–40° N latitudinal band of the North Pacific (Polovina *et al.*, 2000). To investigate the spatial and interannual variability of this springtime feature, Leonard *et al.* (2001) conducted multi-platform surveys (23–33° N, 158° W) during April 1998 and 1999. In this study, shipboard and satellite measurements were employed to characterise the hydrographic conditions associated with this large Chl *a* gradient. Irradiance and pigment data collected during the cruise periods were used in an optical model to compute depth-integrated rates of primary production. The southern portion of both transects resembled the climatological conditions at Station ALOHA, while the northern portion was strongly influenced by the SSTF and STF. The SSTF was located at 27–28° N during 1998 (El Niño conditions) and 32.5° N during 1999 (La Niña conditions), while the STF was positioned nearby at 32° N during 1998 and 34° N during 1999. Integrated Chl *a* and primary productivity both increased at the frontal locations in both years. HPLC pigment analysis was performed on select samples collected during the 1998 cruise, and revealed a change in phytoplankton composition at both fronts. Specifically, Leonard *et al.* (2001) observed distinct increases in peridinin (dinoflagellates) and fucoxanthin (diatoms) at the SSTF. It is not clear at this time whether these interannual variations in frontal positions are a consequence of the El Niño Southern Oscillation (ENSO) forcing or reflect a new phase of the Pacific Decadal Oscillation (Polovina *et al.*, 2000).

Case Study No. 3: Decadal variations in primary production and community structure

In 1987, Venrick *et al.* (1987) reported that the average euphotic zone (0–200 m) Chl *a* concentration in the oligotrophic North Pacific Ocean during summer (May–October) had nearly doubled from 1968 to 1985. Their data collected from the Climax region (26.5–31.0° N and 150.5–158.0° W) was of insufficient sampling frequency to determine whether the Chl *a* increase had been continuous over time or whether there had been a 'step-function' increase between 1973 and 1980 (Venrick *et al.*, 1987). The authors hypothesised that decade-scale changes in the open-ocean habitat, caused by large-scale atmosphere–ocean interactions, had resulted in significant long-term changes in the carrying capacity of the ecosystem.

Data from the ongoing HOT program which also includes occasional sampling in the Climax region, extend this previous record of euphotic zone Chl *a* for nearly another decade (Fig. 9.12; Karl *et al.*, 2001a). The combined Climax–ALOHA data set documents that the mean euphotic zone Chl *a* concentration from the period October 1988 to December 1997 (HOT program results) is also significantly greater than the average, pre-1976 Climax program Chl *a* concentration. In addition to changes in the inventories of Chl *a*, there also appears to have been an increase in primary production in the NPSG. Compared to the 17-year Climax program mean of about 200 mg C m^{-2} d^{-1}, the average rate of primary production measured during the HOT program (484 ± 129 mg C m^{-2} d^{-1}) is significantly higher (Fig. 9.12). It is possible that methodological improvements may be responsible for some or even all of the increase that has occurred. However, if this were the case, then we would still need to explain the Chl *a* increase which is independent of the ^{14}C measurements. If past production in the NPSG were equal to the contemporaneous rate, but with less than half the present-day Chl *a* inventory, then assimilation numbers would have approached or exceeded the theoretical maximum value. Based on all information that is currently available, we would predict a change in primary production rate coincident with a secular change in Chl *a*.

Examination of the Climax and ALOHA depth profiles of Chl *a* and pheophytin

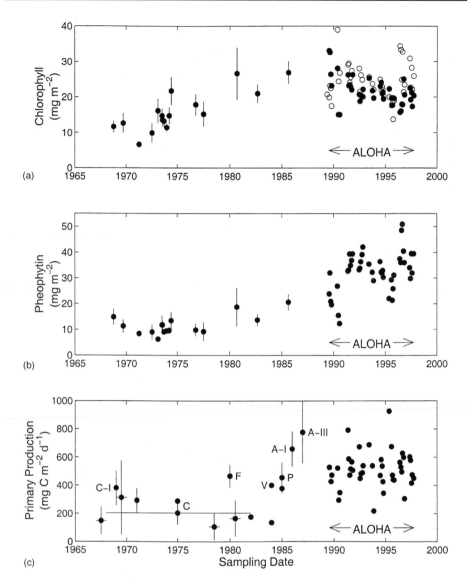

Fig. 9.12 Composite time-series analysis of phytoplankton community parameters for samples collected in the NPSG during the period May–October. Measurements include: (a) euphotic zone depth-integrated chlorophyll concentrations as determined by fluorometer (●) and HPLC (○), (b) euphotic zone depth-integrated pheophytin (FL-pheo) concentrations, and (c) euphotic zone depth-integrated rates of primary production obtained from numerous sources and locations but derived largely from oceanographic investigations at or near the Climax region and Station ALOHA, all derived using the ^{14}C method. Where appropriate, mean values are presented ± 1 standard deviation for multiple observations for a given cruise, season or study. Where shown, horizontal bars indicate the duration of multi-cruise data sets. Data sets (see Karl and Lukas, 1996) from selected expeditions are noted by the following abbreviations: AI/AIII = Asian dust inputs to oligotrophic seas (ADIOS) I and III cruises, C = Climax time-series, C-I = Climax-I and F = FIONA (two separate cruises in the Climax region), G = GOLLUM, P = plankton rate processes in oligotrophic oceans (PRPOOS) cruise, V = vertical transport and exchange (VERTEX-5) cruise. Redrawn from Karl *et al.* (2001a).

based on 'standard' fluorometric procedures also reveals secular increases in these parameters since 1968, especially within the DCML (Fig. 9.13); the most significant change is for pheophytin concentration. It is now recognised that the standard fluorometric method used for pigment determinations over the past 40 years does not provide an accurate measurement of pheophytin, and more reliable determinations using HPLC have generally failed to detect any significant concentrations of pheophytin in the water column near Station ALOHA (Ondrusek *et al.*, 1991). In order to minimise interferences caused by the overlapping excitation and emission wavebands of chlorophylls *a*, *b*, *c* and pheophytin, Turner Designs (Sunnyvale, CA) introduced the multi-spectral TD-700 fluorometer. This instrument was recently beta-tested using samples collected from the euphotic zone at Station ALOHA. While consistent relationships were obtained for HPLC- and spectro-fluorometrically-determined Chl *a*, *b*, and *c* concentrations, the TD-700 did not detect pheophytin in any of the samples analysed (Trees *et al.*, 2000). Karl *et al.*

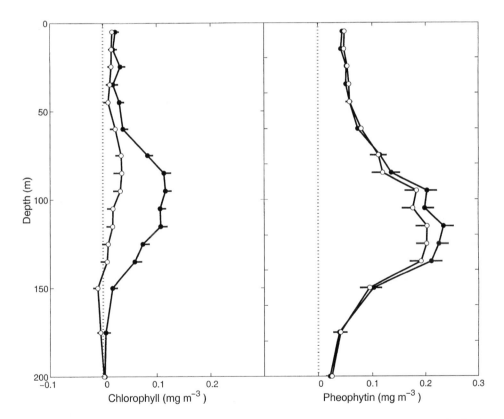

Fig. 9.13 Comparisons of the concentration versus depth profiles for Chl and pheophytin (both determined by fluorometric techniques) in the NPSG for the periods 1968–1975 (Climax data set A), 1976–1986 (Climax data set B) and 1988–1997 (ALOHA data set). Mean (plus or minus 1 standard deviation, as shown) differences for Chl (left) and pheophytin (right) calculated as: ALOHA minus Climax A (●) and ALOHA minus Climax B (○). Redrawn from Karl *et al.* (2001a).

(2001a) have suggested that the increase in pheophytin is actually a manifestation of increasing Chl *b* caused by an enhancement in *Prochlorococcus* abundance. This nearly twofold increase in pheophytin (e.g. in reality, Chl *b*) concentration in the lower portion of the euphotic zone (Fig. 9.13) may have resulted from an increased frequency and duration of El Niño-favourable conditions since 1980 (Karl, 1999) or other climate anomalies such as the Pacific (inter) Decadal Oscillation (Karl *et al.*, 2001a). These large-scale climate variations can lead to major biogeochemical changes in the NPSG including enhanced oligotrophication, altered food web structure and fundamental changes in nutrient fluxes and N:P stoichiometry (Fig. 9.14). Regardless of the ultimate cause, the ecological effects on the NPSG appear to be substantial, including changes in community structure, altered nutrient

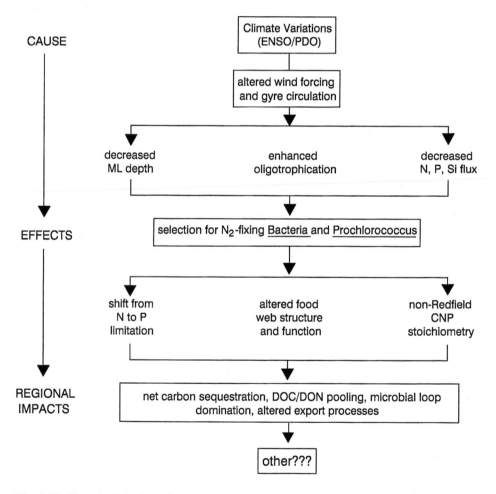

Fig. 9.14 Hypothetical view of the effects of climate variability on ecosystem structure and function in the NPSG based, in part, on results obtained during the decade-long HOT research program. Redrawn from Karl (1999) and Cullen *et al.* (2001).

dynamics and enhanced rates of carbon and, perhaps, nitrogen fixation (Karl *et al.*, 2001b).

Significant variations in phototrophic microbial community structure are also apparent over subdecadal time-scales. For example, zeaxanthin concentration in the euphotic zone increased approximately twofold during 1991–1999, suggesting a doubling of cyanobacteria biovolume during this period (Fig. 9.9(a)). Parallel changes in the zeaxanthin:Chl ratio indicate that this change was caused by a selective enrichment in the cyanobacteria component rather than an increase in all phototrophic (i.e. Chl *a*-containing) microbes. Furthermore, the cyanobacterium, *Prochlorococcus*, appears to account for much of the increase during the first half of this time-series as indicated by changes in *Prochlorococcus*-carbon (Pro-C) biomass during 1992–1996 (Fig. 9.9(b)), but not for the subsequent rise (post-1996). While the source(s) for zeaxanthin increase during 1996–1999 have yet to be identified, *Trichodesmium* and novel, as yet unidentified, N_2-fixing micro-organisms (Zehr *et al.*, 2000) are likely candidates. ^{14}C-determined production rates in the mixed layer at Station ALOHA show large seasonal variations with a mean value of 167 mg C m^{-2} d^{-1} (Fig. 9.9(c)). On average, *Prochlorococcus* production may account for between 50% (assuming $\mu = 0.5\,d^{-1}$) and 70% (assuming $\mu = 0.7\,d^{-1}$) of the ^{14}C-measured rates during 1991–1999, with highest contributions during the period 1994–1997. The rate of *Prochlorococcus* production in the mixed layer shows a gradual, linear increase during 1991–1999, which reflects the secular changes in cyanobacteria biomass described above. The ^{14}C-determined rates, by comparison, show distinct seasonal maxima that may be associated with increases in the abundance of eukaryotic phototrophs.

The NPSG may oscillate, on multi-decadal time-scales, between sink and source habitats for eukaryotic versus prokaryotic photoautotrophic populations. The relative fitness of each subpopulation would be determined by the habitat, which is ultimately under climate control. In this regard, the NPSG probably represents a two-state phototrophic system, with permanent eukaryote and pro-karyote presence, but aperiodic multi-decade long prokaryote dominance as presently occurs. Event-scale phenomena, such as those presented in Case Study No. 2 could provide a short-lived source of new inorganic nutrients, allowing the ecosystem structure to temporarily revert back to one where eukaryotic phytoplankton and herbivorous crustacean zooplankton are selected. Particulate matter export from the euphotic zone and other biogeochemical processes may be more dependent on these short-term shifts in plankton species composition than on the mean ecosystem state. These stochastic biodiversity 'flashbacks' might also help to explain why the >5 μm phytoplankton species list and species rank order abundance have remained relatively constant over the past several decades (Venrick, 1997, 1999), despite the major ecosystem changes that we report in this chapter.

9.4   Summary and prospectus

Primary production in the sea is inextricably linked to physical forcing, including climate variability (Fig. 9.14). However, no simple cause-and-effect relationships and no universally applicable ecological predictive theory has been developed. For example, both enhanced turbulent mixing and enhanced water column stratification can lead to increased rates of primary production, but for fundamentally different reasons. Short-term (< 1 week) changes in the local physical forcing in a given habitat can lead to a production pulse without affecting community structure, whereas longer duration (> 1 year) habitat change tied to climate variability may select for or against organisms with specific physiological capability or ecological function (e.g. N_2-fixing micro-organisms). Both pulsed and sustained changes in photosynthesis have been observed for the NPSG, one of the more stable habitats on Earth, so they are likely to occur in most other regions of the world ocean. In this regard, high-frequency time-series measurements obtained either by approximately monthly repeat vessel occupation or by continuous autonomous mooring observations will be required to capture both anticipated (e.g. seasonal) and aperiodic processes. The study of interannual variability (especially potential ecological response to climate variations) that might occur on decade-to-century time-scales demands a well-designed core management program, dedicated personnel and ample resources. For these and other reasons, long-term time-series measurements of plankton processes are rare despite their obvious importance in ecology.

The ^{14}C method for field estimation of rates of photosynthesis has provided an invaluable database for studies of the marine carbon cycle. However, nearly 50 years after its introduction to the discipline of biological oceanography, the original limitations of this approach continue to stand in the way of progress towards a comprehensive understanding of carbon and energy flow. New approaches for the routine and accurate estimations of GPCP, NPCP, NCCP and total energy flux are needed. It is the tradition in ecology to use carbon as the basic unit for biomass inventories and fluxes; a more systematic and coordinated study of associated major and trace bioelements is also needed, including careful analyses of ecological stoichiometries and controls thereof. Only then can we begin to unravel the complex biogeochemical linkages that we seek to understand.

Finally, it is impossible to predict or even fully comprehend things yet unknown. For plankton assemblages in the NPSG, the unknowns are likely to eclipse the extant knowledge base. In just the past two decades, *Synechococcus* and *Prochlorococcus* were identified as the most abundant and probably most important photoautotrophs, and within the past year alone several new groups of organisms (Zehr *et al.*, 2000; Karner *et al.*, 2001) and novel energy capture/processing pathways (Béjà *et al.*, 2000; Kolber *et al.*, 2000; Zehr *et al.*, 2001) have been described. Additional surprises from the open sea are likely to follow.

Acknowledgements

The HOT program data sets summarised herein would not exist without the dedicated and skilled efforts of a large cadre of scientists and technicians, including K. Björkman, S. Christensen, J. Christian, J. Dore, L. Fujieki, T. Houlihan, M. Latasa, U. Magaard, D. Sadler, F. Santiago-Mandujano, J. Snyder, L. Tupas, C. Winn, and especially D. Hebel who conducted many of the at-sea ^{14}C productivity experiments. From the beginning, R. Lukas has led the companion hydrographic-physical oceanography sub-component of HOT; the biogeochemical data sets described herein would not have been possible without this invaluable collaboration. L. Lum, C. Leonard and L. Fujieki provided additional technical assistance during the preparation of the manuscript. Professor Celia Smith of the U.H. Botany Department provided access to M. Doty's personal archives. This research was supported by grants from the National Science Foundation (OCE96-17409 (DMK and RRB), OCE98-11921 (R. Lukas), OCE00-01407 (R. Letelier)) and the National Aeronautics and Space Administration (NAG5-7171 and NAG5-9757 (both to RRB)), and by the State of Hawaii. US JGOFS contribution No. 667 and SOEST contribution No. 5357.

References

Aabye Jensen, E. & Steemann Nielsen, E. (1952) A water-sampler for biological purposes. *Journal du Conseil International pour l'Exploration de la Mer* **18**, 296–9.

Andersen, R.A., Saunders, G.W., Paskind, M.P. & Sexton, J.P. (1993) Ultrastructure and 18S rRNA gene sequence for *Pelagomonas calceolata* gen et sp. nov. and the description of a new algal class, the Pelagophyceae classis nov. *Journal of Phycology* **29**, 701–15.

Andersen, R.A., Bidigare, R.R., Keller, M.D. & Latasa, M. (1996) A comparison of HPLC pigment signatures and electron microscopic observations for oligotrophic waters of the North Atlantic and Pacific Oceans. *Deep-Sea Research II* **43**, 517–37.

Banse, K. (1987) Clouds, deep chlorophyll maxima and the nutrient supply to the mixed layer of stratified water bodies. *Journal of Plankton Research* **9**, 1031–6.

Beers, J.R., Reid, F.M.H. & Stewart, G.L. (1982) Seasonal abundance of the microplankton population in the North Pacific central gyre. *Deep-Sea Research* **29**, 227–45.

Béjà, O., Aravind, L., Koonin, E.V., *et al.* (2000) Bacterial rhodopsin: Evidence for a new type of phototrophy in the sea. *Science* **289**, 1902–6.

Bidigare, R.R., Prézelin, B.B. & Smith, R.C. (1992) Bio-optical models and the problems of scaling. In: *Primary Productivity and Biogeochemical Cycles in the Sea* (eds P.G. Falkowski & A.D. Woodhead), pp. 175–212. Plenum Press, New York.

Bidigare, R.R., Smith, R.C., Baker, K.S. & Marra, J. (1987) Oceanic primary production estimates from measurements of spectral irradiance and pigment concentrations. *Global Biogeochemical Cycles* **1**, 171–86.

Bidigare, R.R. & Trees, C.C. (2000) HPLC phytoplankton pigments: Sampling, laboratory methods, and quality assurance procedures. In: *Ocean Optics Protocols for Satellite Ocean Color Sensor Validation* (eds J. Mueller & G. Fargion), Rev. 2, pp. 154–61. NASA Technical Memorandum 2000-209966.

Björkman, K., Thomson-Bulldis, A.L. & Karl, D.M. (2000) Phosphorus dynamics in the North Pacific subtropical gyre. *Aquatic Microbial Ecology* **22**, 185–98.

Bricaud, A., Babin, M., Morel, A. & Claustre, H. (1995) Variability in the chlorophyll-specific absorption coefficients of natural phytoplankton: Analysis and parameterization. *Journal of Geophysical Research* **100**, 13 321–32.

Bruland, K., Orians, K.J. & Cowen, J.P. (1994) Reactive trace metals in the stratified central North Pacific. *Geochimica et Cosmochimica Acta* **58**, 3171–82.

Campbell, L. & Vaulot, D. (1993) Photosynthetic picoplankton community structure in the subtropical North Pacific Ocean near Hawaii (station ALOHA). *Deep-Sea Research* **40**, 2043–60.

Campbell, L., Nolla, H.A. & Vaulot, D. (1994) The importance of *Prochlorococcus* to community structure in the central North Pacific Ocean. *Limnology and Oceanography* **39**, 954–61.

Chavez, F.P., Buck, K.R., Service, S.K., Newton, J. & Barber, R.T. (1996) Phytoplankton variability in the central and eastern tropical Pacific. *Deep-Sea Research* **43**, 835–70.

Clements, F.E. (1916) *Plant Succession*. Carnegie Institute, Washington.

Cullen, J.J. (1995) Status of the iron hypothesis after the open-ocean enrichment experiment. *Limnology and Oceanography* **40**, 1336–43.

Cullen, J.J., Franks, P.J.S., Karl, D.M. & Longhurst, A. (2001) Physical influences on marine ecosystem dynamics. In: *The Sea* (eds A.R. Robinson, J.J. McCarthy & B.J. Rothschild), Vol. 12 (in press).

Devassy, V.P., Bhattathiri, P.M.A. & Quasim, S.Z. (1979) Succession of organisms following *Trichodesmium* phenomenon. *Indian Journal of Marine Science* **8**, 89–93.

Doty, M.S. (1961) *Proceedings of the Conference on Primary Productivity Measurement, Marine and Freshwater*, US Atomic Energy Commission, Washington, DC.

Doty, M.S. & Oguri, M. (1958) Selected features of the isotopic carbon primary productivity technique. *Rapports et Procés-Verbaux des Réunions Conseil International pour l'Exploration de la Mer* **144**, 47–55.

Falkowski, P.G. (1981) Light–shade adaptation and assimilation numbers. *Journal of Plankton Research* **3**, 203–16.

Fitzwater, S.E., Knauer, G.A. & Martin, J.H. (1982) Metal contamination and its effect on primary production measurements. *Limnology and Oceanography* **27**, 544–51.

Gordon, H.R. & Morel, A.Y. (1983) *Remote Assessment of Ocean Color for Interpretation of Satellite Visible Imagery*. Springer-Verlag, New York.

Hayward, T.L. (1987) The nutrient distribution and primary production in the central North Pacific. *Deep-Sea Research* **34**, 1593–1627.

Hayward, T.L., Venrick, E.L. & McGowan, J.A. (1983) Environmental heterogeneity and plankton community structure in the central North Pacific. *Journal of Marine Research* **41**, 711–29.

Heinbokel, J.F. (1986) Occurrence of *Richelia intracellularis* (Cyanophyta) within the diatoms *Hemiaulus haukii* and *H. membranaceus* off Hawaii. *Journal of Phycology* **22**, 399–403.

Johnson, P.W. & Sieburth, J.McN. (1979) Chroococcoid cyanobacteria in the sea: a ubiquitous and diverse phototrophic biomass. *Limnology and Oceanography* **24**, 928–35.

Karl, D.M. (1999) A sea of change: Biogeochemical variability in the North Pacific subtropical gyre. *Ecosystems* **2**, 181–214.

Karl, D.M., Bidigare, R.R. & Letelier, R.M. (2001a) Long-term changes in plankton community structure and productivity in the North Pacific Subtropical Gyre: The domain shift hypothesis. *Deep-Sea Research II* **48**, 1449–70.

Karl, D.M., Björkman, K.M., Dore, J.E., *et al.* (2001b) Ecological nitrogen-to-phosphorus stoichiometry at Station ALOHA. *Deep-Sea Research II* **48**, 1529–66.

Karl, D.M., Christian, J.R., Dore, J.E., *et al.* (1996) Seasonal and interannual variability in primary production and particle flux at Station ALOHA. *Deep-Sea Research II* **43**, 539–68.

Karl, D.M., Hebel, D.V., Björkman, K. & Letelier, R.M. (1998) The role of dissolved organic matter release in the productivity of the oligotrophic North Pacific Ocean. *Limnology and Oceanography* **43**, 1270–86.

Karl, D.M., Letelier, R., Hebel, D.V., Bird, D.F. & Winn, C.D. (1992) *Trichodesmium* blooms and new nitrogen in the north Pacific gyre. In: *Marine Pelagic Cyanobacteria: Trichodesmium and Other Diazotrophs* (eds E.J. Carpenter, D.G. Capone & J.G. Rueter), pp. 219–37. Kluwer Academic Publishers, the Netherlands.

Karl, D.M. & Lukas, R. (1996) The Hawaii Ocean Time-series (HOT) program: Background, rationale and field implementation. *Deep-Sea Research II* **43**, 129–56.

Karner, M.B., DeLong, E.F. & Karl, D.M. (2001) Archaeal dominance in the mesopelagic zone of the Pacific Ocean. *Nature* **409** 507–510.

Kolber, Z.S., Van Dover, C.L., Niederman, R.A. & Falkowski, P.G. (2000) Bacterial photosynthesis in surface waters of the open ocean. *Nature* **407**, 177–9.

Kromkamp, J. & Walsby, A.E. (1990) A computer model of buoyancy and vertical migration in cyanobacteria. *Journal of Plankton Research* **12**, 161–83.

Laws, E.A. (1984) Improved estimates of phytoplankton carbon based on ^{14}C incorporation into chlorophyll *a*. *Journal of Theoretical Biology* **110**, 425–34.

Laws, E.A. & Bannister, T.T. (1980) Nutrient and light-limited growth of *Thalassiosira fluviatilis* in continuous culture with implications for phytoplankton growth in the ocean. *Limnology and Oceanography* **25**, 457–73.

Laws, E.A., Karl, D.M., Redalje, D.G., Jurick, R.S. & Winn, C.D. (1983) Variability in ratios of phytoplankton carbon and RNA to ATP and chlorophyll *a* in batch and continuous cultures. *Journal Phycology* **19**, 439–45.

Laws, E.A., Landry, M.R., Barber, R.T., Campbell, L., Dickson, M.-L. & Marra, J. (2000) Carbon cycling in primary production bottle incubations: inferences from grazing experiments and photosynthetic studies using ^{14}C and [18]O in the Arabian Sea. *Deep-Sea Research II* **47**, 1339–52.

Legendre, L. & Le Fèvre, J. (1989) Hydrodynamical singularities as controls of recycled versus export production in oceans. In: *Productivity of the Ocean: Present and Past* (eds W.H. Berger, V.S. Smetacek & G. Wefer), pp. 49–63. Wiley-Interscience, New York.

Leonard, C.L., Bidigare, R.R., Seki, M.P. & Polovina, J.J. (2001) Interannual mesoscale physical and biological variability in the North Pacific Centre Gyre. *Progress in Oceanography* **49**, 227–44.

Letelier, R.M., Bidigare, R.R., Hebel, D.V., Ondrusek, M., Winn, C.D. & Karl, D.M. (1993) Temporal variability of phytoplankton community structure based on pigment analysis. *Limnology and Oceanography* **38**, 1420–37.

Letelier, R.M., Dore, J.E., Winn, C.D. & Karl, D.M. (1996) Seasonal and interannual variations in photosynthetic carbon assimilation at Station ALOHA. *Deep-Sea Research II* **43**, 467–90.

Letelier, R.M. & Karl, D.M. (1998) *Trichodesmium* spp. physiology and nutrient fluxes in the North Pacific subtropical gyre. *Aquatic Microbial Ecology* **15**, 265–76.

Letelier, R.M., Karl, D.M., Abbott, M.R., *et al.* (2000) Role of late winter mesoscale events in the biogeochemical variability of the upper water column of the North Pacific Subtropical Gyre. *Journal of Geophysical Research* **105**, 28 723–39.

Liu, H., Campbell, L. & Landry, M.R. (1995) Growth and mortality rates of *Prochlorococcus* and *Synechococcus* measured with a selective inhibitor technique. *Marine Ecology Progress Series* **116**, 277–87.

Liu, H., Nolla, H.A. & Campbell, L. (1997) *Prochlorococcus* growth rate and contribution to primary production in the equatorial and subtropical North Pacific Ocean. *Aquatic Microbial Ecology* **12**, 39–47.

Lumpkin, C.F. (1998) *Eddies and currents of the Hawaiian Islands.* PhD thesis, University of Hawaii.

Mackey, M.D., Mackey, D.J., Higgins, H.W. & Wright, S.W. (1996) CHEMTAX – a program for estimating class abundances from chemical markers: application to HPLC measurements of phytoplankton. *Marine Ecology Progress Series* **144**, 265–83.

McGowan, J.A. (1974) The nature of oceanic ecosystems. In: *The Biology of the Oceanic Pacific* (ed. C.B. Miller), pp. 9–28. Oregon State University Press, Corvallis.

Moore, L.R., Goericke, R. & Chisholm, S.W. (1995) Comparative physiology of *Synechococcus* and *Prochlorococcus*: influence of light and temperature on growth, pigments, fluorescence and absorptive properties. *Marine Ecology Progress Series* **116**, 259–75.

Ondrusek, M.E. & Bidigare, R.R. (1997) Measurements of photo-physiological parameters and primary production rates in the central North Pacific Ocean. *Proceedings of SPIE Ocean Optics XIII* **2963**, 874–9.

Ondrusek, M.E., Bidigare, R.R., Sweet, S.T., Defreitas, D.A. & Brooks, J.M. (1991) Distribution of phytoplankton pigments in the North Pacific Ocean in relation to physical and optical variability. *Deep-Sea Research* **38**, 243–66.

Ondrusek, M.E., Bidigare, R.R., Waters, K. & Karl, D.M. (2001) A predictive model for estimating rates of primary production in the subtropical North Pacific Ocean. *Deep-Sea Research II* **48**, 1837–63.

Patzert, W.C. (1969) *Eddies in Hawaiian waters.* Hawaii Institute of Geophysics Technical Report No. 69–8, University of Hawaii, Honolulu.

Petzold, T.L. (1972) *Volume scattering functions for selected ocean waters.* SIO Reference 72–78, Scripps Institution of Oceanography, San Diego.

Polovina, J.J., Seki, M.P. & Howell, E. (2000) Sensors detect biological change in mid-latitude North Pacific. *EOS, Transactions of the American Geophysical Union* **81**, 519.

Pope, R.M. (1993) *Optical absorption of pure water and sea water using the integrating cavity absorption meter.* PhD thesis, Texas A & M University.

Riley, G.A. (1953) Letter to the Editor. *Journal du Conseil International pour l'Exploration de la Mer* **19**, 85–9.

Ruben, S., Hassid, W.Z. & Kamen, M.D. (1939) Radioactive carbon in the study of photosynthesis. *Journal of the American Chemical Society* **61**, 661–3.

Ruben, S. & Kamen, M.D. (1940a) Radioactive carbon of long half-life. *Physics Review* **57**, 549.

Ruben, S. & Kamen, M.D. (1940b) Photosynthesis with radioactive carbon. IV. Molecular weight of the intermediate products and a tentative theory of photosynthesis. *Journal of the American Chemical Society* **62**, 3451–5.

Sakshaug, E., Bricaud, A., Dandonneau, Y., *et al.* (1997) Parameters of photosynthesis: definitions, theory and interpretation of results. *Journal of Plankton Research* **19**, 1637–70.

Scharek, R., Latasa, M., Karl, D.M. & Bidigare, R.R. (1999a) Temporal variations in diatom abundance and downward vertical flux in the oligotrophic North Pacific gyre. *Deep-Sea Research I* **46**, 1051–75.

Scharek, R., Tupas, L.M. & Karl, D.M. (1999b) Diatom fluxes to the deep sea in the oligotrophic North Pacific gyre at Station ALOHA. *Marine Ecology Progress Series* **182**, 55–67.

Seki, M.P., Polovina, J.J., Brainard, R.E., Bidigare, R.R., Leonard, C.L. & Foley, D.G. (2001a) Biological enhancement at cyclonic eddies tracked with GOES thermal imagery in Hawaiian waters. *Geophysical Research Letters* **28**, 1583–6.

Seki, M.P., Polovina, J.J., Kobayashi, D.R., Bidigare, R.R. & Mitchum, G.T. (2001b) An oceanographic characterization of swordfish longline fishing grounds in the subtropical North Pacific. *Fisheries Oceanography* (in press).

Smith, R.C. & Baker, K.S. (1981) Optical properties of the clearest natural waters (200–800 nm). *Applied Optics* **20**, 177–84.

Steemann Nielsen, E. (1951) Measurement of the production of organic matter in the sea by means of carbon-14. *Nature* **167**, 684–5.

Steemann Nielsen, E. (1952a) Production of organic matter in the sea. *Nature* **169**, 956–7.

Steemann Nielsen, E. (1952b) The use of radio-active carbon (C^{14}) for measuring organic production in the sea. *Journal du Conseil International pour l'Exploration de la Mer* **18**, 117–40.

Steemann Nielsen, E. & Aabye Jensen, E. (1957) Primary oceanic production: The autotrophic production of organic matter in the oceans. In: *Galathea Report. Volume 1: Scientific Results of the Danish Deep-Sea Expedition Round the World 1950–52* (eds A.F. Bruun, S. Greve & R. Spärck), pp. 49–135. Galathea Committee, Copenhagen.

Strickland, J.D.H. (1960) *Measuring the Production of Marine Phytoplankton*. Fisheries Research Board of Canada, Ottawa.

Strickland, J.D.H. (1965) Production of organic matter in the primary stages of the marine food chain. In: *Chemical Oceanography* (eds J.P. Riley & G. Skirrow), pp. 477–610. Academic Press, New York.

Tanré, D., Deroo, C., Duhaut, P., *et al.* (1990) Description of a computer code to simulate the satellite signal in the solar spectrum: The 5S code. *International Journal of Remote Sensing* **11**, 659–68.

Trees, C.C., Bidigare, R.R., Karl, D.M. & Van Heukelem, L. (2000) Fluorometric Chlorophyll *a*: Sampling, laboratory methods, and data analysis protocols. In: *Ocean Optics Protocols for Satellite Ocean Color Sensor Validation*, Revision 2 (eds J. Mueller & G. Fargion), pp. 162–9. NASA Technical Memorandum 2000–209966.

Vaulot, D., Marie, D., Olson, R.J. & Chisholm, S.W. (1995) Growth of *Prochlorococcus*, a photosynthetic prokaryote, in the equatorial Pacific Ocean. *Science* **268**, 1480–2.

Venrick, E.L. (1988) The vertical distribution of chlorophyll and phytoplankton species in the North Pacific central environment. *Journal of Plankton Research* **10**, 987–98.

Venrick, E.L. (1990) Phytoplankton in an oligotrophic ocean: Species structure and interannual variability. *Ecology* **71**, 1547–63.

Venrick, E.L. (1993) Phytoplankton seasonality in the central North Pacific: The endless summer reconsidered. *Limnology and Oceanography* **38**, 1135–49.

Venrick, E.L. (1995) Scales of variability in a stable environment: Phytoplankton in the central North Pacific. In: *Ecological Time-series* (ed T.M. Powell & J.H. Steele), pp. 150–80. Chapman and Hall, New York.

Venrick, E.L. (1997) Comparison of the phytoplankton species composition and structure in the Climax area (1973–1985) with that of station ALOHA (1994). *Limnology and Oceanography* **42**, 1643–8.

Venrick, E.L. (1999) Phytoplankton species structure in the central North Pacific, 1973–1996: variability and persistence. *Journal of Plankton Research* **21**, 1029–42.

Venrick, E.L., McGowan, J.A., Cayan, D.R. & Hayward, T.L. (1987) Climate and chlorophyll *a*: Long-term trends in the central North Pacific Ocean. *Science* **238**, 70–72.

Villareal, T.A., Pilskaln, C., Brzezinski, M., Lipschultz, F., Dennett, M. & Gardner, G.B. (1999) Upward transport of oceanic nitrate by migrating diatom mats. *Nature* **397**, 423–5.

Walsby, A.E. & Bleything, A. (1988) The dimensions of cyanobacterial gas vesicles in relation to their efficiency in providing buoyancy and withstanding pressure. *Journal of General Microbiology* **134**, 2635–45.

Waterbury, J.B., Watson, S.W., Guillard, R.R.L. & Brand, L.E. (1979) Widespread occurrence of a unicellular, marine, planktonic, cyanobacterium. *Nature* **277**, 293–4.

Waters, K.J., Smith, R.C. & Marra, J. (1994) Phytoplankton production in the Sargasso Sea as determined using optical mooring data. *Journal of Geophysical Research* **99**, 18 385–402.

Williams, P.J.leB. (1993) On the definition of plankton production terms. *Proceedings of the ICES Marine Science Symposium* **197**, 9–19.

Williams, P.J.leB., Berman, T. & Holm-Hansen, O. (1972) Potential sources of error in the measurement of low rates of planktonic photosynthesis and excretion. *Nature* **236**, 91–92.

Williams, P.J.leB. & Purdie, D.A. (1991) *In vitro* and *in situ* derived rates of gross production, net community production and respiration of oxygen in the oligotrophic subtropical gyre of the North Pacific Ocean. *Deep-Sea Research I* **38**, 891–910.

Winn, C.D., Campbell, L., Christian, J.R., *et al.* (1995) Seasonal variability in the phytoplankton community of the North Pacific Subtropical Gyre. *Global Biogeochemical Cycles* **9**, 605–20.

Wu, J., Sunda, W., Boyle, E.A. & Karl, D.M. (2000) Phosphate depletion in the western North Atlantic Ocean. *Science* **289**, 759–62.

Wyrtki, K. (1975) Fluctuation of the dynamic topography in the Pacific Ocean. *Journal of Physical Oceanography* **5**, 450–59.

Zehr, J.P., Carpenter, E.J. & Villareal, T.A. (2000) New perspectives on nitrogen-fixing microorganisms in tropical and subtropical oceans. *Trends in Microbiology* **8**, 68–73.

Zehr, J.P., Waterbury, J.B., Turner, P.J., Montoya, J.P., Omoregie, E., Steward, G.F., Hansen, A. & Karl, D.M. (2001) Unicellular cyanobacteria fix N_2 in the subtropical North Pacific Ocean. *Nature* **412**, 635–8.

Zubkov, M.V., Sleigh, M.A., Tarran, G.A., Burkill, P.H. & Leakey, R.J.G. (1998) Picoplankton community structure on an Atlantic transect from 50°N to 50°S. *Deep-Sea Research* **45**, 1339–55.

Chapter 10
Regional-scale Influences on the Long-term Dynamics of Lake Plankton

D. Glen George

10.1 Introduction

Year-to-year variations in the weather have a profound effect on the physical dynamics of lakes and the growth of phytoplankton (Harris, 1986; Catalan and Fee, 1994; Schindler *et al.*, 1996). Some of these variations are unpredictable but others are quasi-cyclical and are related to regional-scale variations in the atmospheric circulation. The best-known regional effects are those associated with the El Niño Southern Oscillation (ENSO). ENSO events are known to influence the dynamics of lakes in several countries around the Pacific Ocean (Goldman *et al.*, 1989; Anderson *et al.* 1996; Harris and Baxter, 1996) but have little effect on lakes in the Atlantic region. In Europe, the most important climatic effects are those associated with the atmospheric pressure gradient known as the North Atlantic Oscillation (NAO) and the movements of the Gulf Stream in the Atlantic. Variations in the NAO have recently been shown to influence the dynamics of lakes in Germany, Sweden, the UK and Austria (Straile and Geller, 1998; Wehenmeyer *et al.* 1999; George, 2000; Livingstone and Dokulil, 2001). The impact of the Gulf Stream is less wide-ranging but significant effects have been reported in the lakes of the English Lake District (George and Taylor, 1995) and a lake in the south-west of Ireland (Jennings and Allott, pers. comm.).

In this chapter, I describe some of the ways in which these quasi-cyclical variations influence the supply of nutrients and the growth of phytoplankton in four European lakes. The examples are taken from long-term studies in the UK, Sweden and Ireland and have been selected to demonstrate the 'winter' effects of the NAO and the 'summer' effects of the Gulf Stream. In historical terms, phytoplankton ecologists have always been more interested in long-term trends rather than short-term fluctuations. The examples presented here suggest that more attention should be paid to the climatic factors that influence the year-to-year variations superimposed on these long-term trends. The capacity of a lake to support phytoplankton is largely determined by the supply of nutrients from the surrounding catchment but the biomass produced at any given time is strongly influenced by the modulating effects of the weather. The methods currently used to model the trophic status of lakes were developed when weather patterns were different from what they are today (Vollenweider, 1968, 1975). Most are based on statistics acquired in the 1960s when winter temperatures in Europe were lower and summer droughts were relatively rare. In the final section of the chapter (Section 8.3), I discuss some of the ways in which these models can be adapted to meet the requirements of water quality managers in the twenty-first century. By combining the results of historical observations with Monte Carlo simulations of the local weather, we can represent at least some of the 'cascade of uncertainty' associated with the current generation of climate-change scenarios.

10.2   The impact of year-to-year changes in the weather on the dynamics of lakes

The most direct effects of changes in the weather are those associated with the flux of energy across the air-water interface. Other impacts are less immediate, and can include the flushing effects of heavy rain and the entrainment of nutrients by wind-induced mixing. The schematic diagrams in Fig. 10.1 show some of the weather-related factors that influence the seasonal dynamics of lakes and the growth of phytoplankton. In winter (Fig. 10.1(a)), the most important effects are those associated with the supply of nutrients and the seasonal variation in the underwater light. In lakes located at high latitudes, the critical climatic factors are the duration of ice-cover, the accumulation of snow on the lake surface and the influx of water when the snow melts. If there is little snow and the ice is clear, substantial growths of phytoplankton can appear under the ice. In contrast, an accumulation of snow on the lake surface restricts the penetration of light and limits the growth and productivity of the winter phytoplankton. In lakes located at lower latitudes, the critical factors are the number of hours of bright sunshine, the average wind speed and the average rainfall. Variations in the wind speed have a significant effect on the growth and of the winter phytoplankton since they regulate the cells' exposure to light and control the sinking rates of non-motile organisms. The flushing effects of rain are most important in lakes with a short retention time but year-to-year variations in the rainfall can also influence the size of the overwintering 'inoculum' in larger lakes that contain slow-growing species of phytoplankton. In summer (Fig. 10.1 (b)), the most important climatic effects are those associated with changes in the average rainfall and variations in the frequency and intensity of wind-mixing. Changes in the rainfall are particularly important in lakes with a short retention time (Talling, 1993) but they can also influence the dynamics of phytoplankton in larger lakes situated in areas with high rainfall. In lakes that remain thermally stratified for most of the summer, the most important factor influencing the growth of phytoplankton is the frequency and intensity of wind-mixing. Reynolds (1997, 1999) has shown that the

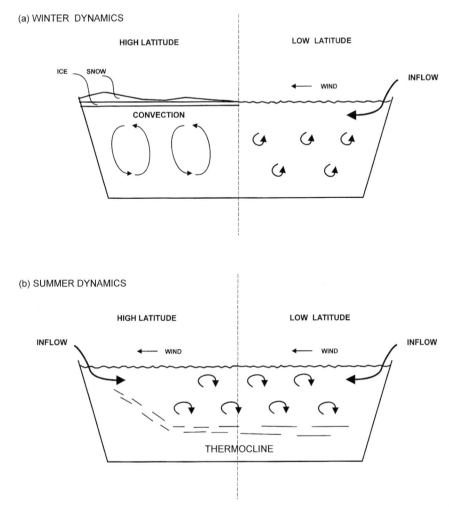

Fig. 10.1 Schematic diagram showing some of the weather-related factors that influence the physical dynamics of lakes and the growth of phytoplankton in: (a) winter and (b) summer.

successional patterns observed in most lakes can be explained by the dynamic interaction of two processes:

(1) An autogenic process, where a number of different 'functional types' dominate the phytoplankton community in an ordered sequence.
(2) An allogenic process, where periodic disturbances disrupt this sequence and return the phytoplankton community to a less stable state.

In most lakes, the key factor influencing the switch from an autogenic to an allogenic pattern of succession is the timing and intensity of wind-induced mixing. Such effects are particularly pronounced in productive lakes where the episodic entrainment of nutrients from the anoxic hypolimnion stimulates the summer growth of phytoplankton.

10.3   Proxy indicators of long-term change in the weather

Over the years, climatologists have devised several ways of characterising regional-scale changes in the global climate. Some, such as those devised by Lamb (1972), are based on the synoptic analysis of daily weather maps. Others, such as those introduced by Walker and Bliss (1932), van Loon and Rogers (1978) and Hurrell (1995) are 'zonal indices' that describe particular features of the atmospheric circulation. In this review, I use two very different 'proxy' indicators to quantify the year-to-year variations associated with long-term changes in the weather.

The first indicator is the NAO index suggested by Hurrell (1995) which is based on the atmospheric pressure gradient that develops in winter between Stykkishol-mur in Iceland and Lisbon in Portugal. Positive values of this index are associated with a predominantly westerly airflow, and milder, wetter winters over most of Europe. Negative values of the index are recorded when the direction of the air-flow is reversed and the winters over much of Europe are colder and drier.

The second indicator is the Gulf Stream Index (GSI) devised by Taylor and Stephens (1998a) which is based on the north–south movements of this current in the Atlantic. Monthly charts showing the position of the northern boundary of the Gulf Stream have been produced from surface and satellite measurements since 1966 (Miller, 1994). In the procedure described by Taylor and Stephens, the position of the north wall is read from these chart at six longitudes – 79° W, 75° W, 72° W, 70° W, 67° W and 65° W – and an index of position constructed using principal components analysis. The first principal component typically accounts for a high proportion of the recorded variation and provides the best measure of its average latitude. North–south movements of the Gulf Stream appear to influence the dynamics of lakes in western Europe by regulating the movement of storm tracks across the Atlantic. In the English Lake District, wind speeds tend to be lower when the Gulf Stream is positioned well to the north (positive GSI) and there is a corresponding reduction in the depth of the early summer thermoclines.

Four case studies now follow – two concerning the impact of the NAO and two concerning the influence of the Gulf Stream.

10.4   Case Study 1: the impact of the NAO on the winter concentration of dissolved reactive phosphorus in Blelham Tarn

Blelham Tarn is a small productive lake situated about 2 km west of Windermere, in the English Lake District (54° 18′ N; 2° 54′ E). The lake covers an area of approximately $0.1 \, \text{km}^2$ and has a mean depth of 6.8 m and a maximum depth of 14.5 m. Samples of water for chemical analyses and phytoplankton counts have been collected from the lake at regular intervals since the late 1940s using a 5 m length of hose of the type described by Lund and Talling (1957). For most of this time, the samples were collected at weekly intervals but fortnightly sampling was introduced during the winter of 1989 and extended to cover the full year in 1992. The data analysed here cover the period between 1961 and 1997, a time when there were large year-to-year variations in the severity of the winter weather. Dissolved reactive phosphorus (DRP) concentrations were measured using the method described by Proctor and Hood (1954) between 1961 until 1964, and thereafter by the method of Stephens (1963). Daily measurements of the air temperature, rainfall, wind speed and the number of hours of bright sunshine were recorded at a site a few kilometres from the lake and assembled into a harmonised time-series that extends from January 1961 to December 1997.

In the lakes of the English Lake District, the average concentration of DRP recorded during the first ten weeks of each year provides a reasonable measure of the trophic status of each lake (Sutcliffe *et al.*, 1982). At that time, there was a high flux of nutrients from the surrounding land and the biological uptake of nutrients in the lake basins was very low. Fig. 10.2(a) shows the year-to-year variations in the winter DRP concentrations recorded in Blelham Tarn between 1961 and 1997. There is some indication of an upward trend, but a linear regression fitted to the time-series showed that this trend was not statistically significant at the 95% confidence level. The most striking feature of the time-series is the short-term variation in the measured concentrations of DRP. The most likely explanation for this variation is the effect that changes in the winter rainfall have on the leaching of phosphorus from the surrounding land. When the winters are relatively dry, much of the DRP released by the decay of organic matter is re-sorbed when the water percolates through the soil (Sharpley and Syers, 1979). When the winters are very wet, the resulting overland flows contain higher concentrations of DRP which are then channelled into the lake through a network of drains and open ditches.

Figure 10.2(b) shows the extent to which the year-to-year variations in the winter concentrations of DRP in Blelham Tarn were related to the average rainfall. The figure shows that there is a strong positive correlation ($r = 0.61$) between the concentrations of DRP and the average rainfall, and the fitted regression is statistically significant at the 99.9% confidence level. One of the most important climatic factors influencing the winter rainfall experienced in western Europe is the NAO. When the NAO is positive, winds tend to blow from the sea and the average rainfall is relatively high. When the NAO is negative, there is a strong easterly component to the flow of air and there is a corresponding reduction in the average rainfall. Figure 10.2(c) shows the extent to which the DRP concentrations recorded in Blelham Tarn were influenced by the year-to-year variations in the NAO. The figure shows that there is a strong positive correlation ($r = 0.53$) between the two variables, and the fitted regression is statistically significant at the 99.9% confidence level.

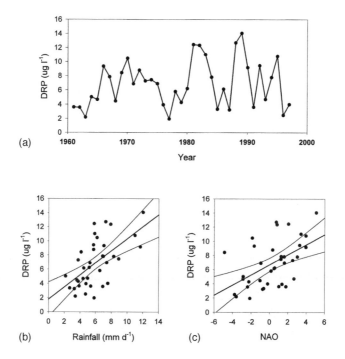

Fig. 10.2 (a) The year-to-year variations in the winter concentrations of DRP recorded in Blelham Tarn between 1961 and 1997; (b) the influence of the winter rainfall on the concentration of DRP; and (c) the influence of the NAO on the concentration of DRP.

The potential impact of these regional-scale variations on the productivity of the lake becomes even clearer when we examine the temporal variations in the DRP and the NAO (Fig. 10.3). This figure shows that the concentrations of DRP in Blelham Tarn can increase by a factor of five when the winters are very mild. In the early 1960s, when the NAO was strongly negative, the average winter concentration of DRP was less than $4 \, \mu g \, l^{-1}$. In the late 1980s, when the NAO was strongly positive, the average winter concentration of DRP exceeded $12 \, \mu g \, l^{-1}$ in two consecutive years.

10.5 Case Study 2: The impact of the NAO on the summer biomass of phytoplankton in Lake Erken

Lake Erken is a moderately eutrophic lake situated in the south of Sweden ($59° 25' N$, $18° 15' E$). It has a surface area of $23.7 \, km^2$, a mean depth of 9 m and a maximum depth of 21 m. The lake is covered with ice for at least four months every year but there are large year-to-year variations in the timing of ice formation and break-up. Water samples for chemical analyses and chlorophyll measurements have been collected from the lake at regular intervals since the 1960s using a Ruttner

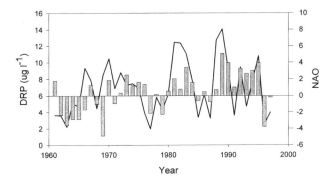

Fig. 10.3 Time-series relating the variations in the winter concentration of DRP in Blelham Tarn (solid line) to the NAO (vertical bars).

bottle lowered to different depths. Nutrient and chlorophyll concentrations have been measured by the methods outlined in Ahlgren and Ahlgren (1975) and sub-samples of water preserved in Lugol's iodine for microscopic examination. Detailed accounts of the phytoplankton populations found in the lake have been given by Pettersson (1985) and Weyhenmeyer *et al.* (1999). The spring phytoplankton is usually dominated by the diatom *Stephanodiscus hantzschii* but substantial growths of the dinoflagellate *Peridinium aciculiferum* also appear from time to time. One of the key factors influencing the relative abundance of the two genera is the time when the ice starts to break in the spring. When the winter is mild and the ice melts early, the spring crop is dominated by diatoms that behave as passive contaminants of the physical flow. When the winter is cold and the spring thaw is delayed, the diatoms are replaced by dinoflagellates that are better able to regulate their position in the ice-covered water column. The factors influencing the growth of phyto-plankton later in the year are more complex, but the timing of ice-break still has a significant effect on the qualitative composition of the community.

Figure 10.4(a) shows the long-term variation in the average concentration of chlorophyll *a* recorded in Erken during the summer in relation to the average concentration of DRP recorded the previous winter. The figure shows that there has been a marked increase in the summer biomass recorded in the 1990s but there is no corresponding increase in the winter concentration of DRP. A *t*-test on the DRP concentrations measured in the 1970s and the 1990s confirmed that there was no significant difference between the two periods but the measured increase in the concentration of chlorophyll *a* was statistically significant at the 95% confidence level. Figure 10.4(b) shows the long-term variations in the summer chlorophyll in relation to the length of the ice-free period. This figure shows that the extension of the ice-free period has had a significant effect on the summer biomass of phyto-plankton and the calculated correlation ($r = 0.49$) is statistically significant at the 95% level. Figure 10.4(c) shows the long-term variations in the summer chlorophyll in relation to the NAO. There was a weak positive relationship between the two

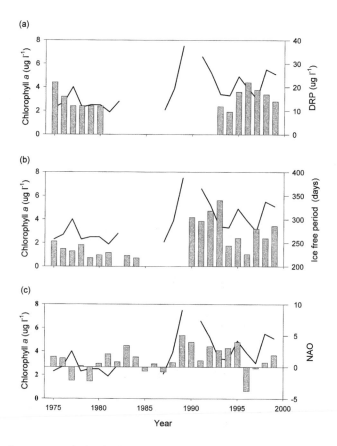

Fig. 10.4 The long-term variation in the concentration of chlorophyll *a* recorded in Erken during the summer in relation to: (a) the concentration of DRP measured the previous winter; (b) the length of the ice-free period; and (c) the NAO. The solid line shows the variation in the summer chlorophyll.

time-series but the calculated correlation ($r = 0.43$) is only statistically significant at the 90% confidence level.

A more detailed analysis of the phytoplankton data acquired from Erken suggests that the main factor responsible for the recent increase in biomass was the enhanced growth of the blue-green alga *Gloeotrichia* in late summer. In the 1970s, this species accounted for less than 10% of the late summer biomass but this proportion increased to an average of 35% in the 1990s. Figure 10.5 shows the relationship between the proportion of cyanobacteria present in Erken during the summer (expressed as a percentage of total biovolume) and the length of the ice-free period. The figure shows that there is a strong positive correlation between the two variables ($r = 0.71$) and the fitted regression is statistically significant at the 99.9% confidence level. Results of this kind demonstrate that quite subtle changes in the physical characteristics of a lake can have a pronounced effect on the long-term

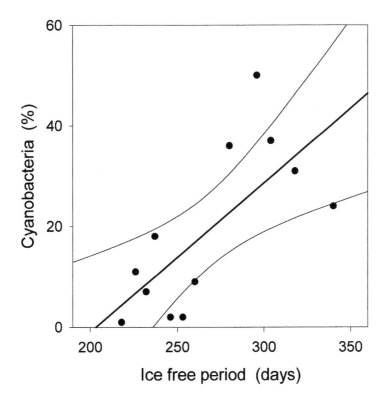

Fig. 10.5 The relationship between the proportion of cyanobacteria (% biovolume) found in Erken in summer in relation to the length of the ice-free period.

development of the phytoplankton community. Here, a change of just a few weeks in the date of ice break-up has resulted in quantitative and qualitative changes that mimic those normally associated with cultural eutrophication. Freshwater ecologists frequently underestimate the importance of time as a controlling factor. Most bloom-forming species of algae grow very slowly, so an extra cell division in mid-summer can have a dramatic effect on the perceived 'productivity' of a lake.

10.6 Case Study 3: The influence of the Gulf Stream on the summer concentrations of nitrate-nitrogen in Lough Leane

Lough Leane is a morphometrically complex lake situated in the centre of the Killarney National Park in the south-west of Ireland (52° 05′ N; 9° 35′ E). The lake has a surface area of 20 km², a mean depth of 13.4 m and a maximum depth of 60 m. The water in the lake is soft and slightly coloured by water-draining areas of peat. Water samples for chemical analysis have been collected from the lake at fortnightly intervals since the 1970s. The only results reported here are those for the winter concentrations of nitrate-nitrogen. These were measured using the Lovibond nessleriser procedure (Mackereth, 1963) between 1976 and 1983 and by a spectrophotometric method (Department of the Environment, 1981) between 1983 and 1998.

The nitrate concentrations recorded in many freshwater systems have increased in recent years due to the increased use of nitrogenous fertilisers (Casey and Clark, 1979; Burt *et al.*, 1988). Comparable increases have been recorded in Lough Leane but there were also large year-to-year variations superimposed on this long-term trend. The results summarised here are taken from an unpublished manuscript by Jennings and Allott and show that these year-to-year variations are closely correlated with the position of the Gulf Stream. Figure 10.6(a) shows the long-term variations in the concentration of nitrate measured in Lough Leane during winter in relation to the position of the Gulf Stream the previous spring. The nitrate concentrations were measured in February and the GSI is the average of the monthly values calculated for March, April and May. There is no direct correlation between the two time-series but there is a strong positive correlation ($r = 0.76$, $p < 0.001$) between the nitrate concentrations measured in February and the GSI for the previous spring.

The climatic factors influencing the leaching of nitrates from grassland soils are now known to be quite complex (Scholefield *et al.*, 1993; Stronge *et al.*, 1997). Winter rainfall can have a direct effect on the flux of nitrate in areas that are relatively dry (Trudgill *et al.*, 1991) but in wetter areas the dominant effects are those connected with the uptake of nitrate by vegetation. In the west of Ireland, June is the peak month for grass growth but the quantity of nitrate assimilated during this period depends on the moisture content of the soil. Figure 10.6(b) shows the effect that year-to-year variations in the spring GSI had on the moisture content of the soil in early summer. Daily soil moisture deficits have been calculated from meteorological measurements in the Lough Leane catchment and the results averaged for the month of June. The results demonstrate that the spring GSI has a delayed effect on the moisture content of the soil ($r = 0.47$) and the regression fitted to this relationship is statistically significant at the 95% confidence level. Figure 10.6(c) shows the effect that these inter-annual variations in the soil moisture deficit had on the residual concentration of nitrate recorded in the lake the following February. In this figure, the nitrate time-series has been detrended using a simple linear regression and the residuals used as a measure of relative change. The results demonstrate that there is a significant positive relationship between the two variables ($r = 0.68$) and the fitted regression is statistically significant at the 95% confidence level. We do not know what effect these variations had on the subsequent growth of phytoplankton but results from elsewhere (Maberly, pers. comm.) have shown that the supply of nitrogen is often a limiting factor in moderately productive lakes.

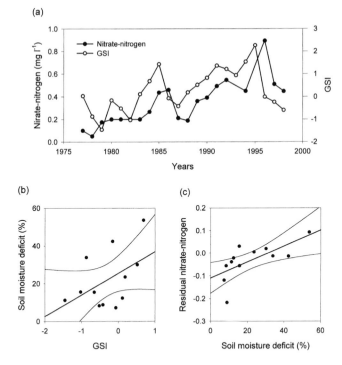

Fig. 10.6 (a) The year-to-year variation in the concentration of nitrate-nitrogen recorded in Lough Leane during the winter and the position of the Gulf Stream in the spring of the previous year; (b) the relationship between the soil moisture deficit recorded in the Lough Leane catchment in June and the position of the Gulf Stream in the spring of the same year; and (c) the relationship between the residual concentration of nitrate-nitrogen in Lough Leane in February and the soil moisture deficit recorded in the June of the previous year.

10.7 Case Study 4: The influence of the Gulf Stream on the summer biomass of phytoplankton in Esthwaite Water

Esthwaite Water (54° 18′ N; 2° 54′ E) is one of the most productive lakes in the English Lake District and has been the subject of intensive scientific study for more than 50 years. The lake has a surface area of 1.01 km², a mean depth of 6.4 m, a maximum depth of 14 m and usually remains thermally stratified from the end of May to the middle of October. Samples of water for chemical analyses and phytoplankton counts have been collected from the lake at regular intervals since the early 1950s using the methods already described for Blelham Tarn. Routine chlorophyll *a* measurements were, however, only started in the early 1960s. The pigment was first extracted using cold methanol and then by the hot methanol procedure described by Talling and Driver (1963–8). Representative samples of phytoplankton were also preserved in Lugol's iodine and examined under an inverted microscope using the methods described by Lund *et al.* (1958). The pattern of phytoplankton succession in Esthwaite Water has remained much the same throughout the period of study (Lund, 1972; Reynolds, 1984; Talling, 1993). Early in the year, the phytoplankton community is dominated by the diatoms *Asterionella formosa* Hass and *Aulacoseira subarctica* (O. Müller). When the diatoms decline at the onset of thermal stratification, they are usually replaced by a variety of small flagellates such as *Rhodomonas* and *Chlorella*. Later in the summer, large, slow-growing species that Reynolds (1984) describes as 'stress tolerators' become more abundant. In the early 1970s, the dominant species were the dinoflagellates *Ceratium hirundinella* and *Ceratium furcoides* but these were replaced by the blue-green algae *Aphanizomenon* and *Microcystis* in the 1980s.

Figure 10.7 shows the seasonal variation in the concentrations of chlorophyll *a* recorded in Esthwaite Water between 1966 and 1989. The solid line shows the 24-year average and the broken lines the 95% confidence intervals around this mean. The seasonal variation follows the classic diacmic pattern described by Talling (1993) with a well-defined spring peak and a more variable summer maximum. Figure 10.8(a) shows the year-to-year variations in the concentration of chlorophyll *a* recorded in Esthwaite Water in summer in relation to the DRP concentrations measured the previous winter. The chlorophyll *a* concentrations are the average of measurements taken between the middle of July and the end of September, i.e. the

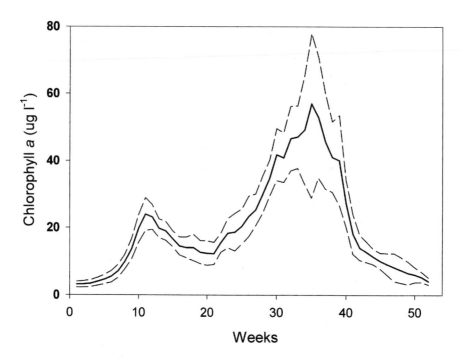

Fig. 10.7 The seasonal variation in the average concentration of chlorophyll *a* recorded in Esthwaite Water between 1966 and 1989. The solid line shows the 24-year average and the broken lines the 95% confidence intervals around this mean.

period that coincides with the summer maximum in Fig. 10.7. The DRP concentrations are the average of measurements taken in the first ten weeks of each year, i.e. the time when the biological uptake of nutrients is known to be very low. The DRP time-series shows that the quantity of nutrients reaching the lake increased substantially in the 1980s but this increase had no effect on the biomass of phytoplankton recorded during the summer.

One of the key factors influencing the summer growth of phytoplankton in a productive lake such as Esthwaite Water is the entrainment of nutrients from the anoxic hypolimnion. In early summer, very high concentrations of DRP accumulate in the upper layers of the hypolimnion and can then be entrained by wind-induced mixing. Figure 10.8(b) shows the effect that year-to-year variations in the intensity of mixing had on the biomass of phytoplankton present in summer. The depth of the early summer thermocline has been used as a general measure of mixing and an average value derived by a visual inspection of the temperature profiles recorded in June. The results show that the highest concentrations of chlorophyll *a* levels were always recorded in years when the early summer thermocline was relatively deep.

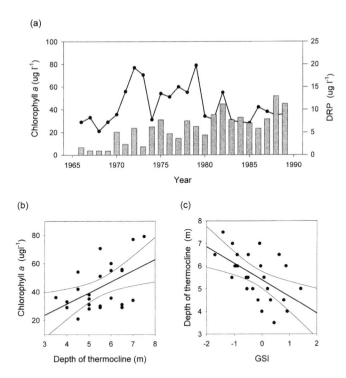

Fig. 10.8 (a) The year-to-year variation in the concentration of chlorophyll *a* recorded in Esthwaite Water in summer (solid line) in relation to the concentration of DRP measured the previous winter (vertical bars); (b) the relationship between the concentration of chlorophyll *a* and the depth of the early summer thermocline; and (c) the relationship between the depth of the early summer thermocline and the GSI.

There is a significant positive correlation ($r = 0.49$) between the two variables and the fitted regression is statistically significant at the 95% confidence level.

Recent studies in the English Lake District have shown that the mixing characteristics of the lakes in early summer are strongly influenced by the latitudinal movements of the Gulf Stream in the Atlantic (George and Taylor, 1995; George, 2000). Figure 10.8(c) shows the effect that year-to-year variations in the annual GSI had on the depth of the early summer thermocline. The results show that there is a strong negative correlation ($r = -0.56$) between the two variables and the fitted regression is statistically significant at the 95% confidence level. The potential impact of this 'teleconnection' becomes even clearer when we compare the inter-annual variations in the chlorophyll and the GSI (Fig. 10.9). This figure shows that highest concentrations of chlorophyll *a* are always recorded when the GSI index is negative ($r = -0.69$) and suggests that much of the recent decrease in biomass can be attributed to changes in the position of the Gulf Stream.

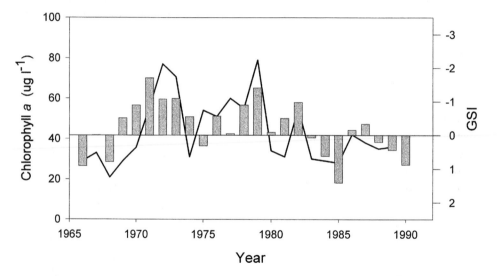

Fig. 10.9 Time-series relating the variations in the summer concentrations of chlorophyll *a* in Esthwaite Water (solid line) to the GSI (vertical bars).

10.8   Discussion

The growth of phytoplankton in lakes is controlled by the combined effects of the 'external' supply of nutrients from the catchment and the 'internal' recycling of nutrients within the basin. In this chapter, I have shown that year-to-year variations in the weather can have a significant effect on both these processes and demonstrated that the key climatic 'drivers' frequently vary from lake to lake.

In the first case study (Blelham Tarn), the proximate driving variable was the winter rainfall but the ultimate factor was the year-to-year variation in the NAO.

This example also shows the extent to which the physical characteristics of a lake can influence its response to changes in the weather. Blelham Tarn is a small lake with a relatively short retention time and is therefore particularly sensitive to changes in the rainfall. Similar effects have also been recorded in Esthwaite Water (George *et al.*, 2001) but the two basins of Windermere are too large for these flushing events to have any measurable effect.

In the second case study (Lake Erken), the proximate driving variable was the duration of ice-cover but the ultimate factor was again the inter-annual variations in the NAO. Many limnologists believe that changes in the freeze–thaw date of lakes could well provide the first unequivocal evidence of global warming. In Lake Erken, the date of ice break-up has hitherto only changed by a few weeks but this has already had a substantial effect on the perceived trophic status of the lake. This case study also demonstrates the complex nature of the changes that we might expect in a warmer world. Most reports of the limnological effects of the NAO emphasise the changes recorded during the winter. The 'productivity' changes in Lake Erken were, however, recorded much later in the year and were the result of a sustained qualitative change in the composition of the phytoplankton.

In the third case study (Lough Leane), the most surprising feature was the strength of the correlation established between the winter concentration of nitrate and the position of the Gulf Stream the previous spring. The strength of this correlation is even more remarkable when we consider the variance associated with the collection of a single water sample and the empirical nature of the GSI. Similar 'lagged' correlations have, however, been recorded in the composition of the vegetation in southern England (Willis *et al.*, 1995) where the proximate factor was again the moisture content of the soil. In the lakes of the English Lake District, George *et al.* (2001) found that the most important factor influencing the winter concentration of nitrate was the average air temperature. When the winters were relatively mild, more nitrate was assimilated by the plants and micro-organisms in the soil and there was a fourfold reduction in the nitrate concentrations reaching the lakes. The nitrate effects noted in Lough Leane were, in contrast, related to the uptake of nitrate by rooted vegetation in the previous summer. The most likely explanation for the different leaching patterns recorded in the two catchments was the regional variation in the management of land. In the English Lake District, most of the land surrounding the lakes is unimproved pasture which is not usually fertilised. In Lough Leane, much of the catchment is intensively grazed and is regularly treated with large quantities of inorganic fertiliser. In such a situation, it seems reasonable to assume that much of the fertiliser applied in the spring would eventually reach the lake if the summer was dry and there was a pronounced reduction in the growth of grass.

In the fourth case study (Esthwaite Water), the north–south movements of the Gulf Stream had a direct effect on the physical characteristics of the lake growth of phytoplankton. Here, as in the marine studies reported by Taylor (1995), the biological effects recorded in the plankton were much more obvious than the physical effects recorded in the lake. Taylor suggests that this 'amplifying' effect reflects the inherent non-linearity of biological systems but it could also be produced by the integrating effects of the GSI. We do not yet know why changes in the position of the Gulf Stream have such a pronounced effect on the dynamics of lakes located on the west coast of Europe. The only established physical effects are those connected with the movement of storms across the Atlantic. Taylor (1996) found that the number of storms crossing the North Atlantic were significantly lower when the Gulf Stream was located well to the north. A physical model of the heat fluxes associated with the Gulf Stream (Taylor, 1996) produced very similar effects and suggested that these movements could have a significant effect on the atmospheric circulation.

Quasi-cyclical variations of this kind have important implications for the way we model lakes and manage catchments in the early years of the new century. In the remaining sections of the discussion, I will consider three issues that need to be addressed if we are to quantify the impact of long-term changes in the weather on the dynamics of lake phytoplankton:

(1) The extent to which proxy 'climatic indicators' can be used to identify systematic patterns of change.
(2) The extent to which regional-scale changes in the weather can generate synchronous patterns of variation in lakes located in the same geographic area.
(3) The extent to which new modelling techniques will have to be devised to assess the impact of future changes in the weather on the trophic status of lakes.

Using 'climatic indicators' to analyse long-term patterns of change

Proxy indicators of change, such as the NAO and the GSI, are now widely used to identify patterns of change in terrestrial, marine and freshwater ecosystems. For example, Willis *et al.* (1995) used the GSI to identify the factors influencing the composition of vegetation in the Cotswolds whilst Beamish *et al.* (1999) used a number of atmospheric indices to analyse regime shifts in Pacific salmon. In the case studies presented here, the NAO and the GSI have been treated as independent measures of the long-term variation in the global climate. Recent studies have, however, shown that the two indices are related and may even be linked to climatic events in the tropical Pacific.

In 1998a, Taylor and Stephens showed that the NAO has a delayed effect on the north–south movements of the Gulf Stream. When the NAO and the GSI are compared on a year-to-year basis, the correlation between the two indices is not statistically significant. In contrast, when GSI is correlated with the NAO index recorded two years previously, there is a strong positive correlation ($r = 0.56$, $p < 0.05$) between the two variables (Fig. 10.10). The most likely explanation for this lagged response is the time taken by the ocean to respond to atmospheric forcing. A similar delayed response has been observed in models that simulate the point at

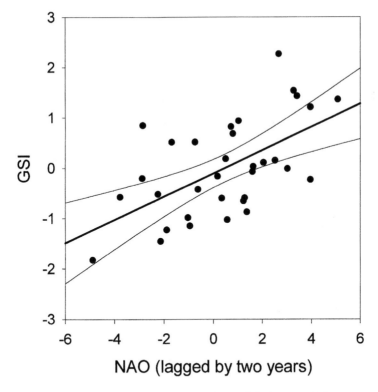

Fig. 10.10   The relationship between the GSI and the NAO index recorded two years earlier.

which the north wall of the Gulf Stream moves away from the US coast. Gang-opdhyay *et al.* (1992) suggest that this lag is related to the time taken by long pla-netary waves to cross the ocean. These waves have a length-scale of about 1000 km and a dominant period of about six months. They travel at a speed of several kilometres per day so the time taken to cross the Atlantic is consistent with the two-year delay.

More recently, Taylor *et al.* (1998b) have noted an unexpected correlation between the position of the Gulf Stream and ENSO events. Once the 'lagged' effect of the NAO has been removed from the GSI time-series, much of the remaining variation can be related to year-to-year variations in the Southern Oscillation Index (SOI). The SOI is based on the difference in atmospheric pressure recorded between Tahiti and Darwen and is widely used as an indirect measure of ENSO events. Here again, there is no direct correlation between the two time series but there is a significant negative correlation ($r = -0.53$) between the residuals in the GSI/NAO regression and SOI recorded two years previously. Results of this kind currently pose more questions than answers but they highlight the need to adopt a more holistic approach to the analysis of long-term change. Any causal inter-pretation must ultimately be based on a rigorous analysis of the critical driving variables. The correlations observed with these proxy indicators simply demonstrate that there are some systematic patterns present that we cannot yet explain.

The influence of regional changes in the weather on the temporal coherence of lakes

Lakes located in the same geographic area frequently show synchronous patterns of inter-annual variability (Kratz *et al.*, 1998; Baines *et al.*, 2000). Magnuson *et al.* (1990) pictured the development of regional coherence as a filtering process where lakes modulate the long-term signal imposed by the weather. Recent studies in the English Lakes (George *et al.*, 2000) have, however, shown that our ability to detect such 'coherent' patterns depends on the amplitude of the variations and the spatial heterogeneity of the driving variables. In this section, I demonstrate the extent to which systematic changes in the weather can influence the dynamics of lakes within a particular district and between districts located in the same climatic region.

The 'within district' example is taken from a 36-year study of the winter biomass of phytoplankton in two English lakes. The two lakes (Esthwaite Water and Blelham Tarn) have already been described and are both sites where year-to-year changes in the rainfall have a significant effect on the winter biomass of phytoplankton. The time-series in Fig. 10.11 shows the extent to which the long-term variations in the residual concentration of chlorophyll *a* in Blelham Tarn mirrored those recorded in Esthwaite Water. Both time-series have been de-trended using a linear regression of the measured concentration of chlorophyll versus the year of sampling. The figure shows that there was a high degree of synchrony between the de-trended measurements at the two sites, and the calculated correlation ($r = 0.56$)

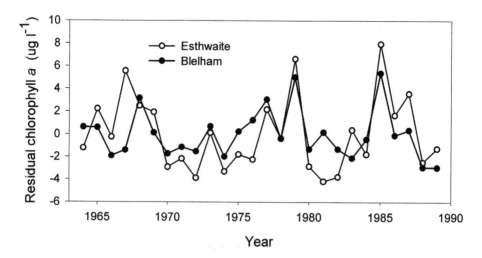

Fig. 10.11 Time-series showing the relationship between the residual winter concentrations of chlorophyll *a* recorded in Esthwaite Water and Blelham Tarn.

is statistically significant at the 95% confidence level. In both cases, the proximate driver was the flushing effect of the winter rains but the ultimate driver was the NAO.

The 'between district' example is taken from a 36-year study of winter water temperatures in two lakes in the English Lake District and one in the west of Ireland (Jennings *et al.*, 2001). Figure 10.12(a) shows the extent to which the winter temperature variations recorded in Blelham Tarn (maximum depth 12 m) mirrored those recorded in the south basin of Windermere (maximum depth 40 m). The results show that the winter temperatures measured at the two sites were strongly correlated ($r = 0.78$) but there were periods when the average difference in temperature exceeded 2°C. Figure 10.12(b) shows the extent to which the winter temperatures measured in Lough Feeagh mirrored those recorded in the south basin of Windermere. The two lakes are more than 300 km apart but the maximum depth of Lough Feeagh is very similar to the south basin of Windermere. The results show that the variations recorded in Lough Feeagh are very closely correlated ($r = 0.87$) with those recorded in Windermere and the maximum difference in temperature is now only 1°C.

Comparisons of this kind show the extent to which the physical characteristics of a lake influence its response to regional changes in the weather. Variations in the winter temperature have little effect on the growth of phytoplankton in these low-latitude lakes but can have a significant effect on the seasonal succession of phytoplankton.

Modelling the trophic responses of lakes to future changes in the weather

The fixation of organic carbon by lake phytoplankton depends on the dynamic interaction of four factors:

(1)   the external supply of nutrients,
(2)   the internal recycling of nutrients,
(3)   the external supply of mechanical energy,
(4)   the external supply of light energy and its attenuation in the water column.

Some of the site-specific factors that regulate these processes are fixed, but others are variable and regulated by regional-scale variations in the weather. Over the years, a variety of methods have been devised to classify lakes and characterise their response to changes in the environment. Most of these methods are, however, based on the analysis of historical data and may not be applicable if there is a major change in the boundary conditions imposed by the regional weather. One of the most effective methods hitherto designed to quantify the trophic status of lakes is that suggested by Vollenweider (1968). This scheme was originally based on nutrient loadings but was later extended to include response variables such as phytoplankton biomass (Anon., 1982). The Vollenweider models are most effective when used to

(a)

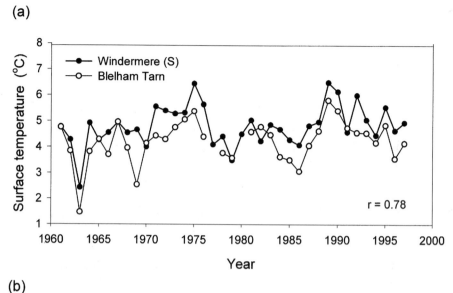

(b)

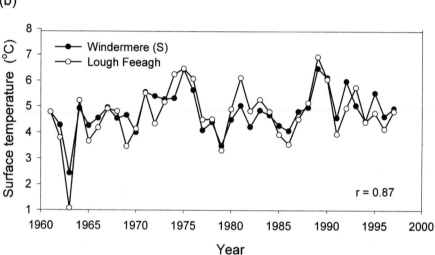

Fig. 10.12   (a) Time-series showing the relationship between the winter surface temperatures of the south basin of Windermere and Blelham Tarn; and (b) time-series showing the relationship between the winter surface temperatures of the south basin of Windermere (UK) and Lough Feeagh (Ireland).

characterise the response of different types of lakes to defined change in the external loading. They are not designed to simulate the precise response of an individual lake but to characterise the general responses of lakes in a particular geographic region (Reynolds, 1992; Harris and Baxter, 1996).

The solid lines in Fig. 10.13 shows the probability distributions for three of the categories used by the Organisation for Economic Cooperation and Development

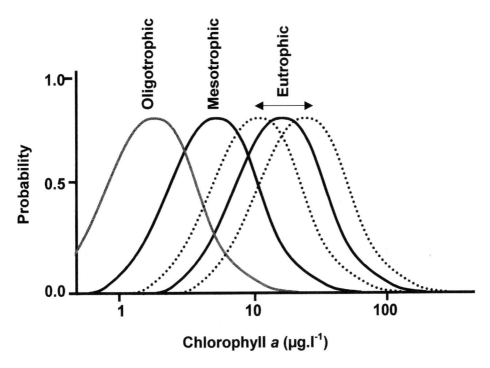

Fig. 10.13 Schematic diagram showing the possible effects of climate change on the classification system used to assess the trophic status of lakes. The solid curves show the historical limits defined by the OECD. The broken curves show the climate-related changes that could occur in lakes that are currently classified as eutrophic.

(OECD) to characterise the trophic status of lakes in Europe. Although there is a considerable overlap between the different categories, the classification is essentially static and assumes that there is no change in the physical factors regulating the productivity of the lakes. The broken lines in the figure show the uncertainties that can arise when eutrophic lakes are subjected to systematic changes in the weather. The frequency distribution to the left of the curve shows the changes that could arise in a lake like Esthwaite Water if there were a systematic reduction in the entrainment of nutrients and the subsequent growth of phytoplankton. The frequency distribution to the right of the curve shows the converse situation where an increase in the wind speed has resulted in an increased rate of nutrient entrainment and an enhanced growth of phytoplankton.

Most of the models that we currently use to manage lakes are based on statistics acquired in the 1960s when the weather patterns differed from how they are today. In this period of rapid change, the only sound approach is to recalibrate the models with new data and develop a probabilistic rather than a deterministic approach to the classification of lakes. A discussion of the different sources of uncertainty associated with climate modelling can be found in a number of recent publications

(Hulme and Carter, 1999; Katz, 1999). The most reliable scenarios are those that present the results of General Circulation Modelling as a frequency distribution and not a regional average.

In the final section of the discussion, I present a simple example to illustrate what can be achieved by combining the results of a Monte Carlo based weather generator with the historical observations described in Case Study 4. The weather generator used was developed by the Climatic Research Unit in East Anglia and is based on the 'medium-high' output of the HADCM2 General Circulation Model (Hulme and Jenkins, 1998). The solid line in Fig. 10.14(a) shows the result of using this weather generator to simulate the early summer wind speeds recorded at Ambleside between 1966 and 1989. The broken line in the figure shows the results of using the same method to generate a frequency polygon for the wind speeds projected for 2050. These simulations suggest that the early summer wind speeds in the area will have decreased by 10% by 2050 and that there will be an associated change in the form of the frequency distribution.

Figure 10.14(b) shows the result using these wind speeds to drive the empirical model described in Case Study 4. In this model, the dependent variable is the summer concentration of chlorophyll and the regressor variables the average wind speed measured in early summer. The results of these simulations have to be treated with some caution, but they imply that the average summer concentration of chlorophyll in Esthwaite Water will have decreased substantially by the middle years of this century.

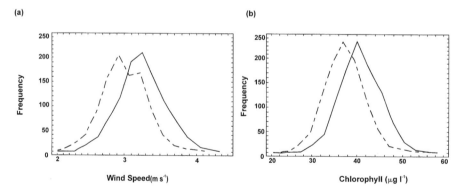

Fig. 10.14   (a) Frequency polygons showing the potential impact of climate change on the wind speeds recorded at Esthwaite Water in early summer. The solid lines show the current situation and the broken lines the frequency distribution projected by the HadCM2 climate-change simulation for the year 2050. (b) Frequency polygons showing the potential impact of climate change on the average concentration of chlorophyll *a* recorded in Esthwaite Water in summer. The solid lines show the current frequency distribution and the broken lines the frequency distribution projected by a HadCM2 climate-change simulation for the year 2050. Both predictions are based on the simple wind-mixing model described in Case Study 4.

Acknowledgements

Most of the studies reported here were supported by the European Union Environment and Climate project REFLECT (Responses of European Freshwater Lakes to Environmental and Climatic Change; contract ENV4-CT97-0453). I would like to thank Ms Diane Hewitt for help with processing data and Ms Yvonne Dickens for producing several diagrams. I am also very grateful to Dr Kurt Pettersson, Thorsten Blenckner, Dr Norman Allott, Dr Eleanor Jennings and Ian Harris for allowing me to use data that had not been published when this chapter was submitted to the editors.

References

Ahlgren, I. & Ahlgren, G. (1975) *Methods of Water-chemical Analyses Compiled for Instruction in Limnology*. Institute of Limnology, Uppsala.

Anderson, W.L., Robertson, D.M. & Magnuson, J.J. (1996) Evidence of recent warming and El Niño-related variations in ice breakup of Wisconsin lakes. *Limnology and Oceanography* **41**, 815–21.

Anon. (1982) *Eutrophication of Waters, Monitoring Assessment and Control*. Organisation for Economic Cooperation and Development, Paris.

Baines, S.B., Webster, K.E., Kratz, T.K., Carpenter, S.R. & Magnuson, J.J. (2000) Synchronous behaviour of temperature, calcium, and chlorophyll in lakes in northern Wisconsin. *Ecology* **81**, 815–25.

Beamish, R.J., Noakes, D.J., FcFarlane, G.A., Klyashtorin, L., Ivanov, V.V. & Kurashov, V. (1999) The regime concept and natural trends in the production of Pacific salmon. *Canadian Journal of Fisheries and Aquatic Sciences* **56**, 516–26.

Burt, T.P., Arkell, B.P., Trudgill, S.T. & Walling, D.E. (1988) Stream nitrate levels in a small catchment in south west England over a period of 15 years (1970–1985). *Hydrological Processes* **2**, 267–84.

Casey, H. & Clark, R.T. (1979) Statistical analysis of nitrate concentrations from the River Frome (Dorset) for the period 1965–1976. *Freshwater Biology* **9**, 91–7.

Catalan, J. & Fee, E.J. (1994) Interannual variability in limnetic ecosystems: origin, patterns, and predictability. In: *Limnology Now: A Paradigm of Planetary Problems* (ed. R. Margalef), pp. 81–94. Elsevier, Amsterdam.

Department of the Environment (1981) The spectrophotometric determination of nitrate. In: *Oxidised Nitrogen in Waters: Methods for the Examination of Waters and Associated Materials*, pp. 31–5. HMSO, London.

Gangopadhyay, A., Cornillon, P. & Watts, R.D. (1992) A test of the Parsons–Veronis hypothesis on the separation of the Gulf Stream. *Journal of Physical Oceanography* **22**, 1286–1301.

George, D.G. (2000) The impact of regional-scale changes in the weather on the long-term dynamics of *Eudiaptomus* and *Daphnia* in Esthwaite Water, Cumbria. *Freshwater Biology* **45**, 111–21.

George, D.G., Maberly, S.C. & Hewitt, D.P. (2001) The influence of the North Atlantic Oscillation on the winter characteristics of lakes in the English Lake District. *Freshwater Biology* (in press).

George, D.G. & Taylor, A.H. (1995) UK Lake Plankton and the Gulf Stream. *Nature* **378**, 139.

George, D.G., Talling, J.F. & Rigg, E. (2000) Factors influencing the temporal coherence of five lakes in the English Lake District. *Freshwater Biology* **43**, 449–61.

Goldman. C.R., Jassby, A. & Powell, T. (1989) Interannual fluctuations in primary production: meteorological forcing at two sub-alpine lakes. *Limnology Oceanography* **43**, 310–23.

Harris, G.P. (1986) *Phytoplankton Ecology: Structure, Function and Fluctuation*. Chapman and Hall, London.

Harris, G.P. & Baxter, G. (1996) Interannual variability in phytoplankton biomass and species composition in a sub-tropical reservoir. *Freshwater Biology* **35**, 545–60.

Hulme, M. & Carter, T.R. (1999) Representing uncertainty in climate change scenarios and impact studies. In: *Representing uncertainty in climate change scenarios and impact studies. Proceedings of the ECLAT-2 Workshop*, Helsinki, pp. 11–37. Climatic Research Unit, Norwich.

Hulme, M., & Jenkins, G.J. (1998) *Climate change scenarios for the United Kingdom: scientific report*, 80 pp. UKCIP Technical Report No. 1, Climatic Research Unit, Norwich.

Hurrell, J.W. (1995) Decadal trends in the North Atlantic Oscillation: regional temperature and precipitation. *Science* **269**, 676–9.

Jennings, E., Allott, N., McGinnity, P., Poole, R., Quirke, B., Twomey, H. & George, D.G. (2001) The North Atlantic Oscillation: implications for freshwater systems in Ireland. *Biology and Environment* **100B** (3), 149–57.

Katz, R.W. (1999) Techniques for estimating uncertainty in climate change scenarios and impact studies. In: *Representing uncertainty in climate change scenarios and impact studies. Proceedings of the ECLAT-2 Workshop, Helsinki*, pp. 38–53. Climatic Research Unit, Norwich.

Kratz, T.K., Sorrano, P.A., Baines, S.B., *et al.* (1998) Interannual synchronous dynamics in north temperate lakes in Wisconsin, USA. In: *Management of Lakes and Reservoirs during Global Climate Change* (eds D.G. George, J.G. Jones, P. Punchochar, C.S. Reynolds & D.W. Sutcliffe) pp. 273–87. Kluwer, Dordrecht.

Lamb, H.H. (1972) British Isles weather types and a register of the daily sequence of circulation patterns. *Geophysical Memoirs* **16**, 85 pp.

Livingstone, D.M. & Dokulil, M.T. (2001) Eighty years of spatially coherent Austrian lake surface temperatures and their relationship to regional air temperatures and to the North Atlantic Oscillation. *Limnology and Oceanography* **46**, 1220–27.

van Loon, H. & Rogers, J.C. (1978) The see-saw in winter temperatures between Greenland and Northern Europe, Part 1: General description. *Monthly Weather Review* (US Department of Agriculture) **106**, 296–310.

Lund, J.W.G. (1972) Changes in the biomass of blue-green and other algae in an English lake from 1945–1969. *Proceedings of the Symposium on Taxonomy and Biology of Blue-green Algae* (ed. T.V. Desikachary), pp. 305–27. University of Madras Press.

Lund, J.W.G., Kipling, C. & Le Cren, E.D. (1958) The inverted microscope method of estimating algal numbers and the statistical basis of estimations by counting. *Hydrobiologia* **11**, 143–70.

Lund, J.W.G. & Talling, J.F. (1957) Botanical limnological methods with special reference to the algae. *Botanical Review* **23**, 489–583.

Mackereth, F.J. (1963) *Some Methods of Water Analyses for Limnologists*. Freshwater Biological Association, Ambleside, Scientific Publication No. 21.

Magnuson, J.J., Benson, B.J. & Kratz, T.K. (1990) Temporal coherence in the limnology of a suite of lakes in Wisconsin, USA. *Freshwater Biology* **23**, 145–59.

Miller, J.L. (1994) Fluctuations of Gulf Stream position between Cape Hatteras and the Straits of Florida. *Journal of Geophysical Research* **99**, 5057–64.

Pettersson, K. (1985) The availability of phosphorus and the species composition of the spring phytoplankton in Lake Erken. *Internationale Revue der gesamten Hydrobiologie und Hydrographie* **70**, 527–46.

Proctor, C.M. & Hood, D.W. (1954) Determination of inorganic phosphate in sea water by an iso-butanol extraction procedure. *Journal of Marine Research* **13**, 122–32.

Reynolds, C.S. (1984) *The Ecology of Freshwater Phytoplankton*. Cambridge University Press, Cambridge.

Reynolds, C.S. (1992) Eutrophication and the management of planktonic algae: what Vollenweider couldn't tell us. In: *Eutrophication: Research and Application to Water Supply* (eds D.W. Sutcliffe & J.G. Jones) pp. 4–29. Freshwater Biological Association, Ambleside.

Reynolds, C.S. (1997) *Vegetation Processes in the Pelagic: a model for Ecosystem Theory*, 371 pp. Ecology Institute, Oldendorf/Luhe.

Reynolds, C.S. (1999) With or against the grain: responses of phytoplankton to pelagic variability. In: *Aquatic Life-cycle Strategies* (eds M. Whitfield, J. Matthews & C.S. Reynolds), pp. 15–43. Marine Biology Association, Plymouth.

Schindler, D.L., Bayley, S.E., Parker, B.R., *et al.* (1996) The effects of climatic warming on the properties of boreal lakes and streams at the Experimental Lakes Area, northwest Ontario. *Limnology and Oceanography* **41**, 1004–17.

Scholefield, D., Tyson, K.C., Garwood, E.A., Armstrong, A.C., Hawkins, J. & Stone, A.C. (1993) Nitrate leaching from grazed grassland lysimsters: effects of fertilizer input, field drainage, age of sward and patterns of weather. *Journal Soil Science* **44**, 601–13.

Sharpley, A.N. & Syers, J.K. (1979) Phosphorus inputs into a stream draining an agricultural watershed: II Amounts and relative significance of runoff types. *Water, Air and Soil Pollution* **11**, 417–28.

Stephens, K. (1963) Determination of low phosphate concentrations in lake and marine waters. *Limnology and Oceanography* **8**, 361–2.

Straile, D. & Geller, W. (1998) The response of *Daphnia* to changes in trophic status and weather patterns, a case study from Lake Constance. *ICES Journal of Marine Science* **55**, 775–82.

Stronge, K.M., Lennox, S.D. & Smith, R.V. (1997) Predicting nitrate concentrations in Northern Ireland rivers using time series analysis. *Journal of Environmental Quality* **26**, 1599–1604.

Sutcliffe, D.W.S., Carrick, T.R., Heron, J., *et al.* (1982) Long-term and seasonal changes in the chemical composition of precipitation and surface waters of lakes and tarns in the English Lake District. *Freshwater Biology* **12**, 451–506.

Talling, J.F. (1993) Comparative seasonal changes, and interannual variability and stability in a 26-year record of total phytoplankton biomass in four English lake basins. *Hydrobiologia* **268**, 65–98.

Talling J.F. & Driver D. (1963) Some problems in the extraction of chlorophyll *a* in phytoplankton. In: *Proceedings on Primary Production measurement, Marine and Freshwater* (ed. M. Doty), pp. 142–6. US Atomic Energy Engineering Commission, Hawaii.

Taylor, A.H. (1995) North–south shifts of the Gulf Stream and their climatic connection with the abundance of zooplankton in the UK and its surrounding seas. *ICES Journal of Marine Science* **52**, 711–21.

Taylor, A.H. (1996) North–south shifts of the Gulf Stream: ocean atmosphere interactions in the North Atlantic. *International Journal of Climatology* **16**, 559–83.

Taylor, A.H., Jordan, M.B. & Stephens, J.A. (1998) Gulf Stream shifts following ENSO events. *Nature* **393**, 638.

Taylor, A.H. & Stephens, J.A. (1998a) Latitudinal displacements of the Gulf Stream (1966 to 1977) and their relation to changes in the temperature and zooplankton abundance in the NE Atlantic Ocean. *Oceanologica Acta* **3**, 145–9.

Taylor, A.H. & Stephens, J.A. (1998b) The North Atlantic Oscillation and the latitude of the Gulf Stream. *Tellus* **50A**, 134–42.

Trudgill, S.T., Burt, T.P., Heathwaithe, A.L. & Arkell, B.P. (1991) Soil nitrate sources and nitrate leaching losses, Slapton, South Devon. *Soil Use and Management* **7**, 200–206.

Vollenweider, R.A. (1968) *Water management research: scientific fundamentals of the eutrophication of lakes and flowing waters, with particular reference to nitrogen and phosphorus as factors in eutrophication.* Technical Report DAS/CSI/68.27, OECD, Paris.

Vollenweider, R.A. (1975) Input–output models with special reference to the phosphorus loading concept in limnology. *Schweizerische Zeitschrift für Hydrologie* **37**, 53–84.

Walker, G.T. & Bliss, E.W. (1932) World weather V. *Memoirs of the Royal Meteorological Society* **4**, 53–84.

Wehenmeyer, G.A., Blenckner, T. & Pettersson, K. (1999) Changes of the plankton spring outburst related to the North Atlantic Oscillation. *Limnology and Oceanography* **44**, 1788–92.

Willis, A.J., Dunnett, N.P., Hunt, R. & Grime, J.P. (1995) Does Gulf Stream position affect vegetation dynamics in Western Europe? *Oikos* **73**, 408–10.

Chapter 11
Ecosystem Function and Degradation

Wilhelm Ripl and Klaus-Dieter Wolter

11.1 Introduction and overall aims

The sustainability of ecosystem function is now the proper concern of all human society. A stabilised climate, even runoff into surface waters, low material losses, and reliable soil fertility are essential features of an enduring, functional landscape. Through human interference, however, particularly regarding water balance and vegetation, the sustainability of our ecosystems is today very much threatened. We cannot expect to secure improved sustainability without first acquiring a proper understanding of the biospheric functioning and its inherent, self-regenerating processes. What is needed is a coherent and comprehensive recognition of the functions of intact ecosystems, on the one hand, and a knowledge of the responses to anthropogenic impacts that result in their loss of function and degradation, on the other.

The present chapter deals with how we can rationalise complex systems to a series of interdependent processes. The basic principles of ecosystem development were articulated in the classical, though not universally adopted, paper of Odum (1969). Fundamental to the development of all ecosystems is the extent to which they regulate the through-flow of energy and water. The Energy–Transport–Reaction (ETR) model (Ripl, 1995a) is a model of energy dissipation, based on the water balance of whole landscapes. In this model, the underlying role of water exchanges – and their energy-dissipative properties – in the function and the development of natural systems is explicitly recognised. These energy-dissipating processes regulate the ecological dynamics within the earth's biosphere in such a way that the development of natural systems is never allowed to proceed in an undirected or random way. A basic characteristic of unconstrained development in natural systems is the increasing role of cyclical processes while loss processes are correspondingly reduced. This gives the coincidental increase in the system's sustainability. This might also be viewed as an increase of ecological efficiency, which is applicable at all spatio-temporal levels – from small (microscopic) up to large (whole landscapes). This hierarchy of spatio-temporal levels can be described as a 'fractal' series, the levels being self-similar in their ordering and in their functional tendency towards the proportionate increases of cyclic over loss-making processes.

Energy flow through ecosystems has feedbacks to all spatio-temporal (fractal) levels. The lower levels (smaller in space, localised, shorter in time) always exhibit higher dynamics than upper or higher fractal levels (more space, longer in time). Thus, for example, longer-wave processes, such as forest-generation cycles or the successional development of ecosystems, only appear static from a short-term viewpoint. From a functional viewpoint, however, they should be considered rather as phases, whose harmonic patterns are simply revealed over longer time periods than those of shorter-wave subsystems.

In this chapter, we attempt to bring this philosophical framework to the behaviour of aquatic ecosystems, including their hydrological catchments, and how an appreciation of their function should underpin their sustainable management. We begin by developing the concept of the ecological efficiency that characterises the development of all natural ecosystems. Then we show how the same principle is manifest at different spatio-temporal (fractal) levels, especially in the ways that intact ecosystems proceed in their development and, equally, how they respond to human impacts that bring about their degradation. At the end, we commend the functional approach as a prescriptive requirement to the design of the measures that are now necessary to restore landscapes and freshwaters to sustainability.

11.2   The efficiency principle

Energy dissipation and efficiency

In this section, we expand the premise that natural systems are energy-dissipative and, thus, self-optimising. As a grounding, the following definitions are fundamental:

- Energy is the interaction between two or more objects (e.g. between molecules, or between the sun and the earth) that results in spatial and/or temporal change between them (e.g. acceleration or delay in a chemical reaction or change of temperature).
- Energy dissipation (the lowering of energy flux) is the process whereby energy gained by an entity as a pulse (interaction phase and/or from an energy potential) is released, out of phase, both in time as well as in space. In this way, the energy pulse is diminished, tending to a dampened or attenuated value. Thus, the energy potential is reduced, eventually to zero (Ripl *et al.*, 1996).
- Water plays a special significance in global energy dissipation, existing as it does in such large quantities, in solid, liquid and gaseous form, and showing high enthalpy leaps between the various phases (see section 11.3).

An important aspect of energy-dissipative systems is that under constraints of space (i.e. they are spatially and/or materially-limited) and time (i.e. they are temporally-structured, or are limited by the energy available), they develop more closed-cycle processes and thus higher levels of persistence. They are thus self-optimising systems, characterised by a tendency to retard the lowering of energy flux (Ripl, 1995a; Ripl *et al.*, 1996). Equally, this tendency increases the energy-efficiency of the system.

Natural systems that experience unimpaired self-structuring development always approach a condition of lowest energy flux. During this development, energy dissipation is directed increasingly within cyclic processes, so that the proportion of irreversible losses are thus being minimised. The system functions more sustainably and its longevity is extended. Given a sufficient period of observation, it has a higher probability of being found than one beset by heavy energy losses (Ripl *et al.*, 1996). In this way, the growth and the attrition of the system become better controlled.

The efficiency of structures may be judged by the extent of their investment in cyclic processes as opposed to loss processes, that is, by their relative closure. It is important, therefore, that delineation of the spatio-temporal units is oriented towards the minimisation of materials transported across their borders. To this end, functional units with minimal openness may be readily recognised at every fractal level:

- atoms with electrons distributed in orbitals about the nucleus,
- molecules with several atomic nuclei and electrons in molecular orbitals,
- molecular aggregates from water (H_2O, H^+ and OH^-), salts and macromolecules, with clusters stabilised by electrical interaction,
- cells with their functional organelles,
- organisms with their different cell aggregates and organs,
- the dissipative ecological unit (DEU; see section 11.3), being the smallest functional unit of a functional ecosystem (after Ripl, 1995a; Ripl *et al.*, 1995),
- ecosystems in aggregate, contributing to global processes.

At maximum stability (permanency), these functional units develop in accord with the efficiency principle. At the levels of atoms, molecules and molecular complexes, they reveal a collective pattern – standing waves, oscillations with the same phase, coherency and coupled (cyclic) processes. At the levels of cells, organisms, up to sustainable ecosystems, the functional units are characterised by comparatively stable patterns and extensive cyclic processes, with minimal material leakage to the environment. Thus, in all functional units, independently of their structural level, there is a common pattern. As the proportion of parallel processes (i.e. close, high-feedback production and decomposition) is increased, so the sequential processes at the same level are reduced, so that, overall, energy flux is optimally lowered. Fully functional units dissipate energy almost completely within local cyclic processes. Moreover, irreversibility falls from hierarchically organised levels in a cascade-like fashion.

In contrast, degraded functional units are marked by higher randomness in their dynamics, which is accompanied by linear losses (irreversible losses, reduced cyclic processes, more spatio-temporally separated); the longevity of the structure is lowered. Following the second law of thermodynamics (entropy theorem), open systems (i.e. all known systems) are inevitably destined to 'irreversible' breakdown, with unordered movement and structural decay of structure, towards thermo-dynamic equilibrium. The increase in entropy of a system can be estimated from its irreversible (unusable) energy and/or its material losses. These are also measures of openness; for instance, loss from a hydrological catchment might be quantified in the energy unit 'proton flow', derived from the total dissolved load of base cations removed in the stream or river, expressed in proton equivalents.

Efficiency of ecosystems

The permanency of natural ecosystems is limited by the available supply of nutrients and minerals in the soil. With a directed water flow, material is transported from the land, into surface waters, and thence to the sea, where the mineral ions thus leached away can only be made available again within geological time periods (i.e. an overall loss process within conceivable time periods). Therefore, only ecosystems that maximise the proportion of matter cycled in relation to that lost irreversibly in the water flow, can be considered sustainable.

In this way, the efficiency of an ecosystem can be measured on the basis of its water and matter balance. Efficiency may be assessed in sub-catchments as well as in entire catchment areas, so long as the minimum relative interchanges with the surroundings are low and the output discharges are known. Two separate measures of efficiency are applicable to ecosystems – one thermal, the other chemical (Ripl *et al.*, 1995; Hildmann, 1999; Ripl and Hildmann, 2000). Although developed origin-ally for use at the landscape and catchment levels, both coefficients may be applied to other spatio-temporal levels, including those of the organelle, the cell, the organism and the biosphere. Both efficiency coefficients yield results on a scale 0 to 1. Although calculated using different parameters, the two measures of landscape efficiency are directly interlinked – through the deductive reasoning of our concept model – and may therefore be compared directly.

The *thermal efficiency* of land systems (the land surface, with its various uses) may be derived from the degree of damping of the daily, and seasonally-modulated, solar-energy pulse. After a preliminary heuristic phase of investigation, this can be measured from regular satellite observations (using the thermal channel) on a regional basis.

The *chemical efficiency* of an ecosystem may be estimated by the extent of cyclical metabolism that is based on the production of fixed carbon atoms. This chemical efficiency (E) can be calculated as an annual ratio between gross production (P, as a proton yield, 1 mol carbon representing 2 mol H^+) and the irreversible loss of cations in the drainage (L, also expressed as 'lost' proton-equivalents as exported base cations). For a given catchment:

$$E = \frac{P - L}{P} \qquad (11.1)$$

Both thermal and chemical measures of landscape efficiency may be used in areas where water is the main medium of the system dynamics; both measures provide valuable inputs to land-use management and planning for sustainability.

Although its fundamental basis is well-known, the efficiency principle has not, to our knowledge, been applied so far to the different fractal levels of various systems. Yet the advantage of this approach lies in its consistent applicability in many other fields of science. By focusing on water and its structuring role as indicators of efficiency, it is possible to deal with the complexity of entire ecosystems, as well as with the full hierarchy of structural levels. We believe that the method is well-suited to meet the urgent requirement for the determination of manageable development and the management of processes towards sustainability.

For more complete accounts about the functional description of ecosystems and the efficiency principle, as founded on the ETR model, see Ripl (1992, 1995a) and Ripl and Hildmann (2000). The ETR model is based on the dissipative-processor properties of water, the main energy dissipative medium on earth. The model is in accord with the concept of self-organising dissipative structures (e.g. Prigogine, 1980).

11.3   Ecosystem function at various fractal levels

Dissipative processes at the molecular and pre-cellular level

Liquid water is a fundamental prerequisite for all life processes. In addition, the energy-dissipative reactions in which water is involved, from the interaction between sun and earth down, and its physical and chemical changes of state (e.g. evaporation, dissociation) make it the most important dynamic medium on the earth. Together with organic macromolecules, water is the primary constituent of organismic cells (Cope, 1972), where its ionic structure and its mole fractions are predetermined. For instance, the manifold vibrational possibilities of the water molecule and its molecular associations (clusters) give it an unusually high heat capacity. Interactions involving its molecular, atomic and sub-atomic behaviours are subject to unpredictabilities (Heisenberg's uncertainty principle) but the consequences are sufficiently probabilistic for quantitative statements about the functionality of aqueous solutions to be made, even at this level.

Each energy interaction becomes effective and observable in the acceleration or deceleration of matter. By way of electromagnetic radiation and/or mechanical oscillations, relaxing (dissipating) systems interact with absorbing systems, changing their states of motion in the process. In liquids, a higher coherency arises from these localised interactions, harmonic oscillations replacing random movement. Unlike the gas phase, wherein each particle moves in a straight line until it strikes another particle, in liquids, especially in polarised liquids such as water, particles are substantially coupled. Through electromagnetic interactions, neighbouring particles move together spatially in the same direction, and temporally, with the same frequency. Within such a coherent structure, the probability of a collision with another particle is therefore much reduced. Only within the proximity of phase boundaries, or other structuring of system material, does some modulation of the oscillation process occur. Here, transitions between the coherent vibrations of clusters and the vibrations of ions and macromolecules lead to a local increase in the likely reactions. As a function of amplitude of the oscillations, there results an intensified dissociation of the water. This applies equally to living cells: the same efficiency principle, demonstrable at the molecular level, is repeated and amplified at successive fractal levels.

In aqueous solutions, oscillation patterns are modified substantially by the presence of ions and macromolecules. In the close vicinity of a positively- or negatively-charged ion, the polarised structure (dipoles) of the water molecules causes their relatively rigid alignment. Through changes in bond lengths and angles, local increases in molecular densities also occur in the vicinity of hydrophilic ions or molecules (Gutmann, 1978). In the immediate environment of ions, therefore, the aqueous solution becomes strongly structured. The electrical field associated with individual water molecules is substituted by mutual hydrogen bonding and developing hydrate coverings and clusters. At the interface between these organised and coherently oscillating clusters, there is a zone of 'distortion' within which fewer hydrogen bonds arise, through the disturbance of orientation by the electrical field. Because of the lowered, 'looser' structuring, this zone is particularly fluid. Within a salt solution, discretely distributed differentiated areas arise from the structured changes in the distribution pattern of charge and in the vibration pattern. Thus, due to the oscillation of water molecules, charge density distributions within the salt solution are altered. These then have a sorting effect on other molecules, including macromolecules. Such sorting of polymerisation products would have been critical in the evolution of life: the present-day function of macromolecules in living cells must have evolved by a similar mechanism.

Macromolecules have regions of greater or less polarisation, as well as areas with stronger and weaker vibrations. In an aqueous solution, therefore, these macromolecules are sorted precisely in space and time by the differentiated dynamics of modulated water vibrations. Such dynamic dissipative structures comprise the 'active centres of enzymes'. As the essential medium for life processes, water must continue to host these structures. This is why, irrespective of whether the cells came from terrestrial, marine or freshwater organisms, cellular salt concentrations must be regulated within a relatively narrow range. It is also the reason why the temperature optimum of cellular processes also lies within a relatively narrow temperature range, around 37°C.

The reactivity of water encompasses acid-base reactions (H^+ and OH^-), and thus the dissolution and precipitation of salts, as well as the redox reactions (reduction and oxidation); alternately splitting and joining of water (in, respectively, photosynthesis and respiration). Reactivity is particularly high at phase boundary surfaces (liquids/solids, gas/liquids, etc.). Owing to differences in the natural frequencies of the materials interfaced at these state-change boundaries, individual water molecules experience a form of mechanical stress, which leads to a higher dissociation rate. Accordingly, the electrical energetics (electron distribution in different chemical bonds) and the mechanical energetics (vibrations of atomic nuclei, atoms and molecules) become separated. In this way, the point of reduction of a proton by an electron is reached. By chemical convention, this point is assigned, under normal conditions, the pH value of zero. Thus, the probability of finding mono-atomic hydrogen (i.e. pH = 0) reaches 1. In correspondingly small, delimited areas, the pH value is precisely steered in space and time, by way of this modulated oscillation of water molecules. The reactivity of the water and/or radicals, based as it is on the mobility of electrons and protons, subject to the abundances of carbon dioxide, organic molecules (e.g. enzymes) or metals (Cunningham *et al.*, 1987), determines its chemical and biochemical reactions. Far from being an inert solvent, water is an essential component in the structure, transport and reactivity of life processes. Phase boundaries represent the preferential reaction sites for all biochemical life processes.

On the same conservative principle, energy received by a system (e.g. a light energy pulse – a photon) must be dissipated. As a consequence of its function, the system partitions the pulse among chemical energy and heat.

Conceptual models of elementary, energy-dissipative life processes

Because of its rather unique characteristics, water probably also played an important role during the emergence of life on earth. Energetically, the processes described above, at the molecular level of water, explain satisfactorily the emergence of photosynthesis. Conceptually, life processes owe their origins from the structure of water, as outlined below.

Light – the electromagnetic interaction between the earth and the sun – leads to an excitation of terrestrial electrons. Modulation of coherent cluster oscillations is achieved by transformation of absorbed radiation into mechanical movement of the absorbing molecules. The quantity of such molecules is regulated by equilibrium between the rates of formation and decay and, therefore, is controlled by the coherency of molecule movement. Bicarbonate ions or carbon dioxide are integrated into the water structures described above, according to their mole fractions. Reduction to a carbohydrate radical ($-CH_2O$) will take place at reaction centres activated through resonance by mechanical energy, when pH = 0 (formation of a hydrogen radical, as above) is reached. Such spatially minute and temporally short-lived reaction centres, it may be assumed, are not in areas of 'free' water, but rather in the proximity of phase boundaries (e.g. sub-microscopic fissures in rocks, or light-absorbing, macromolecular modulating oscillations). The selection of the best-suited macromolecules is controlled by evolution according to the efficiency principle. This explanation of the origin of photosynthesis would also account for the fact that the oxygen so liberated originates from the water and not from the carbon dioxide.

According to the efficiency principle, the carbohydrate radicals formed should be transported directly to sites with low energy state, where their far-from-thermo-dynamic-equilibrium state would be most stable (i.e. where coherent oscillations of molecules and ions are lowest). When meeting with other radicals, carbohydrates (e.g. glucose) are formed and further polymerisation leads to the formation of, for example, starch and cellulose, further lowering their kinetic energy. Since the vibrations of macromolecules are of longer wavelengths (because of their higher molecular weight) than are those of water molecules, a discretely distributed, non-Gaussian spatial structure develops at the level of molecular associations. Likewise, their dynamics are not normally distributed, resulting in the sorting and selection of the structures and processes involved (i.e. the probability of reaction) through resonance coupling. This process is sustained until the rates of assembly and dis-assembly of these substances achieve a dynamic equilibrium. Thus, a higher overall efficiency for the structure is obtained when the openness of the subsystem is lowered and its material losses are minimised.

A further possibility for increasing the efficiency of the overall structure would be the establishment of better matter cycles at the next higher fractal level (in the present case, in the cell association). This 'jump' is the mainspring for self-organi-sation, bifurcation and for the development towards gradually more complex and larger-scale structures represented by the evolution of organisms and the devel-opment of ecosystems on earth. It must be assumed that the necessary reactions could not have arisen simultaneously, neither were they normal in distribution but, equally, they must have appeared very early in the development of living systems. This means, in turn, that a thermodynamically-based selection procedure arose in the lowest levels of molecular self-organisation. The system had turned from that of 'Boltzmann space' (random, chance distributions of processes) to that of 'Hilbert space', in which chance events were eliminated by process coupling – and therefore a substantially higher degree of coherency and patterns would have arisen.

Today, it is generally accepted that self-structuring and self-organising dissipative structures arose early in geological time and that, eventually, these led to the emergence of ever more complex structures and life itself (e.g. Lehninger, 1978; Ebeling and Feistel, 1982). Teleological and comprehensive explanations for the self-organising phenomena of nature (e.g. Resch and Gutmann, 1987) have so far been classified as philosophical and of having no scientific basis. On the other hand, the stabilisation of simple biochemical cycles (e.g. hypercycles, after Eigen and

Schuster, 1979) and smaller biochemically active particles (e.g. 'coaservates' and 'micro-spheres') could hardly be explained with the usual selection criteria that had been considered up to that time (Ebeling and Feistel, 1982). The suggestion would therefore seem to be to demand that the more traditional explanation for development – by way of random selection and trial and error – should give way to the outcome of the efficiency principle operating at a series of small to large scales. Thus, chemical and biological evolution would be governed by the same functional principle. Structures that are optimised in standing-wave patterns, or spatio-temporally closed cyclic processes, survive for longer than structures subjected to heavy losses. The latter are structures exhibiting a more random dissipation of energy with a higher proportion of irreversible processes. Accordingly, structures with a higher efficiency and higher stability with regard to these cyclic processes (efficiency optimisation), will almost certainly be selected for over longer periods of time (see also Eigen and Winkler, 1993). In this way, biochemical evolution and the development of ecosystems can be just as readily explained – as a thermodynamic necessity (Ripl and Hildmann, 1997).

Use of energetic patterns by cells and organisms

At the level of cells and organisms, the efficiency principle operates in the organisation of material flows through low-loss, coupled, cyclic structures. The cell components (structured water, organic and inorganic substances contained in water, as well as the organelles, membranes and DNA which they constitute) are arranged in, to a large extent, stationary positions. Nevertheless, they are closely involved in cyclic operational sequences. The cell membrane governs the exchange of material with the outside environment and in so doing minimises it. The materials needed in cell metabolism (water, ions, low-molecular organic substances and gases) are transported selectively at rates regulated by the cell membrane. Since all biochemical reactions and transported materials are carried in an aqueous medium enriched with organic and inorganic components, organisms could be just as well regarded as 'structure-optimised water' (Ling, 1984; see also Cope, 1972). Controlled energy dissipation keeps the processes far from chemical equilibrium, while their efficiency is optimised harmonically with regard to the dynamic surroundings. Coherent, resonant, vibration patterns and short-circuited material cycles are generated.

For single-celled organisms, the efficiency principle already applies: the more effectively the cell can turn over the material running through structured cyclic processes and the fewer losses it makes, then the more stable and thus more survivable it becomes. Thus, the intake of food needed for operating processes is minimised. In multicelled organisms, this efficiency criterion is supplemented by the synergetic division of labour between cells and cell tissue – and by the organism's internal transport systems (e.g. phloem and xylem in plants, blood and lymph circulatory systems in animals). In addition, multicellular organisms minimise a part of their irreversible losses by cycling material internally.

Once again, use of the energy-dissipative efficiency concept for understanding processes at the level of organisms proves helpful. In the distribution of an energy potential, organisms always possess several degrees of freedom for their use of this energy (i.e. to partition it into chemical reactions or allow a rise in temperature). Examples of energy use by organisms include: absorbed electromagnetic radiation (light) and the mechanical transport of reactants. Some organisms use the energetics of their own environment for themselves – and thus operate vital functions with smaller amounts of internally-produced energy (i.e. from $NADH + H^+$, $NADPH + H^+$ or ATP), rather than by the exclusive use of any one of these energy forms.

Some examples of these functions, hitherto almost ignored, are noted. They include strategic mechanisms to overcome supply deficits, such as CO_2 deficiency or a lack of light. They also include feedback processes for rectifying an unwanted surplus, such as of oxygen evolved in excess under bright light. In this manner, the possibilities for organisms to attenuate an existing energy pulse to a mean value are greatly extended; matter cycles are more readily short-circuited and closed.

The Dissipative Ecological Unit (DEU) – the interdependence of organisms

At the next higher fractal level – between organisms – close coupling between two organisms can become more effective, as is seen, for example, in symbiotic nitrogen fixation or in the use of algal exudates by bacteria. However, the more familiar interactions among organisms cover such familiar topics as competition, predation and coexistence. We wish to progress our consideration of the fractal hierarchy by following the same principles and adopting the Dissipative Ecological Unit (DEU) as the next fractal level of organismic interaction.

The idea of a minimum unit ecosystem was developed during the 'Stör' river project, carried out in northern Germany, and conceived the context of the ETR model (Ripl, 1992; Ripl, 1995a; Ripl *et al.*, 1995, Ripl and Hildmann, 2000). Here again, efficiency may be understood as the criterion for selection – favouring closure by cyclic processes, minimisation of those that are irreversible and the greatest retention of resources. Higher efficiency in a compound dissipative structure means that it can spread itself more rapidly than an adjacent structure with lower efficiency. It is more stable and it is more sustainable.

At the level of ecosystems, the DEU represents the smallest unit in which evolution and succession takes place and for which a thermodynamic efficiency can be determined. The structure comprises five functional components, capable of mutual coupling in such a way that water and material cycles can be internalised and, to a large extent, closed. The five components are as follows:

(1) *Primary producers*, having the dual function of manufacturing material and providing the energy needed for all heterotrophic structures, and of pumping the second functional component – water – as coolant, transport and reaction medium, through the process of evapotranspiration.

(2) *Water* – through evapotranspiration – provides a feedback control of production.

(3) The *detritus* buffer – the store of organic matter, the fine capillary structure of which controls the entrance of air (including oxygen) for the fourth component.

(4) The *decomposers* (consisting of bacteria and fungi). Nutrients and minerals are recovered from the detrital store of material and energy when water-saturation is sufficiently alleviated for the microbial mineralisation to proceed. Thus, the store can be used efficiently with a low level of losses. The first four components are subject to a fifth, processing component, regulating energy dissipation through localised structures incurring minimised losses.

(5) This fifth processing component is the *food chain*, the network of consumers consisting of lower and higher fauna, creating space by the management of primary producers and decomposers and, thus, also maintaining the important process of reproduction and increasing overall dissipative efficiency.

Within the DEU, there is regulation through a potentially perfect feedback, wherein the 'bottom-up' and 'top-down' controls collapse into one another, in favour of a basic 'control loop'. Its stability and durability may be assessed from the temporal efficiency of local resource use. This control loop implies that there is a management span, in which management by the food chain is regulating energy efficiency in such a way that the daily energy pulses, for example, are smoothed to give an increased efficiency of energy use. With resource cycling at a maximum, residual energy potentials can be better deployed. Expressed in ecological terms, relating to the processes of production and respiration, this means the maximisation of gross productivity but with a reduction in net productivity towards zero. Energy pulses are more completely dissipated and energy flux is maximally attenuated. In a typical DEU, attenuation of the energy flux proceeds by way of: (1) evaporation of water; (2) development of biomass; (3) development of short (and thus more closed) evaporation–condensation cycles; (4) with internal matter cycles; and (5) by the warming of the soil.

As an example, we can consider the spatio-temporal components of processes comprising a functional DEU. We may imagine an experiment with two treatments, both including water and identical quantities of algal and bacterial cells. In the first treatment, algae and bacteria are randomly distributed throughout the water. Switching a light source on and off results in fluctuations in, for example, the levels of oxygen, dissolved phosphate, pH, etc. The exchange of materials will always be made by the relatively long distance of open water separating algae and bacteria with fluctuations of the above determinands arising as a function of the light/dark phases.

In the second treatment, the same quantity of algae and bacteria are closely juxtapositioned, interconnected and interdependent as they might be in periphyton. With the lighting switched on, the oxygen produced by the algae can be respired immediately by the bacteria in a short, localised cycle. Moreover, nutrients and carbon dioxide set free by the bacteria may be taken up substantially more rapidly by the algae. For the same given conversion of material, the second treatment experiences distinctly smaller fluctuations in the determinands than would be measureable in the first treatment. For the same energy flux, the more interconnected community in the second experimental treatment has operated more efficiently and has minimised its material losses.

From this simple reconstruction of a DEU, it may be hypothesised that, through recursive feedbacks among all components over time, harmonic coherent patterns can develop, involving appropriate, mid-range abundances of functionally necessary organisms (primary producers, decomposers and consumers). Moreover, a well-structured DEU should be capable of surviving a pulse of external, disruptive forcing. Provided their frequency is low, such forcing events may be absorbed (the system is resilient). However, disruptive forcing at high frequencies or intensity may exceed the resilience of the DEU, leading directly to its degradation.

The efficiency of a DEU can be determined from the rate that matter is lost relative to the incoming energy flux. Rising matter losses are indicative of poorer attenuation of the energy flowing through the DEU. Conversely, DEUs having high efficiency are longer-lasting than DEUs of a lower efficiency: available minerals (base cations) tend to be retained more effectively than they are in lower-efficiency structures. The most effective DEUs will have all five components well-represented and functioning well.

An indication of the comparative efficiencies of various habitats (DEUs) can be shown by continuous, automatic temperature records, lower-efficiency units showing wider temperature fluctuations than those in units with higher efficiencies (Ripl *et al.*, 1995). Through the time-course of an ecological succession, DEUs are restructured in favour of a strengthening representation of components that gain the closest coupling between organisms and their environment, increasing efficiency as they do so. In a succession, as in evolution, additional species can be introduced into the maturing ecosystem, only if they are able to add to the system's efficiency and coherence and so broaden the communal interdependence. The structuring potential and speed of succession are set by the difference between the maximum, usable energy potential and the proportion actually used. Then, the spatio-temporal dynamics move the DEU from the spatially strong but temporally weak establishment phase to the spatially weaker but temporally stronger optimisation phase of later succession. To put it in another way, rapidly increasing organisms are favoured during the establishment phase of a succession because development is constrained only by energy (i.e. time) and progress depends only on the capabilities of the organism (*r*-strategy). During the optimisation phases, the constraint on further development is set primarily through spatial limitation (e.g. shortage of nutrients).

Closely interdependent organisms that contribute to the minimisation of material losses (*K*-strategists) are preferentially selected. For the optimisation and self-organisation in nature, as in human society, spatial (resource) limitation is an indispensable condition and characteristic feature.

Ecosystem efficiency of landscapes

How completely a landscape dissipates its daily input of solar energy is a measure of the efficiency of the entirety of its ecosystem(s). The extent to which the energy income to a delimited land surface (e.g. a catchment, subcatchment, or the domain of aggregated DEUs) and over a given time period (e.g. a year, or the complete lifecycle of longer-lasting primary producers such as forest trees) is dissipated is a measure of the efficiency of the whole ecosystem. However, ecosystem efficiency, too, has inherent limits. If the energy pulse of a certain frequency were fully damped within the given boundaries of a system (suppressing all spatio-temporal variability in energy flow and material supply, or matter budget, across the given surface), then matter-cycling would be perfect and loss-free, and the efficiency of the system would equal unity. The second law of thermodynamics, however, precludes this limit ever being reached (perpetual motion machines do not exist) and a certain degree of subsystem openness persists.

   The notion of measurable ecosystem efficiency can be applied only to defined partially closed spatio-temporal systems, like a DEU or larger unit, such as a sub-catchment. Where the limits cannot be so defined, the functional interrelationships (such as matter flow) will be arbitrarily separable and the system's efficiency will be unquantifiable. The larger the unit considered, the easier the system delineation becomes. Estimates of matter losses – from a single plant, from a forest stand, from a subcatchment or a larger river basin – are self-evidently scale-dependent. According to this logic, the time-scale for assessment needs to be set appropriately to accommodate the generation times of its characteristic components (that of a tree in some instances) when interpreting ecosystem processes.

11.4   Development of ecosystems – self-organisation and degradation

In this section, we seek to demonstrate, by reference to particular types of ecosystem, how the efficiency that is transmitted through the fractal sequence of biological function may be assessed. For a series of functional biospheric compartments, we contrast two extremes of organisation:

(1)   intact, enduring systems that are self-structured and self-optimised,
(2)   transient, degraded systems that are losing structure.

There are, in reality, many intermediate conditions but the 'intact' and 'transient' categories are adequate to delimit the range of ecosystem function. They also make it possible to see generalised targets for the reorganisation and restoration of ecosystems and the achievement of sustainability.

As discussed in the preceding sections, all ecosystems intercept energy that is essentially pulsed and material resources (water, resources) that are essentially also in flux. Intact ecosystems optimise the dissipation of the energy and minimise material losses. Transient or degraded systems fail to retain energy or material resources and are ultimately unsustainable. Self-organisation works towards making open systems more intact, by damping fluctuations in energy income, minimising temperature fluctuations and water loss and by increasing the cycling of resources.

Under each of the following section headings, we contrast the functions of intact and transient compartments of the biosphere by reference to the ways in which water transformations are regulated. Thus, optimising functions characterise the self-organisation of intact systems; degrading functions distinguish transient systems. Differences are emphasised in the boxed tabulations. The text provides some quantitative amplification where appropriate.

Atmosphere

The atmosphere is the capacious medium for the long-range cycling of water between the oceanic global material sink and the continents. The hydraulic discharge from the land masses averages $\sim 250\,\mathrm{mm\ a^{-1}}$ in total, against a mean precipitation over the continents of $\sim 670\,\mathrm{mm\ a^{-1}}$ (Zachmann, 1976). The difference ($420\,\mathrm{mm\ a^{-1}}$) corresponds to the conventional estimate of the mean evaporation from the land to the air. However, this computation obscures the exchanges due to high-frequency, local-scale cycling by evaporation and condensation (e.g. dew and hoar-frost formation on cooled surfaces) which make up a substantial part of the water cycle. Exchanges of water exceeding the estimated precipitation but lost invisibly in a larger evaporation are indicative of a short-circuited water circulation in an intact atmosphere.

The consequences for optimisation of this compartment at the continental scale are noted in Box 11.1. The near-ground cooling by evaporation enhances the presence of vapour near the ground surface where it provides a 'thermal-protection filter' against accelerated solar heating.

Atmospheric degradation may result from destruction of the permanent vegetation covering and impairment of the water relations. Reduction in evaporation over large areas allows wider diel and seasonal variations in air temperatures and larger differences between the drier, overheated surfaces on the one hand, and, on the other hand, the better-cooled, damper areas (high spatial temperature gradients). Near-ground thermostatic buffering is impaired by the loss of wet surfaces, while intensified periods of dryness lead to apparent increases in warming over large areas. Dusts, possibly toxin-laden, carbon dioxide, nitrogen oxides, methane, ozone

---

Box 11.1   Self-organisation and degradation in the atmospheric water compartment.

Optimising functions
- Frequency of evaporation-condensation in small-scale cycles is increased
- Capacity of local diel exchanges is raised
- Large-scale, non-local water cycling is reduced
- Dust minimised

Degrading functions
- Temperature damping weakened.
- Thermostatic buffering of near-ground water vapour is lost.
- Heating, drying, air movements, dust generation in large-scale cycles are intensified.

---

formed near the surface and chlorofluorocarbons (CFCs) are more readily transported into the upper atmosphere. Their interference with global heat exchanges is already well-known. Meanwhile, the reduction in evaporative cooling and the absorption of heat further degrades the atmosphere. The increasing reduction of the surface area that retains its dissipative evaporation–condensation cycles seems to be a contributory factor in modern climate change.

Vegetation

A well-developed, tree-dominated vegetation intercepts a large proportion of the incoming solar energy which is invested in the further production of biomass. An intact vegetation is in close feedback with the local water regime of its location (Box 11.2). During the course of succession, the vegetation develops a humus cover that improves water retention and, through evapotranspiration, further damps the daily energy pulse. The cooler surface has a feedback in favouring more condensation of water. Localised cycling turns the well-vegetated landscape into a dynamic water reservoir. Added to the storage capacity of the soil's humus layer, runoff to rivers is progressively moderated. Water retention also contributes to lower redox levels in the litter layer, whence spontaneous mineralisation losses are minimised. The slow release of ions is more effectively cycled back to the rooting zone. In this way, decomposition and production become closely coupled (in parallel), benefiting the retention of water, matter and limiting nutrients.

Vegetation is degraded principally through reduction or removal of natural, diverse woodland cover, its replacement with agriculture or commercial monocultures and through accelerated drainage of wetlands. The risk of desiccation is increased, followed by humus oxidation and the loss of temperature damping. On sunny days, unregulated heating may raise the temperatures of vegetation-free surfaces (arable fields, city streets) by more than 10°C above the comparable forest

> **Box 11.2**   Self-organisation and degradation in the vegetation water compartment.
>
> *Optimising functions*
> - Plant succession maximises solar energy harvest.
> - Stores of water and humus increased.
> - Temperature fluctuations are minimised.
> - Dissolution and irreversible losses of materials are minimised.
> - Lowered redox levels resist local loss of minerals.
> - Retention and internal cycling of resources are increased.
>
> *Degrading functions*
> - Water exchanges are less controlled: inundation and desiccation are more common.
> - Humus is increasingly oxidised.
> - Mineral salts and nutrients leaked to drainage or removed irreversibly in agricultural crop.

(Ripl *et al.*, 1995; Hildmann, 1999; Ripl and Hildmann, 2000). The control of short-circuited circulation exerted by vegetation is lost.

Soil

Unhindered development of vegetation is accompanied by the accumulation of a humus layer in the upper horizon of the soil. Its presence cuts evaporative water losses from below, so helping to maintain the water-retentive capacity and moderate soil temperature (Box 11.3). In heavily saturated soils, the nitrates and sulphates formed in soil water near the surface are rapidly diminished by denitrification and desulphurisation. Liberated hydroxyl ions precipitate base cations that are retained as hydroxides. The resistance to rainwater that humus layers provide encourages lateral water movement (runoff), also minimising the leaching of nutrients beyond the root zone. With the arrangement of the processes of matter production and decomposition in parallel, long-range translocation and misalignment of material with the water flow are avoided.

Ascribing such importance to soil water level, we prefer a more robust distinction than is usually made by hydrologists, between the dynamic, near-surface soil water and the deeper store below the water table and where dynamic exchanges are generally weak. The true groundwater of intact soils supports a small, relatively even discharge with a low mineral and nutrient content.

At the soil surface, with optimal cooling of the surface, the gradients of important parameters (e.g. moisture content, temperature, substrate) are reduced at ecotone boundaries. At major habitat boundaries with sharp gradients in material turnover (e.g. the land–water ecotone), an optimisation and stronger interconnectedness of organisms develops through the process of succession and leads to a highly diverse and efficient community.

Box 11.3   Self-organisation and degradation in the soil-water compartment.

Optimising functions

- Moisture content and water content are moderated by accumulating humus.
- Mineralisation, especially denitrification and desulphurisation, is retarded.
- Root-zone translocation and resource misalignment is resisted.
- Runoff discharges are moderated.
- Ecotone gradients (with atmosphere, vegetation and groundwater) are flatter; more interconnectedness of organisms is facilitated.

Degrading functions

- Drainage of agricultural, forested and wetland areas diminishes temperature damping.
- Increase in unsaturated zone and mineralisation with associated acid formation impairs feedback inhibiting material flow.
- Groundwater abstraction increases desiccation and permeability of upper soils.

With the intensification and industrialisation of agriculture, many soils have suffered degradation. Drainage and continuous cultivation usually result in the increasing mineralisation of soil humus and in the loss of water-retention capacity. The dynamics of near-surface water are accelerated, while deeper groundwater is left unprotected. Gradients are sharpened and some microhabitats are diminished. On the broader scale, water relations are greatly modified.

The symptoms, summarised in Box 11.3, are those of much accelerated water exchanges. Episodes of drying and water logging of the humus-depleted soil also accelerate material losses, respectively through mineralisation and leaching. Structure and, eventually, fertility is lost.

Before the industrial era, agricultural production was largely confined to drained, nutrient-rich, organic soils (e.g. of calcareous fens). Lowering of the level of water saturation would lead to aeration and oxidation of organic matter and to wider variation in soil moisture and temperature. With improved drainage, the formerly closed water and material cycles would be opened. With the oxidation of reduced nitrogen and sulphur compounds, strong acids (nitric and sulphuric acid) would be more prevalent. In an analogous manner, the higher partial pressure of CO_2, resulting from the decomposition of organic substances would contribute to acidification with the formation of carbonic acid.

Since the industrial era, these leaching processes have accelerated with the raising of net productivity, itself made possible by the increased use of additional (mostly fossil and nuclear) energy. Today, the main cause of degradation is, ironically, 'land improvement' for agriculture and forestry (soil drainage, deep ploughing), multiplied by expansion in land area affected. Soils cease to act as an intact interface between bedrock and vegetation. Rather, they are reduced to an industrial substratum for the maximisation of agricultural production. The plants (in particular, trees used in commercial forestry) have had to adjust to the increased losses of base cations (Ca, Mg, K) and phosphate by producing acids in their root zones, which, in turn, leads to greater acidification of drainage water. Apart from the incidence of highly acidic atmospheric depositions, this process has provided a significant contribution to the acidification of soils and water.

Material losses have greatly accelerated as a consequence of this industrialisation of the landscape. Dissolved loads to river drainage have increased by factors of between 5 and 100 (Ripl, 1995a). In terms of calcium, magnesium and potassium alone, German streams transport cations from the land at rates of between 185 and 367 kg ha^{-1}a^{-1} (Ripl *et al.*, 1996; Hildmann, 1999).

Running waters

An intact catchment area mediates time-equalised output discharges, in which nutrient and mineral ions are simultaneously diluted. With unidirectional drainage, material losses tend to be greater from upper catchments than from the lowlands: the naturally structured catchment imposes an increasing downstream trend to capture and retain matter and minerals transported in surface or soil-water flows.

Through their structure (morphology, colonisation by organisms) and discharges, rivers mirror the water and matter balances of their catchment areas. From an intact catchment, generating an optimally equalised discharge throughout the year (low amplitude between high and low water) and shedding only minor leachate loads, the structure of the drainage network should also be expected to be as stable, diverse and self-organising as possible. In fact, natural river flows conspicuously fashion their channels and adjacent riparian features which, together with the persistent sorting of bedload materials throughout the stream, contribute to a maximisation of flow-resistance and hydraulic retention. Along these self-modified channels, flowing water is subjected to numerous small-scale accelerations and delays. Examples are provided by the gradients of velocity over single pebbles through to the acceleration and delay zones associated with meandering. The transport capacity of flowing water decreases in favour of small-scale flow resistance. The small-scale flow velocity gradients favour not only a high variety in sorted material, and thus of ecological niches, but at the same time the development of a diversified community. A highly-structured and mainly stable running water is the result – i.e. a long-lasting functioning stream or river.

Sustainable rivers carry only small amounts of entrained material. Depending on their position within the catchment area and their geomorphology, stream banks tend to be rather gently sloped, so that small increases in discharge are sufficient to inundate the riparian zone. In flat and broad streams, where water is disrupted in areas of high flow resistance (riffles), there is enhanced gas exchange with the atmosphere, bringing stabilisation of O_2 and CO_2 levels.

We deduce that rivers also go through conspicuous self-optimisation processes, favouring the establishment of stable, part-closed biological systems, tending to retain materials rather than void them in the flow (Box 11.4).

Box 11.4   Self-organisation and degradation in running waters.

Optimising functions
- Intact catchments damp fluctuations in water flow.
- Catchment structuring captures and retains matter against unidirectional drainage.
- Channel formation and riparian development resist flow and favour retention of water and materials.
- Bankside ecotones are flattened, habitat breadth is increased.

Degrading functions
- Piecemeal catchment management increases amplitude of flow, increased soil erosion and loss of fertility.
- Canalisation and revetting to 'improve' drainage impoverishes fluvial habitats and diminishes riparian function.
- Flood and drought incidents become more severe.

Degradation of flowing waters features opposite trends. The present-day, near arbitrary arrangement of land usage, coupled with the intensive interference to the natural passage of water (drainage, abstraction, irrigation, centralised sewage treatment and waste-water disposal, channel canalisation) have mostly amplified fluctuations in runoff and accelerated the leaching of matter from the land. Moreover, agricultural tillage is often carried out to the stream edge, encouraging erosional soil loss, which is then countered by revetting or other reinforcement. River beds are narrowed and deepened. Natural riparian storages are removed. It is of little wonder that floods are more common or more severe, that plant life and fluvial microhabitats have been lost and that rivers carry more particulate matter, more leachate and more nutrients than heretofore.

Standing waters

Lakes and other impoundments, particularly standing waters with shorter residence times, are subordinate to the intactness and degradation of the water and matter balance within their catchments. During development of the landscape (such as in the post-glacial period) terrestrial matter losses are reduced and the trophic state of the receiving lakes decreases (Digerfeldt, 1972). Overloading the metabolic capacity of the lake with the losses of matter from the catchment will lead to degradation to a lower functional efficiency. By contrast, intact lakes have the metabolic capacity to be able to process their material and nutrient loads in an efficient manner. There are intergrades between the intact and the transient lake type: those whose condition would be described, on the conventional trophic scale, as mesotrophic, approximate to the point of transition between the two types, although any perception of metabolic functionality based only on nutrients and production hardly takes full account of the internal dynamics of the lake system.

The concept of intactness and transience adopted here applies to all kinds of lakes (from arctic to tropical, high to low altitude, very small to very large). Behavioural differences are often greatest between deep and shallow lakes. (Limnologically speaking, shallow lakes are defined as having a maximum depth which is less than 4% of the diameter of a circle equal in area to that of the lake (Wetzel, 1983).) Almost all lakes are plainly open systems – outflowing volumes are balanced against inflowing hydraulic discharge but some or all of the particulate income is retained through sedimentation. They can develop a long-lasting functionality if the catchment loads are relatively modest and if permanent losses of organic matter to outflow and to the sediments are minimised.

In intact lakes, the catchment load is collected and retained largely in the transition zone between land and water, within the subdivision of the system (mesohabitat, or biocoenosis) contributed by littoral vegetation with a well-established network of highly-interconnected organisms, including the periphyton (biofilm) and a well-developed, near self-sufficient food web. Subject to physical constraints, the continuous supply of colonisable material facilitates the expansion of the littoral vegetation, with a positive feedback in terms of matter retention (Björk, 1988; Brinck *et al.*, 1988). A well-developed littoral structure also buffers waves, so that dead and decomposing material accumulates until it is successively amenable to invasion by swamp and marsh plants and, eventually, by terrestrial vegetation. In this case, the lake becomes overgrown starting from the banks, pointing to a littoral type of sedimentation (Lundqvist, 1927).

The open-water pelagic zone of the intact lake functions on very dilute nutrient base, in part because much of the small phosphorus load is rendered unavailable by a stoichiometric excess of P-precipitating materials (including of iron, aluminium and calcium). The impoverished nutrient capacity can support only a very small producer biomass (phytoplankton) but this nevertheless sequesters a significant fraction of the energy flux through the rapid microbial cycling of the carbon fixed. The production to respiration (P/R) quotient in the intact pelagial is close to 1. The sedimentary output from the pelagic comprises little more than refractory organic substances and biominerals (silica, carbonate), accumulating at less than $1 \, \mathrm{mm} \, \mathrm{a}^{-1}$. Conversely, the poverty of the organic carbon supply offers a limited energy source and restricts the habitat suitability of the deep benthos to consumers of detritus and microbial decomposers. Redox levels remain positive. Ageing and silting-up of the lake are slow and longevity (sustainability) is maximised.

Development of an intact lake is constrained by the seasonal pattern of hydraulic and nutrient loading. The theoretical retention time of a lake has a fundamental influence on its overall functioning. In a lake with minimal external loading, closely-connected coupling among the organisms is strongly favoured and within which organisms are increasingly able to regulate their own surroundings (nutrient content, organic substances, final metabolites, light availability, etc). Thus, intact lakes supplied by intact catchments are durable, sustainable structures and they are reckoned to have high value for species diversity (Ripl, 1995b).

Box 11.5 Self-organisation and degradation in standing waters.

Optimising functions
- Although subordinate to their catchment areas, lakes act as capacitators of hydrologic and material inflows, damping reactivity to fluctuating inputs.
- Mesohabitats and biocoenoses are stabilised with internal recycling of resources.
- Succession in littoral increases material retention and flattens ecotones between water and land surfaces.
- In open (pelagic) waters, sparse resources become segregated and stored in the sediments; phosphorus is precipitated with iron or calcium present in stoichiometric excess.
- Modest organic recruitment to sediments offers limited microbial energy sources.

Degrading functions
- Accelerated catchment losses stress the capacity of internal lacustrine processes.
- Material overloads shortcut internal recycles, deregulating pelagic production and increasing the incidence of phytoplankton blooms.
- Increased turbidity diminishes submerged littoral growth.
- Phosphorus-binding capacity is stoichiometrically exceeded, providing a positive productive feedback.
- Sedimentary flux of organic material is increased; mineralisation depletes oxidative capacity and low redox of sediments and deep waters constrains habitat availability and function.

The symptoms of transience in degraded standing waters (summarised in Box 11.5) provide stark contrasts. Degradation may have several drivers but the major generic cause is catchment degradation. Following the deforestation that proceeded from neolithic times (approximately 5000 BP) and, especially, since the start of industrialisation of agriculture the exploitation of non-renewable energy sources (in just the last 150 years), material cycles in the catchment areas of lakes have been increasingly opened. As the first interface is at the surface, the reactivity of the lake is much influenced by the ratio of catchment area to lake surface (Ohle, 1971). Some early responses may be restricted to certain parts of the lake – and be exploited by newly-introduced plants and animals (e.g. Canadian waterweed, *Elodea canadensis*; grass carp, *Ctenopharyngodon idella*) – but symptoms soon proliferate through the communities of the lake. However, it is not the organisms that are necessarily problematic so much as the modified environment provided by enhanced deliveries of nutrients, organic substances and metabolites. Mostly, the problems arise from elevated metabolic capacity and biological overproduction in the water, spilling over in increased sedimentation and ageing. This process is known as eutrophication.

Excess nutrient availability obviates the close, interconnected resource regulation and releases the environmental constraints on fast-growing, opportunist organisms (*r*-strategists). For instance, if larger biomasses of phytoplankton and periphyton can be supported, then the light and, in turn, the space available to macrophytic vegetation is compressed and the diversity of the littoral components is diminished and their function is impaired. The stoichiometric capacity of P-binding minerals is saturated and phosphorus ceases to provide a key limitation on phytoplankton production.

While degradation in shallow lakes (i.e. dominated by littoral) is presented first as a 'bottom-up' effect of enhanced nutrient loading, it is also possible to produce analogous effects from 'top-down' adjustments to the multiple feedback provided by fish-stock levels. Increased stocking of white fish (carp, roach, etc.) or the increased harvesting of predatory fish may alter the structure of the food web, reducing the zooplankton and increasing the phytoplankton (Hrbácek, 1994).

Meanwhile, production (*P*) in the pelagic zone, increases relative to respiration (*R*) and sedimentary output of organic carbon (*P/R* > 1) is enhanced. The oxidative capacity of the deep-water and the sediments is increasingly challenged, and organic sediments accumulate more rapidly than $1\,mm\ a^{-1}$. However, it remains difficult for benthic consumers of detritus and microbial decomposers to exploit the carbon richness because of a lack of oxidant. Specialist anaerobes, such as sulphate-reducing bacteria and methanogens contribute to the degradation of more easily degradable organic substances. The hydrogen sulphide thus formed leads to 'rapid lake-ageing' (Ohle, 1953, 1954): trivalent iron (Fe^{3+}) is reduced (to Fe^{2+}), liberating small amounts of phosphorus bound to the iron into the water by the mechanism described by Mortimer (1941/1942). However, for the release of large quantities of phosphate from the sediment, divalent iron must be scavenged by free sulphide and precipitated as FeS. Thus, P-binding capacity is effectively lost (Ripl, 1978; Ripl and Lindmark, 1979). In this way, the internal metabolism of lakes is matched to processes at the sediment–water boundary layer: degradation of sediment metabolism and the subsequent release of phosphate exerts a positive feedback on production in the pelagic. With the sedimentation of excessive quantities of organic substances, the automatic control loop is opened and substituted by a transient metabolism.

The process of eutrophication is explained conventionally in terms of the increased inputs of phosphorus into surface waters. The basis of this explanatory model was developed by Vollenweider (e.g. 1971), from the statistical analysis of data from a series of large lakes. Originally, most of the additional nutrient was supposed to emanate from point sources, in particular, from sewage purification plants. As a consequence, many sewage purification plants in industrialised countries have been equipped with phosphorus-removal plants, which lowered the phosphorus content in waste-water discharges to far below $100\,\mu g\ P\ dm^{-3}$. However, only in few cases has a return to oligotrophic conditions been achieved. Diffuse, non-point sources and the release of phosphorus from lake sediments seemingly continue to supply sufficient phosphorus to maintain the prevailing eutrophic conditions. The role of diffuse sources of P from inappropriate, maladjusted land management has increased and has compensated, to a large degree, for the measures undertaken by water management to control P.

Whether the nutrients originate from diffuse or point sources is of secondary importance to the view that, in the end, both are symptomatic of landscape degradation consequential upon several interrelated pressures – groundwater abstraction, food production and waste-water collection. 'Point sources' are merely the foci of an altogether degraded, inefficient water and material balance (see Ripl, 1992).

Nevertheless, eutrophy represents only a temporary phase in the development of a lake. A similar condition would have been encountered during the brief period following the last glaciation when a poorly-vegetated landscape suffered with high matter losses. Eutrophic lakes persist only until either the landscape is fully leached or the lake becomes finally terrestrialised. In contrast to a stable system, degraded lakes, together with their catchments, are characterised by a tendency towards accelerated change. This can lead not only to their rapidly silting-up, but also to their acidification, if base cations are also depleted from their catchments.

Nutrients and minerals at the surface of a catchment represent limited resources. If they are leached because of human interference, then both the vegetation and associated water circulation degrades. The processes of temperature-equalisation from functioning evaporation and condensation cycles, a substantial mechanism for normal temperature distribution on earth, is thereby lost. As a function of the continentality of climate, the continued existence of lakes as surface waters is then put into doubt. Under these aspects, furthermore, the eutrophication of lakes can be brought into connection with the desertification process (collapse of the water and matter balance within a landscape), through which, also, many larger lakes in continental areas are already subject to regression (e.g. Aral Sea).

11.5   Summary and conclusions

An integrative understanding of the coherency in the process we call 'nature' is indispensable for sustainable management. In the present text, we have tried to construct, on the basis of the dissipative water cycle, an integral deductive model ('*Leitbild*') for ecosystems. An integrative approach – the Energy–Transport–Reaction (ETR) model – is used to describe the essential processes and functions in ecosystems. Therein, the daily pulse of solar energy is dampened through the dissipative medium of earth–water interaction. In constrained systems, unhindered in their development, external loss processes (away from system borders) are replaced by internal cycles. Losses from systems may be manifest in many ways – such as, for example, leaching from soils, sedimentation in a lake, or the dissolved load in the discharge from a catchment area. Based on the relationship between cyclic and loss processes, the system's ecological efficiency may be assessed. It approaches unity where material cycling is closed (high chemical efficiency) and the damping of the sun's energy pulse is equalised (high thermal efficiency).

Short-circuited cycling is the result of self-optimisation in the ecosystem – by way of a well-interconnected functioning community of organisms (biocoenosis). Self-optimisation can be achieved only with a system that is limited (space or resource-limited and time or energy-limited). The criterion for optimal integration of organisms is the efficiency: all organisms that contribute to the better governing of closely coupled cycles help to stabilise the material existence of the ecosystem and thus its sustainability (K-strategy). An ecosystem optimised in such a way can be regarded as an intact system.

If cycles are disturbed, if the losses of a system increase, and if the (daily) pulsed solar energy is converted more into heat and accelerated random particle movement, then the system degrades: its intactness is lost. Examples given are the loss of minerals and nutrients from soils and the subsequent material overloading of surface water, where it stimulates eutrophication and the promotion of r-strategists. Stability and sustainability are reduced (see Schrödinger, 1967).

We have shown that the efficiency principle is applicable at all the various fractal (spatio-temporal) levels of ecosystems. Harmonious and stochastic processes arise in water, which, even at the molecular level, is structured by ions, macromolecules and clusters. The physical and chemical characteristics of water may be understood rather as oscillation or vibration patterns, which are characterised by coherent (i.e. in-phase) resonances, with acceleration or delay of ions and molecules at phase boundaries. During biochemical evolution, these oscillation patterns affected the development of systems through a process of sorting – as stable long-lasting structures would be selected, according to the efficiency principle. These oscillation patterns are the basis of life processes in cells and organisms.

Continuing to higher fractal levels, we have shown that the integration of closely connected organisms in an ecosystem also accord to the efficiency principle. A simple ecosystem comprises primary producers, the storage/buffer of detritus, consumers, decomposers and water as the transport and reaction medium. The smallest representative and fully functional part of an ecosystem accommodating all five components is called a Dissipative Ecological Unit (DEU). The association of these five functions is necessary for the ability to close a cycle of matter cycles – or, on disturbance, to yield losses. Thus, the DEU, like the ecosystem and landscape of which it is part, is according to the efficiency principle.

The working of the efficiency principle at these higher levels is demonstrated by reference to the intact and degraded (transient) ecosystem 'compartments' of air, vegetation, soil, running waters and lakes. Short-circuited atmospheric-water cycles are characteristic of intact landscape areas carrying vegetation. Since the vegetation develops a powerful humic layer, the water-retention capacity remains high in these areas. Even at times of high temperature, evaporation, and thus the cooling of the land surface, can be maintained. Since larger temperature gradients in the landscape are then much reduced or missing, large spatial climatic events are minimised. Discharges from these intact catchment areas are relatively even and, given good retention capabilities of catchments, are poor in nutrients and mineral ions. Lakes are always subordinate to the functionality of their catchment areas but they will usually develop intact systems under oligotrophic or, at most, mesotrophic conditions. In such lakes, littoral structures, characteristically developing controlled, closed cycles (macrophytes, biofilm), tend to dominate. Production and decompostion are in balance and material losses to the sediments are small. The rate of ageing of an intact lake (i.e. its rate of silting-up) is slowed as long as the functioning of the system continues to self-optimise.

The present impact of humans on landscapes is to degrade their ecosystems in almost every respect. With the removal of vegetation, the cooling function of landscape areas is decreased dramatically and material losses in runoff discharges from catchments are very much increased. Receiving waters become more heavily loaded, so that littoral vegetation and biofilm structures decrease and, in many cases, plankton becomes the dominant controlling component. Through increased sedimentation, the rate of ageing of a lake is accelerated. In addition, lakes may be subject to internal fertilisation from sediments loaded with degradable organic substances, through desulphurisation and ligand exchange of sulphide for phosphate.

Humans do not stand outside the natural system by any means. We still depend on the oxygen in the atmosphere, on a tolerable climate with a generally constant and even precipitation, and on the fertility of soils. Therefore, all management measures and efforts towards reorganisation and restoration of the landscape are subject to exactly the same selection criteria as operate in natural landscapes. With a higher efficiency, the sustainability of the system rises; with high losses, it falls. Without a functional understanding and the adjustment of human activities to conform to the functional conditions of natural systems, extensive breakdowns arising from misuse can hardly be prevented. Against this eventuality, the openness within defined systems should be reduced as far as possible, through intelligent management, where the ageing of the landscape as a whole is slowed down and natural stability is increased. New structures, based on geographical catchments and with humankind integrated as system helmsmen, are necessary. Guided always by the efficiency principle, humankind should be deploying its scientific wherewithal to the development of new, sustainable structures.

Acknowledgements

The basis for this contribution, the ETR model, was developed at the Department of Limnology at the Technical University (Technische Universität) Berlin, under the direction of W. Ripl in cooperation with M. Feibicke, C. Hildmann, K.-D. Wolter and further colleagues. For continuing suggestions and the critical examination of the manuscript we thank I. Otto. Also thanked are S. Ridgill and C.S. Reynolds for improvements to the English language of the text.

References

Björk, S. (1988) Redevelopment of lake ecosystems – a case study approach. *Ambio* **17** (2), 90–98.

Brinck, P., Nilsson, L.M. & Svedin, U. (1988) Ecosystem redevelopment. *Ambio* **17** (2), 84–9.

Cope, F.W. (1972) Structured water and complexed Na^+ and K^+ in biological systems. In: *Water Structure at the Water–polymer Interface* (ed. H.H.G. Jellinek), pp. 14–18. Plenum Press, New York.

Cunningham, K.M., Goldberg, M.C. & Weiner, E.R. (1987) An examination of iron oxy-hydroxide photochemistry as a possible source of hydroxyl radical in natural waters. In: *Chemical Quality of Water and the Hydrologic Cycle* (eds R.C. Averett & D.M. McKnight), pp. 359–63. Lewis, Chelsea.

Digerfeldt, G. (1972) The post-glacial development of Lake Trummen. Regional vegetation history, water level changes and paleolimnology. *Folia Limnologica Scandinavia* **16**.

Ebeling, W. & Feistel, R. (1982) *Physik der Selbstorganisation und Evolution.* Akademie-Verlag, Berlin.

Eigen, M. & Schuster, P. (1979) *The Hypercycle. A Principle of Natural Self-organization.* Springer, Berlin.

Eigen, M. & Winkler, R. (1993) *Laws of the Game. How the Principles of Nature Govern Chance.* Princeton University, Princeton.

Gutmann, V. (1978) *The Donor–acceptor Approach to Molecular Interactions.* Plenum Press, New York.

Hildmann, C. (1999) *Temperaturen in Zönosen als Indikatoren zur Prozeßanalyse und zur Bestimmung des Wirkungsgrades. Energiedissipation und beschleunigte Alterung der Landschaft.* Dissertation Technische Universität Berlin, Fachbereich Umwelt und Gesellschaft, D 83. Mensch & Buch, Berlin.

Hrbácek, J. (1994) Food web relations. In: *Restoration of Lake Ecosystems – a Holistic Approach.* (ed. M. Eiseltová), pp. 44–58, IWRB Publication 32. International Waterfowl and Wetlands Research Bureau, Slimbridge.

Lehninger, A.L. (1978) *Biochemistry. The Molecular Basis of Cell Structure and Function.* Worth Publications, New York.

Ling, G.N. (1984) *In Search of the Physical Basis of Life.* Plenum Press, New York.

Lundqvist, G. (1927) Bodenablagerungen und Entwicklunstypen der Seen. *Die Binnengewässer* **2**, 1–124.

Mortimer, C.H. (1941/1942) The exchange of dissolved substances between mud and water in lakes. *Journal of Ecology* **29**, 280–329 and **30**, 147–201.

Odum, E.P. (1969) The strategy of ecosystem development. *Science* **164**, 262–70.

Ohle, W. (1953) Der Vorgang rasanter Seenalterung in Holstein. *Naturwissenschaften* **40**, 153–62.

Ohle, W. (1954) Sulfat als 'Katalysator' des limnischen Stoffkreislaufes. *Vom Wasser* **21**, 13–32.

Ohle, W. (1971) Gewässer und Umgebung als ökologische Einheit in ihrer Bedeutung für die Gewässereutrophierung. *Wasser – Abwasser (Aachen)* **1971** (2), 437–56.

Prigogine, I. (1980) *From Being to Becoming. Time and Complexity in the Physical Sciences.* Freemann, New York.

Resch, G. & Gutmann, V. (1987) *Wissenschaftliche Grundlagen der Homoeopathie*, 2nd edn. Österreichisches. Verlag, Berg am Starnberger See.

Ripl, W. (1978) *Oxidation of Lake Sediments with Nitrate – a Restoration Method for Former Recipients*. Institute of Limnology, University of Lund.

Ripl, W. (1992) Management of water cycle: An approach to urban ecology. *Water Pollution Research Journal of Canada* **27** (2), 221–37.

Ripl, W. (1995a) Management of water cycle and energy flow for ecosystem control: the energy–transport–reaction (ETR) model. *Ecological Modelling* **78**, 61–76.

Ripl, W. (1995b) Der landschaftliche Wirkungsgrad als Maß für die Nachhaltigkeit. In: *Umwelt und Fernerkundung: Was leisten integrierte Geo-Daten für die Entwicklung und Umsetzung von Umweltstrategien* (eds R. Backhaus & A. Grunwald), pp. 40–52. Wichmann, Heidelberg.

Ripl, W. & Hildmann, C. (1997) Ökosysteme als thermodynamische Notwendigkeit. In: *Handbuch der Umweltwissenschaften* (eds O. Fränzle, F. Müller & W. Schröder), Ch. V-3.1.1. Loseblattsammlung. Ecomed, Landsberg am Lech.

Ripl, W. & Hildmann, C. (2000) Dissolved load transported by rivers as an indicator of landscape sustainability. *Ecological Engineering* **14**, 373–87.

Ripl, W., Hildmann, C., Janssen, T., Gerlach, I., Heller, S. & Ridgill, S. (1995) Sustainable redevelopment of a river and its catchment – the Stör River Project. In: *Restoration of Stream Ecosystems. An Integrated Catchment Approach*, pp. 76–112. (eds M. Eiseltová & J. Biggs), IWRB Publication 37. International Waterfowl and Wetlands Research Bureau, Slimbridge.

Ripl, W., Janssen, T., Hildmann, C. & Otto, I. (eds) (1996) *Entwicklung eines Land-Gewässer Bewirtschaftungskonzeptes zur Senkung von Stoffverlusten an Gewässer (Stör-Projekt I und II)* (in conjunction with F. Trillitzsch, R. Backhaus, H.-P. Blume & P. Widmoser (eds)). Im Auftrag des Bundesministeriums für Bildung, Wissenschaft, Forschung und Technologie (BMBF) und des Landesamtes für Wasserhaushalt und Küsten Schleswig-Holstein, Endbericht.

Ripl, W. & Lindmark, G. (1979) The impact of algae and nutrient composition on sediment exchange dynamics. *Archives of Hydrobiology* **86** (1), 45–65.

Schrödinger, E. (1967). *What is Life?* Cambridge University Press, Cambridge.

Vollenweider, R.A. (1971) *Scientific Fundamentals of the Eutrophication of Lakes and Flowing Waters, with Particular Reference to Nitrogen and Phosphorus as Factors in Eutrophication*. Organisation for Economic Cooperation and Development (OECD), Paris.

Wetzel, R.G. (1983) *Limnology*. Saunders, Philadelphia.

Zachmann, G. (1976) *Lexikothek*. Spektrum der Naturwissenschaften. Bertelsmann, Gütersloh.

Chapter 12
On the Evolution of the Carbon Cycle

Paul G. Falkowski

12.1 Introduction

The thin film of liquid water covering most of its surface has been a continuous, essential, support system for life on earth. Whereas the oceans have existed on this planet since the late heavy bombardment period, approximately 4.2 Ga[1] (Mojzsis *et al.*, 2001), they have been constantly changing. Changes in physical circulation, chemistry and biological activities are recorded in the fossil and lithospheric records, from which we are left to infer the causes and consequences. Some of the inferences are truly remarkable: completely ice-covered oceans and deep oceans with temperatures of 12°C or more; periods of relatively rapid sedimentation and sequestration of organic carbon and numerous, basin-wide anoxic events and outgassing of methane; periods of high carbonate and silicate deposition and periods of carbonate erosion; periods of rapid biological radiation and periods of rapid, massive extinctions. Ultimately, all of these processes are driven by insolation and tectonic processes, yet their interactions with the ocean carbon cycle are profound.

Historically, the interactions have been chronicled and interpreted through the eyes of geochemists and paleoceanographers; however, over the past decade, there has been increasing awareness and discussion amongst biological oceanographers and evolutionary biologists that understanding biological processes is essential to understanding the history of earth's oceans. Here I focus on the evolution of the carbon cycle on earth, and the essential role of the oceans in that process.

12.2 The origin of the elements

The relative abundance of the elements in Earth's crust reveals a saw-toothed distribution of much more abundant light, even numbered elements, relative to rarer,

---

[1] Ga: giga annum (billions of years before present), Ma: mega annum (millions of years), Ka (kilo annum) thousands of years.

odd-numbered, heavier elements (Fig. 12.1). While the two lightest elements, H and He, were formed approximately 16 billion years ago in the 'Big Bang', all the heavier elements result from fusion of ^{2}He nuclei or fusion/spallation (proton or neutron loss) reactions in stars (Williams and Frausto da Silva, 1996). The fusion reactions involving ^{2}He with itself, H, O, C or S tend to form even-numbered atomic nuclei, while spallation and proton capture leads to odd-numbered nuclei (Broecker, 1998). Additionally, as the nuclei of elements with paired protons are slightly more stable than for those with an odd number, there is generally a larger relative abundance of even-numbered elements. Hence, of the six light elements that make organic material, H, C, N, O, S and P, C and O are approximately equal in abundance, whilst N and P are relatively rare in comparison with adjacent elements.

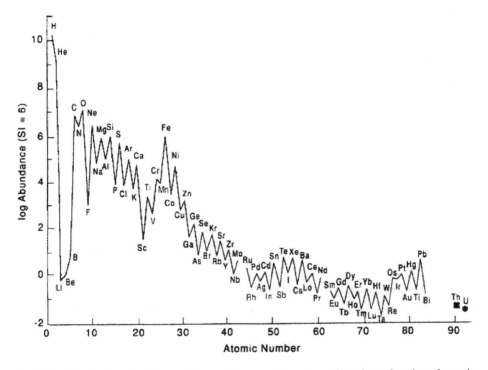

Fig. 12.1 The relative abundance of elements in our solar system plotted as a function of atomic number. Note that abundances are plotted on a semi-logarithmic scale and scaled to silicon = 1×10^6 (reproduced with permission from *Biogeochemistry, An Analysis of Global Change* by W.H. Schlessinger, (1997) Academic Press, New York).

In the origin of our solar system approximately 4.6 billion years ago, elemental composition and planetary accretion was strongly influenced by the gravitational forces of the sun. Planetary bodies closer to the sun contain relatively higher proportions of heavy elements than those further away. The four innermost planets are approximately three times denser than the outer planets and have solid rock surfaces that contain a relatively high proportion of metals, especially iron and aluminium. In the accretion process, a further gravitational distillation occurred within the planets themselves. On earth, the abundant heavier elements, such as nickel and iron, tended to migrate towards the centre of the internal gravitational field, while lighter elements tended to float above the metal core and, upon cooling, accumulated as a solid surface and crust. The lightest elements formed a gaseous phase. Almost all of the two lightest gases, H_2 and He, escaped the gravitational field and diffused into interplanetary space during this initial period of earth's history.

The composition of the gases in earth's atmosphere following accretion is not completely resolved, but almost certainly contained high concentrations of CO_2, N_2, and H_2O, HCl and H_2SO_4 (Holland, 1984).[2] This gas composition is similar to that on Venus, at the present time. Precipitation of minerals and formation of felsic rocks led to the condensation and upward migration of liquid, precipitable water that overlies vast regions of denser, mafic rocks. In addition, water was probably provided by cometary and meteoritic bombardment. Based on thermodynamic equilibrium calculations with crustal elements, the acidic gases or hydrated equivalents (e.g. HCl, H_2SO_4) solubilised mineral cations in the primordial ocean, leading to a sea water dominated by Na^+, K^+, Mg^{2+} and Ca^{2+}. The anion balance was supplied by vulcanism and outgassing from deep crustal sources. The dominant anions were Cl^-, and to a lesser extent SO_4^{2-} and HCO_3^-. PO_4^{2-} was probably present, but to a lesser extent, and fixed inorganic nitrogen (as NH_3 or NO_3) was almost certainly absent, or very rare. The pH of the early oceans was probably close to neutral or slightly acidic (6.8 to 7.0). Additionally, radiogenic heat, high concentrations of greenhouse gases, and geothermal activity would have provided a source of heat to maintain elevated temperatures in earth's early ocean history. This was critical, as solar luminosity was considerably lower than today.

Over the course of earth's history, there has been a continuous trajectory from the mildly reducing conditions that prevailed at the time of origin of the planet to the highly oxidising conditions that prevail in the contemporary geological epoch. This oxidation trajectory has been largely driven by biological processes. The oxidation of earth's atmosphere is the consequence of the biological chemistry of carbon and oxygen, but even in the absence of any life on earth, there still would be a carbon cycle.

12.3   The abiotic carbon cycle

Tectonic processes, driven by the internal heat of the planet, continuously sweep the oceans' sedimentary layers into the mantle of earth, to be later regurgitated by volcanic processes as igneous rocks. In so doing, carbon dioxide is outgassed to the

---

[2] The rate of escape of H_2 to space is critical. If planetary accretion is fast, and the planet cools rapidly, less H_2 escapes and the resulting atmosphere of the planet reduces faster than if the planet cools slowly.

atmosphere by vulcanoes and consumed by chemical weathering reactions (Fig. 12.2). The geochemical reaction schemes are often attributed to Harold Urey (Urey, 1952; Berner *et al.*, 1983; Berner, 1991), and are commonly called the 'Urey reactions'. The primary source of carbon dioxide that vents into the atmosphere is vulcanism. Once in the atmosphere, the gas equilibrates with the oceans, forming carbonic acid. Neutralisation of the excess protons is accomplished by the chemical erosion of alkaline metals, primarily calcium and magnesium.

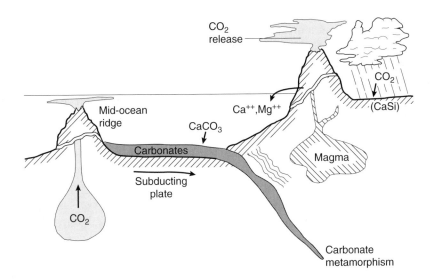

Fig. 12.2   Schematic of the geochemical carbon cycle. CO_2 is outgassed to the ocean and atmosphere from magma chambers feeding vulcanoes and at mid-ocean ridges. The CO_2 exchanges with silicates in soils and rocks, and mobilises Mg and Ca ions, that are carried to the oceans by rivers. In the oceans, the Mg and Ca precipitate as carbonates and are subducted into the upper mantle at plate boundaries. CO_2 is released from the subducting carbonates at high temperatures and pressures, and accumulates in magma chambers, where it feeds vulcanoes.

Uplift of shales on continents during orogenic events exposes silicates, which react with carbon dioxide. The overall reaction is:

$$Ca(Mg)SiO_3 + CO_2 \rightarrow Ca(Mg)CO_3 + SiO_2 \qquad (12.1)$$

The mobilisation of HCO_3^- in the aqueous phase delivers inorganic carbon to the oceans to be precipitated as magnesium and calcium carbonates (dolomites and limestones). On long time-scales, the rate of vulcanism must closely match the rate of weathering, or the atmosphere/ocean system would gain or lose carbon dioxide. How are these two processes connected?

Increased vulcanism leading to a greater rate of carbon dioxide supply, increases the acidity of rain and promotes more weathering. Conceptually, that simple negative feedback stabilises carbon dioxide; however, there were several periods in earth's history when the system went somewhat out of control. During the late Devonian and early Carboniferous, there appears to have been a drawdown in carbon dioxide, corresponding to the evolution and rapid colonisation of terrestrial ecosystems by land plants (Berner, 1997). The decrease in carbon dioxide, inferred from isotopic fractionation in marine and soil carbonates (Ekart *et al.*, 1999), appears to be first related to the chemical and physical exposure of rocks to the atmosphere as roots began to form true soils (Berner, 1997). Thus, by accelerating weathering, terrestrial plants led to a decline in atmospheric carbon dioxide. During this period, organic carbon burial also was high, and in the invasion of terrestrial plants, there is evidence of a transient rise in atmospheric oxygen (Berner *et al.*, 2000); however, carbon burial occurred following the decline in atmospheric carbon dioxide. The decline in carbon dioxide was probably the single most important factor leading to the Permo-Carboniferous glaciation, the longest and most extensive glacial period of the entire Phanaerozoic (Crowley and North, 1991). The escape from this glacial period presumably arose when vulcanic outgassing exceeded the weathering and burial in the late Permian, and ultimately culminated in the massive end-Permian extinction, which marks the end of the Paleozoic.

The second period of extensive drawdown in carbon dioxide occurred at the middle of the Mesozoic, and rapidly accelerated in the Cenozoic (the past 65 Ma; Fig. 12.3). This period in earth's history has also witnessed rapid radiations in terrestrial plants and marine animals (as well as phytoplankton, which we will discuss later) (Rothman, 2001). What factor is responsible for all three processes? One fundamental process that explains the correlation between all three proxies is break-up of Pangea and the opening of the Atlantic Ocean in mid-Mesozoic time, about 180 Ma (Falkowski and Rosenthal, 2001). The tectonic processes during this period exposed large continental surfaces of silicate to the atmosphere, thereby accelerating weathering reactions that led to a long, and apparently, steady depletion of carbon dioxide. This process accelerated throughout the Cenozoic, presumably because orogenic activity increased the rate of mantle exposure or outgassing of carbon dioxide declined (Raymo and Ruddiman, 1992). Simultaneously, however, the same process greatly facilitated the diversification of terrestrial plants through isolation and subsequent genetic drift. With the formation of the new ocean basin, benthic marine invertebrates became increasingly genetically isolated from the seed stock supplied from the Pan-Thalassian ocean, such that new species (but not classes) emerged. Indeed, the tempo of evolution was accelerated by continental drift while the associated geochemical processes helped to deplete a vital resource – carbon dioxide. That depletion, which continued up until the Industrial Revolution, led to a long-term decrease in earth's temperature, such that by 30 Ma, polar ice again became a permanent feature.

While most of our discussion on long-term changes in carbon dioxide has focused on weathering, the source of carbon dioxide in vulcanic outgassing does not have to be constant. Vulcanic source rocks are subducting sedimentary rocks. Since the break-up of Pangea, the greatest area of subduction has been in the Pacific Ocean.

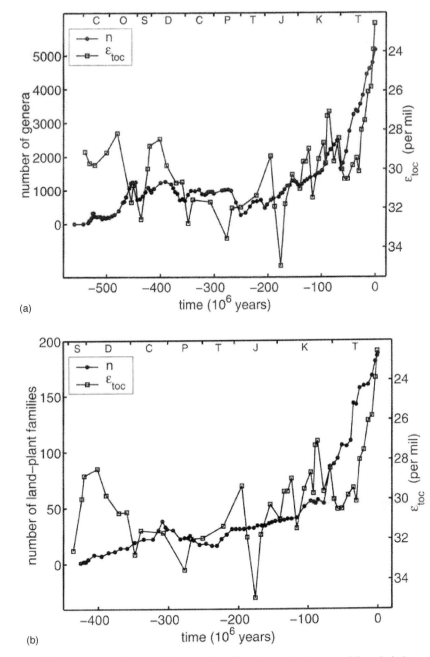

(a)

(b)

Fig. 12.3 (a) The changes in ε_{toc} and the number of marine animal genera (diversity) throughout the Phanaerozoic. ε_{toc} is the difference in the isotopic ratio of $^{13}C/^{12}C$ between carbonates and organic matter. The smaller the value of ε_{toc}, the more scarce CO_2 is in the atmosphere–ocean system. (b) Changes in ε_{toc} and terrestrial plant families throughout the Phanaerozoic. Note that both marine animal and terrestrial plant diversity increased markedly over the past ~ 50 Ma. (Figures kindly provided by Daniel Rothman and are taken from his analysis published in the Proceedings of the National Academy of Science, 2001.)

Because the Pacific is deep, carbonates do not survive long in the sediments,[3] and hence the source sedimentary rocks are carbonate-poor but silicate-rich. In the metamorphic processes that lead to vulcanic outgassing, if the source rock is depleted in carbonates, the carbon dioxide leak will be also depleted. If the Atlantic Ocean were the source of vulcanic rock, or if the Pacific were shallower, almost certainly the carbon dioxide outgassing would exceed or at least match the weathering reactions, thereby maintaining significantly higher carbon dioxide.

The estimated fluxes of carbon dioxide associated with Urey reactions were calculated by Berner *et al.* (1983). The uptake of carbon dioxide from the atmosphere related to carbonate dissolution on land is 11.8×10^{12} mol C per year while that associated with silicate weathering is 11.5×10^{12} mol C per year, yielding a total of 23.3×10^{12} mol C per year. Assuming a steady state, carbon dioxide inputs resulting from calcite (and dolomite) precipitation in the ocean amounts to 17.4×10^{12} mol C per year, while metamorphism involving carbonates and 'low temperature' silicates in the earth's crust, and vulcanism produces 5.90×10^{12} mol C per year. While these global fluxes are relatively small compared with biological fluxes of carbon, which are of the order of 5×10^{15} mol C per year, they are responsible for the sequestration of approximately 5×10^{21} moles of carbon as carbonates and 1.25×10^{21} moles of organic carbon in the earth's crust: that is by far the largest repository of carbon on the planet.

12.4   Biological reactions

As discussed, the abiotic part of the carbon cycle calls on acid-base reactions, i.e. the exchange of protons without electrons. The biological chemistry of carbon is based on redox reactions, i.e. the exchange of electrons with or without atoms.

The oldest rocks on earth come from three formations, in Greenland, northern Australia, and South Africa and date to around 3.8 to 3.4 Ga. The metamorphosed rocks from Greenland are laminated pelagic shales, with interleaving black and green/grey bands. The organic carbon within the black bands comprises as much as 0.5% by weight of the rock and is the oldest datable organic carbon on the planet. That carbon is isotopically enriched in ^{12}C by approximately 19‰ relative to inorganic carbon and its presumptive dissolved forms in the Archean Ocean. A fundamental interpretation of that isotopic enrichment is that as early as 3.7 Ga, an enzymatic process was capable of isotopic fractionation. A key suspect is the enzyme ribulose-1,5-bisphosphate carboxylase/oxygenase (RUBISCO) (Rosing, 1999).

---

[3] The carbonate compensation depth (CCD) is a confluence of temperature and pressure at which carbonates are dissolved. Most of the Pacific sediments are below the CCD, while most of the Atlantic is above the CCD.

The confluence of biochemical studies and molecular phylogenetic reconstruction has helped elucidate the origins of RUBISCO and their remarkable radiation through the microbial world. These analyses reveal that the enzyme has a common ancestor but quickly diverged into four major subgroups (Tabita, 1999). Molecular phylogenetic analysis suggests there are approximately four major subgroups, with radiations in each of the subgroups. The primary role of all forms of RUBISCO is to carboxylate a 5-C substrate, ribulose-1,5bisphosphate, to form two identical three-carbon products, 3-phosphoglycerate. The carbon substrate required is CO_2, not HCO_3^- (Cooper *et al.*, 1969). In so doing, RUBISCO kinetically fractionates against ^{13}C, such that the product is enriched in ^{12}C by up to $\sim 30‰$ relative to the source CO_2 (Park and Epstein, 1961). This isotopic fractionation is a signature of life on earth; the presence of this signature in the oldest rocks suggests that the enzyme was present at or near the beginning of life on the planet. We have not found older rocks that do not have that signature.

At this point it may be instructive to speculate about the origins of the carboxylation reaction itself and the reasons why nature selected a 5-carbon sugar as its major carboxylation substrate. In the subsequent biological chemistry of 3-phosphoglycerate, there is only one reaction where carbon is actually chemically reduced: it is in the formation of the corresponding aldehyde from phosphoglycerate. The rest of the intermediary metabolism in carbon fixation is directed towards regenerating the substrate, ribulose-1,5 bisphosphate. Ribose (and its derivative 5-carbon sugars, deoxyribose and ribulose) is among the simplest sugars capable of forming a ring yet having chiral properties. Ring formation facilitates polymerisation reactions, while chirality is important in enzyme selectivity.

RNA is a primary self-regenerating entity; that is, the polymeric nucleic acid can self-replicate under the proper conditions. One condition for this is a supply of ribose 5-phosphate. Metabolically, ribose is derived from ribulose via the activity of phosphopentose isomerase; and herein lies a chicken and egg problem. Did a primitive metabolic sequence arise that supplied ribose to help a replication machinery, or did a replication machinery direct or select a metabolic sequence that assured a supply of ribose? Whatever the answer to this problem, it would appear more than coincidental that the origins of intermediate carbon chemistry in autotrophs and the requirements for 5-C sugar phosphates in nucleic acid synthesis are intertwined. The marriage of these two pathways assured that carbon fixation and chemical reduction are amongst the earliest of earth's metabolic processes. That particular biological chemistry has continued to supply virtually all organic carbon for the planet to the present day. These reaction sequences, which are sometimes called the Calvin–Benson–Basham cycle, form one half of the wheel on the organic carbon cycle on earth. The other half is the oxidation of organic carbon – respiration in its various forms. Both reaction sequences are mediated via cytochrome complexes (Cramer and Knaff, 1990).

12.5   Engines and power supplies

Of the six light elements required for life as we know it, H, C, N, O, S and P, the first five undergo oxidation-reduction reactions that characterise all metabolic cycles.[4] All redox reactions are coupled. The reduction of inorganic carbon requires electron and proton sources. Water supplies virtually all of the reductant in the contemporary world. The oxidation of water is, however, energetically expensive. If light is the power supply, photons at wavelengths no longer than 735 nm are required to oxidise water. That reaction did not appear to arise until approximately 2.7 Ga; over a billion years after the origins of carbon fixation (Summons *et al.*, 1999).

In the Archean and early Proterozoic oceans, electron sources were relatively abundant in the form of reduced metals such as iron or manganese, and especially reduced sulfur compounds emanating from hydrothermal vents. The reducing power (i.e. electron sources) in the upper mantle could have delayed the onset of the oxidation of earth's atmosphere significantly; Kump *et al.* (2001) propose that mantle 'overturn' led vulcanic outgassing to become increasingly more oxidised at the Archean–Proterozoic transition (around 2.5 Ga ago). The coupling of carbon reduction to the oxidation of metals or sulfur was required to develop the engines of metabolic sequences and electron transfer reactions that subsequently would become prosthetic groups in oxido-reductases. It should be pointed out that, in addition to carbon reduction, a second oxidant was also abundant in earth's atmosphere and oceans: that element is nitrogen.

12.6   Nitrogen fixation and the origins of coupled redox reactions

All synthetic biological chemistry of nitrogen proceeds from ammonium.[5] Fixed inorganic nitrogen is very sparse in the lithosphere, and present models of planetary accretion suggest that there was little or no fixed nitrogen in earth's early atmosphere or oceans (Kasting, 1993). Moreover, as ammonium has a UV cross-section, and the solar output of UV radiation was not attenuated by oxygen or ozone before about 2.2 Ga, any atmospheric ammonium produced, for example by the bolide impact and the subsequent reduction of the nitrogen oxide by-products, would have been photolysed in the atmosphere (Kasting, 1990). Without a continued supply of

---

[4] Arguably, a hallmark of the chemistry of life are redox reactions that are far from thermodynamic equilibrium. Redox reactions involve the transfer of electrons with or without atoms and differ from acid-base reactions (that help drive the purely geological/geochemical carbon cycle) in that in the latter, protons are exchanged without electrons.

[5] Oxides of nitrogen are biological signal molecules, but are not involved in synthetic processes.

fixed nitrogen, life on early earth would have been strongly impeded (Falkowski, 1997).

The paucity of fixed nitrogen almost certainly led to the early emergence of biological nitrogen fixation; indeed, dinitrogen reductases are strictly anaerobic enzymes that utilise iron-sulfur cluster motifs as prosthetic groups. Sequence analysis of all extant nitrogenases suggests a single common ancestor. *In vivo* the enzyme utilises reductant produced via the oxidation of carbohydrates to reduce dinitrogen to ammonium; that is, the fixation of nitrogen essentially is a *respiratory* process that can be summarised by the following balanced reaction:

$$2N_2 + 4H^+ + 3[CH_2O] + 3H_2O \rightarrow 4NH_4^+ + 3CO_2 \qquad (12.2)$$

Prior to the evolution of oxygen, the only inorganic product of the biologically catalysed nitrogen cycle would have been ammonium. The deep ocean would have accumulated ammonium, and the vertical eddy-diffusive flux of the element to the euphotic zone would have sustained all primary production. Over several hundred million years, it is probable that nitrogen fixation would have become limited by the supply of phosphate. This period in earth's history may have been the only time when phosphate rather than fixed nitrogen limited primary production in the oceans. Continued evolution of the nitrogen cycle was coupled to the biological production of oxygen.

12.7   Oxygen evolution

By far, the most important source of oxygen in earth's atmosphere is through the biological oxidation of water. Why is water an electron donor?

Phylogenetic trees, constructed from comparing the nucleotide sequences of 16S RNA molecules in procaryotes, suggest that the earliest organisms to have evolved on earth were non-photosynthetic, thermophylic chemoautotrophs that are placed at the root branch between the Archea and Eubacterial kingdoms (Woese, 1987; Pace, 1997) (Fig. 12.4). These early organisms could have used inorganic substrates such as H_2, H_2S and Fe^{2+} as proton donors to reduce carbon dioxide to carbohydrate. Indeed, such organisms persist and thrive in deep sea vents, volcanic hot springs, deep in earth's crust, and in other 'extreme' environments where liquid water and suitable oxidisable inorganic substrates are available. Chemoautotrophs are almost certainly the precursors of photosynthetic cells (Blankenship, 1992). The evolution of a photosynthetic process in a chemoautotroph forces consideration of both the selective forces responsible (why?) and the mechanism of evolution (how?).

Reductants for chemoautotrophs are generally deep in the earth's crust. In the contemporary ocean, the chemical disequilibria between vent fluids and bulk sea water (which is highly oxidised) provides a sufficient thermodynamic gradient to

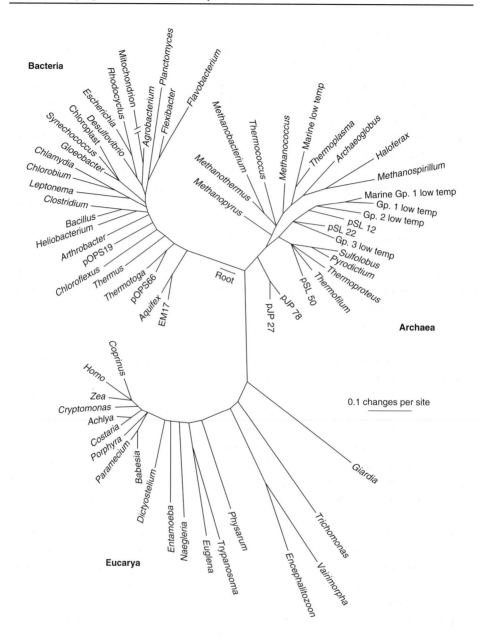

Fig. 12.4 A rooted phylogenetic tree of life based on 16 S ribosomal RNA sequence analyses. The three major domains, Archea, Bacteria and Eucarya, are rooted at a hypothetical thermophilic chemoautotroph that gave rise to both Archea and Bacteria. Eucaryotes are thought of as having arisen from a symbiotic association between an Archean host cell with Bacterial endosymbionts (mitochondria and plastids). (Figure kindly provided by Norman Pace.)

continuously support chemoautotrophic metabolism. However, prior to the evolution of a sustained biologically or chemically mediated reaction that produced a strong oxidant (e.g. oxygenic photosynthesis), the redox gradient in early earth's oceans would not have had a sufficiently large thermodynamic energy potential to support a pandemic outbreak of chemoautotrophy in the ocean basins. Moreover, magma chambers, vulcanism and vent fluid fluxes are tied to tectonic subduction regions, which are transient features of earth's crust and, hence, only temporary habitats for chemoautotrophs. To colonise new vent regions, the chemoautotrophs would have needed to be dispersed throughout the oceans by physical mixing. This same dispersion process would have helped ancestral chemoautotrophs exploit solar energy near the ocean surface. Indeed, phylogenetic analyses suggest that contemporary deep sea vent communities are colonised by organisms living on continental margins or coastal regions (Robert Vrijenhoek, pers. comm.).

Although the processes that selected photosynthetic reactions as the major energy transduction pathway remain obscure, central hypotheses have emerged based on our understanding of photosynthesis, biophysics and molecular phylogeny. The metabolic pathway for the synthesis of porphyrins and chlorins is one of the oldest in biological evolution, and is found in all chemoautotrophs. Mulkidanian and Junge (1997) proposed that the chlorin-based photosynthetic energy conversion apparatus originally arose from the need to screen UV radiation from damaging essential macromolecules such as nucleic acids and proteins. The UV excitation energy could be transferred from the aromatic amino acid residues in the macromolecule to the Soret band of membrane-bound chlorins to produce a second excited state which subsequently decays to the lower energy excited singlet. While non-photochemical dissipation of energy from the lower energy singlet excited state can occur, photochemical energy transduction involving electron transport inward across a membrane is also an energy-dissipating process. This energy-dissipation pathway can be harnessed to metabolism if the photochemically produced charge-separated primary products are prevented from undergoing charge recombination on the time-scale necessary for the reaction pair to interact with the redox catalysts of the chemoautotrophic machinery. In this way, light energy could be used to energise the reduction of carbon dioxide to carbohydrates, using reductants such as S^{2-} or Fe^{2+} which have redox potentials that are too positive to reduce carbon dioxide directly.

The geochemical effects of anoxygenic phototrophs are the reduction of carbon and the oxidation of such reductants as Fe^{2+} (to Fe^{3+}) and S^{2-} (to S^0 or SO_4^{2-}). This synthesis of reduced (i.e. organic) carbon and the oxidised form of the electron donor permit the phototroph to develop 'respiratory' systems that effectively operate in reverse. Thus, the exergonic conversion of photosynthetically reduced carbon to carbon dioxide, coupled to the reduction of Fe^{3+}, S^0 or SO_4^{2-} to Fe^{2+}, S^{2-} or S^0, can support growth and maintenance processes in the dark. However, not all of the reduced carbon and oxidants remain accessible to the photoautotrophs. Cells sink, carrying with them organic carbon. Fe^{3+} precipitates. The sedimentation and subsequent burial of carbon and Fe^{3+} can remove these components from the water column. Without replenishment, the essential reductants for anoxygenic photosynthesis would eventually become depleted in the surface waters. Thus, the necessity to regenerate reductants potentially prevented anoxygenic photoautotrophs from providing the major source of fixed carbon on earth for eternity. Major *net* accumulation of reduced organic carbon in sediments from the Proterozoic implies at least *local* depletion of reductants such as S^{2-} and Fe^{2+} from the photic zone of the ocean and sets the scene for the evolutionary selection pressure to find an alternative electron donor. Such a geochemical limitation of photosynthesis is consistent with $^{13}C/^{12}C$ measurements from the Greenland shale.

Water is a reductant with effectively an unlimited supply on earth. Liquid water contains 50 kmol H_2 per m^3 and there is approximately $10^{18} m^3$ of water in the hydrosphere and cryosphere. Use of water as a reductant for carbon dioxide, however, requires a larger energy input than does the use of Fe^{2+} or S^{2-}. Indeed, for oxygenic photosynthesis to occur, several biochemical innovations were required, including: (a) a blue-shift in the lowest singlet excited state of the chlorophylls, such that the oxidising energy could be poised at >1 electron volt; (b) the development and incorporation of a prosthetic group capable of oxidising water; (c) safety valves that reduce the damage caused by the production of reactive oxygen species. The biological sequences that led to these key innovations are not clearly discernible in the phylogenetic reconstruction; these innovations are not specifically tied to single gene analysis. However, based on topological analysis of the reaction centres in oxygenic photoautotrophs, we can make reasonable inferences about processes that occurred and, based on geochemical analyses, we can make reasonable inferences about the timing of the occurrences.

All oxygenic photoautotrophs contain the same basic reaction centre structure, and almost certainly are derived from a single common ancestor. Two of the key reaction centre proteins responsible for oxidising water have strong sequence homologies with reaction centre proteins found in purple sulfur bacteria. Similarly, the two proteins that comprise photosystem I in oxygenic organisms have strong sequence homology with reaction centre proteins found in green sulfur bacteria. Interestingly, the proteins for both reaction centres are heterodimers containing two polypeptides, each of which has five transmembrane spanning regions. This topological similarity has led to the suggestion that both reaction centres arose from a single common origin; however, sequence homology does not appear to allow direct testing of that hypothesis. Nonetheless, whereas all non-oxygenic photosynthetic bacteria contain a single reaction centre, all oxygenic photoautotrophs contain two reaction centres.[6] There is no clear evidence of how the two types of bacterial

---

[6] The suggestion that mutants of *Chlamydomonas* lacking PSI are photosynthetically competent and can evolve oxygen (Lee *et al.*, 1996), is an artifact. True deletion mutants of *Chlamydomonas* are not capable of oxygenic photoautotrophy (Redding *et al.*, 1999).

reaction centres were merged into a single organism. The only extant procaryotes that are capable of oxygenic photoautotrophy are cyanobacteria, which have clearly undergone significant modifications in the structure of the reaction centres. The earliest dual reaction centre organisms are extinct. A relict metabolism found in the contemporary oceans is represented by anoxygenic aerobic photosynthetic bacteria. These organisms are derived from purple bacteria, but rather than being restricted to anaerobic waters, they comprise approximately 10% of the bacteria in the upper ocean. They appear to be facultative photoheterotrophs – that is, they photo-oxidise organic matter to reduce inorganic carbon.

A major innovation in the oxidation of water is the formation of a tetra-Mn cluster localised within one of the photosystem II reaction centre proteins (Zouni *et al.*, 2001). Two of the Mn atoms are electrically silent; however, the other two participate in two electron transfer reactions, undergoing successive reduction from Mn^{4+} to Mn^{2+}. Each electron transfer is driven by one photon, each with a quantum yield of unity. The resulting 4 electron transfer reaction yields the evolution of oxygen from the oxidation of 2 molecules of water. Dismukes *et al.* (2001) have suggested that the evolution of the Mn cluster was perhaps selected as a mechanism to reduce bicarbonate. Whatever the origins, the oxygen evolving complex is the only 4 photon/4 electron transfer reaction in biological chemistry. The origins of the Mn cluster are obscure. There is little sequence homology with a single Mn electron transfer reaction involving oxygen, an Mn superoxide dismutase, and there are no other 4 Mn cluster prosthetic groups in a protein of which we are presently aware.

Finally, the release of free oxygen poses a severe danger to proteins and especially porphyrins. Singlet oxygen atoms are highly reactive. Cells use carotenoids and other polyunsaturated lipids as antioxidants, and developed escape valves including cyclic electron flow around PSII and superoxide dismutases to help ameliorate against the damage. Nonetheless, nature is not perfect, and the reaction centre proteins in PSII are constantly replaced. In fact, one of these proteins, D1, has a half-life of approximately 30 min *in vivo*.

12.8   When did oxygen evolution appear?

The fossil record suggests that micro-organisms reminiscent of extant cyanobacteria were present at around 3.45 Ga (Schopf, 1993); however, the oldest geochemical evidence for oxygen evolution comes from the presence of methyl-hopanes at around 2.7 Ga. These steroid-based lipids are taken as markers of cyanobacterial membrane lipids, and in contemporary cyanobacteria, require molecular oxygen for synthesis (Schopf, 1993; Summons *et al.*, 1999). A second, concordant signal of oxygen evolving organisms is the isotopic fractionation of sulfur (Canfield and Teske, 1996). However, paleosol records do not show a clear evidence of an oxidising atmosphere until around 2.2 Ga (Holland and Rye, 1998). The paleosols record the oxidation of terrestrial iron-sulfur minerals (pyrites) by oxygen.

Although paleosols are relatively resistant to molecular oxygen, on time-scales of tens of millions of years, with exposure of oxygen and rain containing dissolved oxygen, the reduced iron will become oxidised. The deposition and dating of the iron is taken as a reliable indicator of the timing and duration of the 'great rust' event. That appears to have occurred sometime between 2.2 and 2.1 Ga. Why was there a 500 Ma delay between the origin of oxygenic photoautotrophs and the oxidation of earth's atmosphere?

12.9   The fixed nitrogen crisis of the early Proterozoic

The evolution of oxygen by the first oxygenic photoautotrophs did not lead to the oxidation of the atmosphere for several reasons, but possibly one of the most important was the consequences of this new pathway on the nitrogen cycle (Box 12.1). Whereas there were potentially many electron sinks for the newly derived oxidising agent, including ferrous iron and sulfide, ammonium was certainly an energetically attractive substrate. The oxidation of ammonium to nitrite and nitrate provided a thermodynamically attractive gradient for the reduction of inorganic carbon. All that was needed was a mechanism of oxidising ammonium – in comparison with the oxidation of water, that reaction is relatively trivial, and certainly thermodynamically much more accessible. Once the oxidised fixed nitrogen species were formed in the partially oxidised upper layer of the late Archean or early Proterozoic Oceans, however, they ran into an extermination gallery at the interface with anaerobic microbes inhabiting the ocean interior. These oxidised forms of nitrogen were huge sinks for electrons for respiration, and denitrifiers surely used nitrate and nitrite in that capacity with disastrous results for the oxygenic photoautotrophs. The more oxygen produced in the upper ocean, the more nitrate formed – and the more nitrate formed, the more that fixed nitrogen was lost from the oceans. The oceans almost certainly had to reinvent the nitrogen cycle! As soon as the primordial ammonium in the oceans was exposed to oxygen, it was as good as doomed – to be released back to the atmosphere as N_2. The only possible saviour for this Sisyphean cycle was burial of organic carbon and nitrogen, with the subsequent release of some free molecular oxygen to the atmosphere. The oxidation of earth's atmosphere apparently took several hundred million years. With oxygen in the atmosphere, physical mixing of the ocean could finally ventilate the interior, thereby setting a stable environment for the accumulation of oxidised forms of nitrogen. Whenever the ocean became anoxic, this process had to be repeated; however, unlike in the initial condition, to our knowledge, once it was oxidised, the atmosphere never again lost oxygen. Hence, the duration of a nitrogen crisis in the oceans was never again as long as in the Archean. But the fixed nitrogen crisis is not the whole story.

Box 12.1   The fundamental governing equations for the marine nitrogen cycle.

The formation of particulate organic matter is represented in terms of the Redfield ratio of elements, and is based on the availability of fixed inorganic nitrogen as either nitrate or ammonium. The oxidation state of the nitrogen source influences the ratio of O_2 evolved per CO_2 fixed (i.e. the photosynthetic quotient). Nitrogen fixation is fundamentally a respiratory process, and is a function of both iron availability and O_2 concentration (see Fig. 12.7). Then the resulting fixed nitrogen product, ammonium, can be oxidised by nitrifying bacteria, which are mostly aerobic, to form nitrite and nitrate. This reaction is dependent upon O_2 concentration, with a half-saturation constant of $\sim 20\,\mu M$. Under anaerobic conditions, the oxidised species of nitrogen can be used as electron sinks for respiration by denitrifying bacteria, yielding both ammonium and, ultimately, N_2. This process is also sensitive to O_2, with a half-saturation constant for inhibition of $\sim 5\,\mu M$. Inspection of these equations reveals that the cycles of the three major elements, N, C and O, are biologically coupled through redox reactions with primary feedback based on O_2 concentrations.

Primary governing equations for the nitrogen cycle

Photosynthesis

$$106CO_2 + 16NO_3^- + H_3PO_4 + 122H_2O \Rightarrow C_{106}H_{263}O_{110}N_{16}P + 138O_2$$

$$106CO_2 + 16NH_4^+ + H_3PO_4 + 106H_2O \Rightarrow C_{106}H_{263}O_{110}N_{16}P + 106O_2$$

Nitrogen fixation

$$NF = f(O_2, Fe)$$

$$2N_2 + 4H^+ + 3[CH_2O] + 3H_2O \Rightarrow 4NH_4^+ + 3CO_2$$

Nitrification

$$NI = f(O_2) \qquad i.e. O_2 \geq 20\,\mu m$$

$$NH_4^+ + 2O_2 \Rightarrow NO_3^- + 2H^+ + H_2O$$

Denitrification

$$ON = f(O_2, NO_3^-) \qquad i.e. O_2 \leq 5\,\mu m$$

$$C_{106}H_{263}O_{110}N_{16}P + 84.8HNO_3 \Rightarrow 106CO_2 + 42.4N_2 + 16NH_4^+ + H_3PO_4 + 148.4H_2O$$

$$C_{106}H_{263}O_{110}N_{16}P + 94.4HNO_3 \Rightarrow 106CO_2 + 55.2N_2 + 16NH_4^+ + H_3PO_4 + 177.2H_2O$$

12.10   Phosphorus supply and the role of glaciation in super-charging the ocean

Three billion years ago, the sun was approximately 30% colder than today, and without strong greenhouse forcing, this would have been an ice-covered planet. The greenhouse forcing function can be estimated by the Stephan–Boltzmann relationship:

$$\sigma T^4 = S_o/4(1 - \alpha) \tag{12.3}$$

where σ is a constant (the Stephan constant $= 8.14 \times 10^{11}$ cal cm^{-2} min^{-1}), T is earth's temperature in Kelvin, S_o is the incoming solar radiation at the top of the atmosphere, and α is Earth's albedo. The right-hand side of this equation is the loss of radiation to space, while the left-hand side is the incoming radiation from the sun integrated for the planet.

Assuming a basic planetary albedo of ~ 0.28 for that period, we can estimate that a greenhouse forcing function amounting to at least 38 W m^2 was required to keep the oceans from freezing. One suspect factor in this unfolding and unsolved saga of the 'faint young sun' problem is methane (Kasting, 1997). Methane is a strong greenhouse gas. Once exposed to the atmosphere, the absorption of out-bound IR radiation by methane would have kept the world warm enough to prevent ice from forming all over the oceans. If, however, methane is oxidised to carbon dioxide, the greenhouse forcing is greatly reduced. The atmospheric lifetime of methane decreased from around 500 years in the Archean period (prior to the oxidation of the atmosphere), to around 10 years following the oxidation of the atmosphere. The resulting loss of greenhouse forcing would almost certainly have led to much colder temperatures. Additionally, if ice sheets advance to around 30° of the equator, planetary albedo rapidly increases and an unstable situation occurs – there is nothing to prevent ice from completely covering the oceans.[7] This apparently happened shortly after the oxidation of the atmosphere at around 2.0 Ga, and then at least two and possibly four times in the NeoProterozoic (Hoffman et al., 1998).

The 'snowball' earth phenomenon leads to ocean anoxia and escape requires outgassing of carbon dioxide from the atmosphere. Because the hydrological cycle is basically stopped, the Urey reactions do not operate efficiently, and vulcanically supplied carbon dioxide builds up in the atmosphere until the radiative forcing overcomes the albedo feedback. The net effect is that sufficient radiative forcing must have developed to facilitate ice melt from the equator to the poles. The

---

[7] This ice–albedo feedback is positive once the ice belt gets to low latitudes; that is, as ice sheets cross the 30° latitudes (N and S), there is a tendency (in climate models) for the ice sheets to rapidly expand and cover the globe. Conversely, as the ice sheets melt and ice-free zones expand from the equator toward the poles, there is a positive feedback whereby the ice sheets tend to rapidly melt (Budyko, 1982).

melting process can be rapid – occurring within a few decades (on this time-scale, even if the process took several thousand years, it would be considered almost instantaneous) – and once exposed to the atmosphere, the oceans appear to have undergone massive precipitation of carbonates; indeed, the deposition of the so-called 'cap carbonates' from these periods is a major marker of snowball events (Peryt *et al.*, 1990). The potential outgassing of nitrous oxides, resulting from the denitrification reactions, would have accelerated the warming process.[8]

The oceans at 2.6 Ga appear to have been relatively oligotrophic. It is possible to estimate the equilibrium phosphate concentration in the oceans from banded iron formations. The oxidation of ferrous iron in the upper ocean by either UV or other oxidising agents led to the precipitation of ferric iron. The ferric iron combines with hydroxides and phosphates; the equilibrium reaction can be used to infer the soluble phosphate concentration at the time of precipitation. Based on such calculations, Canfield estimates that the free phosphate concentration was $\sim 0.3\,\mu M$ compared with $\sim 2.3\,\mu M$ in the contemporary ocean (Canfield, pers. comm.). Such a low phosphate concentration would have certainly impeded the oxidation of the atmosphere, even given adequate fixed nitrogen supplies. Glaciations help overcome that problem. By making the ocean interior anoxic, glaciations facilitate mobilisation of phosphate bound with iron or other cations. Hence, glacial periods allowed for the oceans to become supercharged with phosphate. Fixed nitrogen almost certainly was the major element limiting primary production during these periods. In effect, glaciations may have helped stimulate oxidation of the atmosphere.

12.11 Carbon burial and the role of sedimentation on continental margins

The oxidation of earth's atmosphere by oxygenic photoautotrophs requires keeping the reduced species from becoming reoxidised. Once Fe(II) and reduced sulfur species were oxidised, oxidation of the atmosphere required the burial of the photosynthetically reduced (i.e. organic) carbon in the lithosphere. In the Archean and early Proterozoic, all oxygenic photoautotrophs were cyanobacteria. How cyanobacterial carbon is buried is an interesting question.

In the contemporary ocean, the export flux of organic carbon is usually mediated by large cells, such as diatoms, which have high sinking rates (Bienfang, 1992; Goldman, 1992; Dugdale and Wilkerson, 1998). However, in the contemporary ocean, very little of the sedimenting organic carbon is actually buried in the sea floor; almost all of the organic material is remineralised in the water column. Carbon is buried on continental margins, especially in major 'mud belts' of Oceania

---

[8] The warming potential of nitrous oxide is approximately 300 times that of carbon dioxide.

and at north-eastern margin of South America (Aller, 1998). The carbon burial is linearly related to the fluvial supply of sediment (Fig. 12.5). This process must have been critical to the burial of organic carbon in the Archean and Early Proterozoic oceans as well.

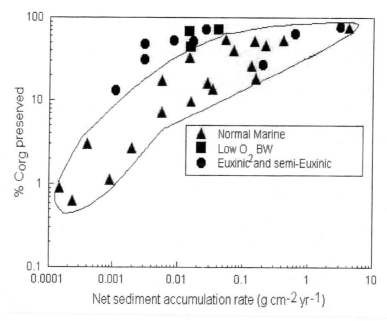

Fig. 12.5 The relationship between organic carbon preserved in marine sediments and sedimentation rates for normal (oxidised) marine systems, low oxygen conditions (BW = Brackish Water), and euxinic conditions. Note that the rate of organic carbon burial is roughly a log normal function of the rate of sedimentation. (Figure kindly provided by Robert Aller.)

In this scenario, cyanobacteria formed mats on continental margins and in shallow seas. These regions were supplied with phosphate and iron from upwelling and fluvial sources, while the organisms themselves fixed both nitrogen and carbon. Seasonal monsoons supplied the shelves and shallow seas with a high sedimentary load that episodically buried the mats.[9] Planktonic cyanobacteria might further have been entrained in the sediments via adsorption onto particle surfaces. The burial processes led to diagenic production of kerogens in shales, the outcrops of which were continuously swept into the deep ocean sediments and ultimately further buried by resupply of continent-derived source sediment.

It should be noted that weathering reactions also occur for organic carbon in

---

[9] Prior to the colonisation of terrestrial ecosystems by higher plants, sediment supply may have limited the rate at which carbon was buried. However, in the absence of widespread terrestrial plants, the sedimentary sources would have probably been scarce.

shales. This reoxidation process appears to be catalysed by heterotrophic microbes within shales (Petsch *et al.*, 2001). In the early Proterozoic, the weathering (or, more accurately, the respiration) of shale organic carbon must have been relatively slow compared with burial. A simple explanation that may have led to the imbalance between these two processes is weak orogenic activity during this period in earth's history.

12.12   The rise of oxygen and its stabilisation

The rate of rise of oxygen in the atmosphere almost certainly was controlled by the rate of burial of organic carbon. As we discussed, the burial process requires a steady supply of sediments. In this period of earth's history, all sources of sediments were derived from physical and chemical weathering: there were no terrestrial plants to accelerate the process. Hence, to first order, even given adequate nutrient sources, and having overcome the Sisyphean nitrogen crisis, sediment supply almost certainly was the primary factor determining the rate of oxidation of the oceans and atmosphere.

There are two basic hypotheses for setpoints on the upper limit of O_2. One is based on the notion that as the mixing ratio of O_2 exceeds ~ 25–28% of the total atmospheric gas content, lignin and other cellulose-based polymers 'spontaneously' combust (Lenton and Watson, 2000). This model, based on experimental data on the combustion of paper, suggests that the frequency of forest fires during the mid-Phanaerozoic helps constrain O_2 at ~ 20–25% (i.e. the present value), and is supported, in part, by the fossil record of charcoal. A second mechanism, proposed here, may have been set by the feedback between nitrogen fixation and oxygen. Being a strictly anaerobic enzyme, nitrogenase activity is repressed at oxygen concentrations above $\sim 50\%$ present atmospheric level (PAL). To overcome this feedback, diazotrophs have evolved several strategies to protect the enzyme. In nitrogen-fixing cyanobacteria, organisms either 'time-share' nitrogen fixation with oxygen evolution, such that a cell fixes nitrogen during some period of the day, while evolving oxygen during another period(s), or they spatially segregate the nitrogen fixation reactions in special cells (heterocysts). Spatial segregation almost certainly evolved long after the 'time-sharing' schemes (Fig. 12.6), and persists as a strategy to this day in most marine diazotrophic cyanobacteria (such as *Trichodesmium*). However, as oxygen concentrations rise, nitrogenase activity is repressed (Fig. 12.7), such that a simple negative feedback results. The more oxygen evolved, the less nitrogen fixed. As nitrogen fixation decreases, more fixed N is removed from the ocean via denitrification, and less oxygen is produced. As oxygen levels decline, nitrogen fixation increases. Both the fire-based feedback and the oxygen regulation of nitrogen fixation continue to the present time.

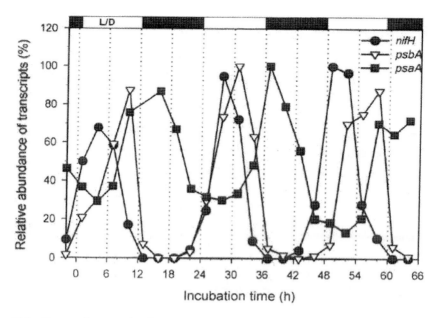

Fig. 12.6 Changes in transcript (i.e. messenger RNA) levels for a nitrogenase subunit (nifH), a core component protein of photosystem II (psbA), and a core component protein of photosystem I (psbB) in the nonheterocystous marine cyanobacterium, *Trichodesmium*. Note that nifH transcript abundance reaches its maximum in the light about 4 h before the maximum abundance of psbA. The organism requires light to fix nitrogen, but the evolution of oxygen strongly inhibits nitrogen fixation. Hence, in *Trichodesmium*, one solution for these competing pathways has been temporal segregation during the photoperiod. (Figure kindly provided by Yibu Chen.)

12.13   The Paleozoic oceans – the rise of prasinophytes and higher plants

The modern ocean carbon cycle evolved over hundreds of millions of years following the oxidation of the atmosphere. Whereas the oxidation itself was primarily due to the photosynthetic and diazotrophic activities of cyanobacteria, the evolution of eucaryotes was critical to the evolution of metazoans.

There are approximately 10 000 species of marine phytoplankton spanning at least eight taxonomic divisions. Plastid ultrastructure, 16 S rRNA sequences, and the conservation of key photosynthetic proteins, suggest a common origin for all plastids, namely a procaryotic oxygen evolving group closely related to extant cyanobacteria (Bhattacharya and Medlin, 1998). In principle, the points of divergence and rates of evolution of the eucaryotic photoautotrophic can be derived by comparing ribosomal RNA sequences between and within taxonomic divisions (Woese, 1987; Medlin *et al.*, 1994). Such results provide clues about probable time of origin and rates of evolution of the host and/or the associated plastid, but often are ambiguous about the timing of the symbiotic event and the rates of evolution of the symbiotic association. Two major plastid lineages can be traced. One is a 'green line'

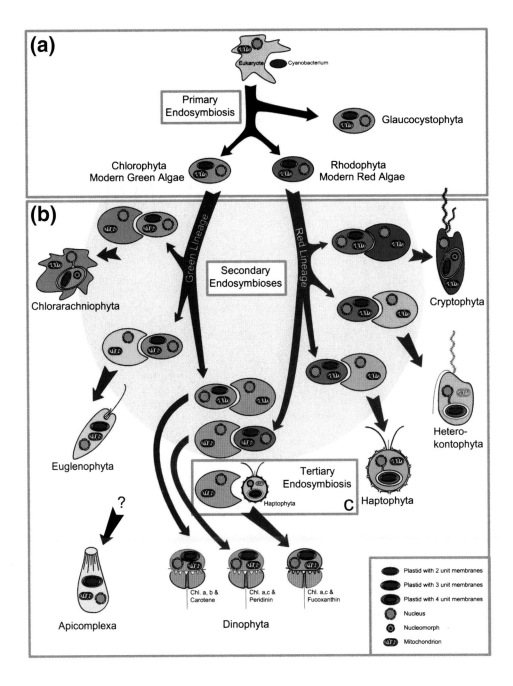

Plate 12.1. The basic pathway leading the evolution of eucaryotic algae. The primary symbiosis of a cyanobacterium with an apoplastidic host gave rise to both Chlorophyte algae and red algae. The Chlorophyte line, through secondary symbioses, gave rise to the 'green' line of algae, one division of which was the predecessor of all higher plants. Secondary symbioses in the red line with various host cells gave rise to all the chromophytes, including diatoms, cryptophytes and haptophytes. (From Delwiche, 2000 with permission.)

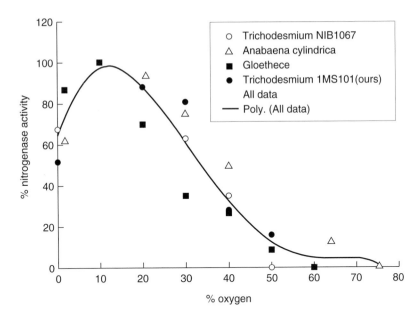

Fig. 12.7 The relationship between O_2 concentrations and nitrogen fixation in four species of marine cyanobacteria. Note that at approximately 25% O_2, the N_2 fixation is half the maximum Michaelis rate constant. The feedback between O_2 and N_2 fixation potentially constrains O_2 concentrations in the earth's atmosphere. (Figure courtesy of Ilana Berman-Frank.)

descended from cyanobacteria and leading to Chlorophytes and other chlorophyll-*b*-containing eucaryotic phytoplankton, especially the Prasinophytes. Chlorophytes and cyanobacteria dominated the Proterozoic and Palaeozoic oceans, but do not dominate contemporary seas. The second, 'red line', is derived from rhodophyte plastids through secondary and, in some cases, tertiary symbioses (Delwiche, 2000), giving rise to chlorophyll-*c*-containing algae, broadly called Chromophytes (Plate 12.1). Since the end-Permian extinction, 250 million years ago, three major groups of chlorophyll-*c*-containing algae – the diatoms, dinoflagellates, and the cocco-lithophorids – developed and arose to ecological prominence that continues to the present period.

The broad emergence of new phytoplanktonic phyla in the Mesozoic is one of the clearest cases of quantum evolution in the Phanaerozoic. Although there is bio-geochemical evidence suggesting Precambrian origins for dinoflagellates (Moldo-wan and Talyzina, 1998), the first identifiable thecate (and presumably, photoautotrophic) dinoflagellates are found in the early Triassic, about 20 Ma after the end-Permian extinction (Fensome *et al.*, 1996) (Fig. 12.8). Curiously, the early Triassic appears to coincide with a widespread ocean anoxic event (OAE) that occurred at the Permian–Triassic boundary (Isozaki, 1997). Coccolithophorids emerged late in the Triassic, under oxic conditions. Both of these groups radiated extensively throughout the Jurassic and into the Cretaceous. The earliest diatoms

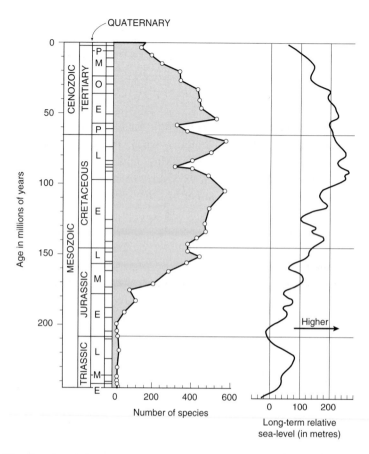

Fig. 12.8 The diversity of dinoflagellates over the past 250 Ma as inferred from fossil cysts. The right-hand side of the figure shows the inferred sea levels. (From Fensome *et al.*, 1996, with permission.)

appear to be neritic, and their radiation appears roughly consistent with regression events.

The recorded first appearances for these three groups represent minimum estimates of the time of origin. For example, not all dinoflagellates produce fossilisable cysts,[10] and preserved biomarker molecules provide evidence for the clade as early as the Neoproterozoic Era (Moldowan and Talyzina, 1998). Nonetheless, dinoflagellate biomarker concentrations increase significantly in Triassic organic matter, in parallel with the radiation recorded by microfossils. Thus, the radiation of dinoflagellates to assume their model role in the ocean system is faithfully chronicled by the geological record. The same is true for coccolithophorids and diatoms. Their earliest representatives may have been lightly skeletonised, decreasing their

---

[10] About 15% of the extant dinoflagellates produce cysts.

probability of preservation. Their observed radiations, however, are marked by clear changes in the marine carbonate and silica cycles, again leaving little doubt that fossils accurately capture their rise to ecological prominence. Indeed, with the emergence of both coccolithophorids and foraminifera in the Mesozoic, for the first time in earth's history virtually all carbonate precipitated in the ocean was due to the activity of plankton.

There were several OAEs in the mid-Cretaceous, that were, from a geological perspective, relatively short-lived (1 million years). Two of these events, the early Aptian Livello Selli (OAE1a, at 120.5 Ma) and the Cenomanian–Turonian boundary Livello Bonarelli (OAE2, at 93.5 Ma) are marked by positive excursions of $\delta^{13}C$ in carbonates, which is presumed to have been associated with the widespread burial of marine organic matter. The causes for the increased burial of organic matter are not resolved: hypotheses include increased upwelling; an accelerated hydrological cycle that increased rock weathering; the evasion of isotopically light, mantle-derived carbon dioxide associated with volcanic activity in the formation of the Ontong-Java Plateau; and the dissociation of isotopically-light methane clatherhates in continental margin sediments. Both events led the formation of extensive black shales, that are found through the then emerging North Atlantic basin and the European Tethys seas. Associated with these events were the reduction in carbonate-forming species, including foraminifera, coccolithophorids, and corals. The nannoconids were a group of heavily calcified nanophytoplankton that became extinct during the Selli event, while planktonic forams and radiolarians were heavily pruned during the Bonarelli. Interestingly, following the latter OAE events, diatoms, which had played a relatively minor role in planktonic community structure previously, erupted, and played a dominant role in the Antarctic basin, finally erupting in the global oceans in the Cenozoic, following the extinction at the K/T boundary (at 65 Ma).

12.14   The oceanic fax machine – the role of mixing

In the general scheme of things, the primary forcing functions for ocean chemistry are earth's radiation budget, continental configuration, sea floor topography, orbital rotational velocity and the trajectory of the earth around the sun. These primary processes determine ocean circulation, which in turn determines, to first order, the distribution of elements in the oceans. In the early Mesozoic, the lack of polar ice and the compressed continental configuration almost certainly reduced the equator-to-pole and the continent-to-ocean heat gradients. The reduced heat gradients led to weak zonal wind velocities which would have led to a weakening in eddy pumping and upwelling along continental margins. On long time-scales, such a physical field can be a selective pressure in phytoplankton evolution. Organisms that are adapted to relatively quiescent or low nutrient regimes would be expected to flourish; indeed, the deposition of calcite throughout the Jurassic and early Cretaceous suggests that coccolithophorids, which lack storage vacuoles and are adapted to low nutrient conditions, were relatively abundant. When the Atlantic basin began to form, at around 175 Ma, the altering continental configuration facilitated a relatively shallow seaway with potentially high turbulent mixing, and freshwater drainage from the continents provided a fluvial nutrient source with high silicic acid loads. This region must have been conducive to diatom blooms. Indeed, the first fossil diatoms are found at around 165–170 Ma, in what was the early Atlantic seaway.

If we follow this basic thought, we realise that one potential mechanism for determining calcite as opposed to opaline silica fluxes may have been nutrient availability. The fossil calcite and opal distributions in the sediments over the last 200 Ma, represent a 'fax' of ocean turbulence. My reasoning is that under highly turbulent conditions, diatoms, which devote a large fraction of their cell volume to vacuoles, are selected over coccolithophorids. When nutrients are pulsed, via eddy pumping or seasonal upwelling, diatoms can rapidly assimilate nitrate and phosphate (note that NH_4^+ is not easily stored in vacuoles), and by storing these nutrients in vacuoles, can rapidly draw down the nutrient fields, leaving other phytoplankton groups at a competitive disadvantage (Olaizola et al., 1993). Under relative quiescent conditions, however, coccolithophorids and dinoflagellates have an advantage over diatoms. The coccolithophores do not have storage vacuoles, but have high affinities for nutrients and, combined with small cell sizes, are relatively successful under low nutrient conditions. Many dinoflagellates are faculative heterotrophs and can assimilate dissolved organics to supplement their phototrophic life styles. Hence, during periods of low turbulent mixing, coccolithophorids and dinoflagellates can outcompete diatoms, and vice versa.

Thus, following the end-Permian extinction, the early Triassic ocean contained lots of dissolved organic carbon and was a heaven for heterotrophic bacteria and eucaryotes, such as the primordial dinoflagellates, that could take advantage of the particulate and organic carbon. These conditions were especially true for the continental margins and inland seas. The free metal availability was probably poised towards the reducing series (Fe and Mn) rather than the oxidised series (Cu and Mo). The open ocean may have been deprived of several nutrients, but if there had been anoxia during the extinction, then fixed N would have been scarce. The postulated high dissolved and particulate organic carbon would further have stimulated denitrification. If the ocean interior was suboxic, metazoan grazers would likely have been scarce. If so, the food chain would have been primarily dominated by a 'microbial food web' with rapid recycling of nutrients but little net community production (i.e. little left for higher trophic levels). As the ocean slowly became aerobic, nitrate levels would have been restored, giving an opportunity to organisms capable of utilising that relatively new source of nitrogen. This probably opened up an opportunity for coccolithophorids in the late Triassic or shortly after the Triassic extinction. I suggest that diatoms, which do well under conditions of high upper ocean turbulence, arose when the equator-to-pole heat gradient strengthened and continental configuration facilitated mixing in the newly forming Atlantic basin. It should be noted that diatom success was not simply due to the availability of silica. Prior to the emergence of diatoms, radiolaria were relatively abundant, and in the Cenozoic there appears to be an inverse relationship between these two groups of silica-requiring organisms (Harper and Knoll, 1975).

How do these patterns cascade into the more modern world oceans?

12.15   The return to the ice house

The beginning of the Cenozoic (65 Ma ago) saw bottom water temperatures of approximately 12°C, and sluggish ocean circulation with no polar ice (Crowley and North, 1991). Over the next 30 Ma, carbon dioxide would slowly become depleted as vulcanism failed to replenish the losses due to weathering and burial of organic carbon (Rothman, 2001). By around 30 Ma, permanent polar ice caps would re-establish, a phenomenon that the earth had not witnessed since the Permo-Carboninferous glaciation, almost 150 Ma earlier. The remarkably long period of no polar ice in the Phanaerozic (encompassing all of the Mesozoic) was almost certainly associated with high levels of carbon dioxide in the atmosphere/ocean system. Again, the primary factors influencing the drawdown on these long time-scales are tectonic and geochemical processes, that are largely, but not entirely, independent of biological responses and feedbacks.

The plunge into the ice house for the past 30 Ma, and the corresponding decline in carbon dioxide has further led to changes in terrestrial plant diversity, with a rise in C_4 plants around 14 Ma ago. Detailed inspection of sedimentary cores over this period reveals high-frequency changes occurring on approximately 40 000 Ka time-scales (Imbrie *et al.*, 1992). These oscillations, in both ^{18}O in marine carbonates (which records both paleotemperature and water volume), and ^{13}C in the carbonates (which records the amount of total inorganic carbon in sea water at the time of carbonate formation), have been suggested to be triggered by minor changes in solar radiation arising from small, but astronomically predictable, variations in earth's orbit around the sun (Berger, 1988). Within the past 1 Ma, the cycles have been dominated by a 100 000-year window. Both the 40 000- and 100 000-year periods are part of the so-called 'Milankovich cycles', small changes in solar radiative forcing, that have been clearly established in polar ice itself (Petit *et al.*, 1999). In reconstructions of ice records from Antarctica, the signatures for the past 420 000 years of four such cycles have been elucidated in isotopic records of the ice and trapped gases, as well as the chemical composition of the gases and aeolian particles in the ice (Petit *et al.*, 1999).

Although we do not completely understand the factors leading to glacial–interglacial states, we believe that the radiative forcings (amounting to $\sim 3\,W$ m^{-2}) are too small to have led to the climate changes without some positive feedbacks (Broecker and Denton, 1989; Sigman and Boyle, 2000). Most of the arguments in the literature are based on the mechanism and importance of the feed-backs. Several issues are certainly related to the glacial–interglacial transitions over the past 420 Ka. First, during glacial periods, terrestrial plant productivity decreases and ~ 600 Pg of carbon is transferred from terrestrial ecosystems to the oceans (Sigman and Boyle, 2000). The inverse occurs at the termination of the glacial periods. Second, continental shelves are flooded during interglacials, but are exposed during glacial periods; the changes in ice volume lead to changes in sea level of the order of 120 m. Third, glacial periods are drier, and the aeolian fluxes of essential elements, such as Fe, are greatly increased during this period.

From the above information, it is reasonable to infer that during glacial periods, the combined inorganic nitrogen inventory of the oceans increases, while during interglacials it decreases (Falkowski, 1997). The changes in nitrogen, combined with increased sea ice extent, and the northern migration of the southern polar front, enhance the uptake and retention of carbon dioxide by the ocean due to both an increase of exported production and a decrease in the leakage of carbon dioxide back to the atmosphere (Falkowski *et al.*, 1998). In effect, for the last million years the ocean has been 'breathing' on 100 000-year cycles; exhaling carbon dioxide to the atmosphere for around 20 000 years during interglacials, and inhaling carbon dioxide from terrestrial ecosystems for some 80 000 years during glacial periods. On these time-scales, biological, rather than tectonic, processes dominate the carbon chemistry of earth (Falkowski *et al.*, 2000).

12.16   The contemporary view of the carbon cycle

The contemporary view of the ocean carbon cycle, where diatoms are the major exporters of silica and organic carbon to the sediments, and coccolithophorids are the major exporters of calcite, is that it has operated for approximately the past 100 Ma. The success and radiations of these groups almost certainly were related to continental drift that provided new habitat for these eucaryotes (Falkowski and Rosenthal, 2001). In the background, however, the most abundant phytoplankton in the sea are cyanobacteria. These organisms have remained the dominant oxygenic photoautotrophs for over 2.5 billion years (Knoll, 1989), and it is both remarkable and humbling to realise that although the fact that their average lifetime in the oceans is only a few days, they have successfully transmitted their genetic information to the successive generations through, and despite, all the events recorded in earth's history. Thus, the procaryotes contribute over 85% of the total net primary production in the oceans and virtually 100% of the fixed nitrogen (Falkowski and Raven, 1997). They are the true unsung heroes of the biological carbon cycle in the sea. It is paradoxical that despite their long history, it is only comparatively recently that we have become aware of their importance in the contemporary ocean.

12.17   Concluding remarks

The carbon cycle is one of the most complicated of the biogeochemical cycles. Its feedbacks with climate and life are hallmarks of the chemistry of this planet, and the cycle evolved over three billion years to arrive at the present state. There are two carbon cycles on earth, and they operate in parallel. The first, an abiotic cycle, driven by vulcanism, weathering, and tectonic subduction, operates slowly, but has extremely large capacity, and over million-year time-scales dominates the carbon chemistry of the planet. This chemistry of the abiotic cycle is driven primarily by acid-base reactions. The biologically driven carbon cycle operates very rapidly, but has significantly lower capacity, and is subjected to wide perturbations, almost all of which are related, either directly or indirectly, to climate forcing. The biological cycle is primarily driven by redox reactions far from thermodynamic equilibrium.

Over the past 200 years, human (mostly energy-related) activities have altered the natural carbon cycle, primarily through an acceleration of release of carbon dioxide into the atmosphere. In the process, redox chemistry is impacted through the accelerated oxidation of organic matter. These activities will have unforeseen consequences; however, in the perspective of earth's history, this is a relatively minor perturbation (Fig. 12.9). While the impacts on humans will undoubtedly be

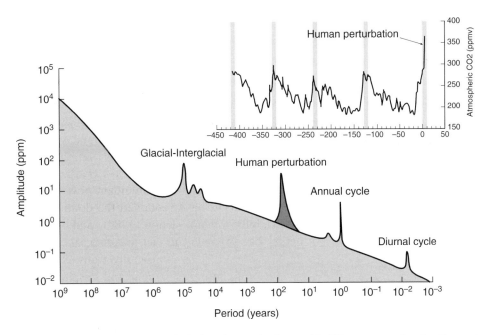

Fig. 12.9   A variance spectrum depicting changes in the earth's CO_2 concentration for the past four billion years. Note that human activities over the past two centuries are relatively short perturbations on the natural cycle; however, the amplitude of the perturbation is high. (From Falkowski *et al.*, 2000, with permission.)

felt within the coming century, on longer time-scales carbon dioxide will decline as weathering reactions continue to outpace natural outgassing processes.

Acknowledgements

I thank the National Science Foundation and the National Aeronautics and Space Administration for support of my research. I thank John Raven, Bob Berner, Norman Pace, Andy Knoll, Dan Schrag, and Chuck Dimukes for their discussions and thoughts.

References

Aller, R.C. (1998) Mobile deltaic and continental shelf muds as fluidized bed reactors. *Marine Chemistry* **61**, 143–55.

Beerling, D., Osborne, C. & Chaloner, W. (2001) Evolution of leaf-form in land plants linked to atmospheric CO_2 decline in the Late Palaeozoic era. *Nature* **410**, 352–4.

Berger, A. (1988) Milankovitch theory and climate. *Reviews of Geophysics* **26**, 624–57.

Berner, R.A. (1991) A model for atmospheric CO_2 over Phanerozoic time. *American Journal of Science* **291**, 339–76.

Berner, R.A. (1997) Geochemistry and geophysics: the rise of plants and their effect on weathering and atmospheric CO_2. *Science* **276** (5312), 544–6.

Berner, R.A., Lasaga, A. & Garrels, R. (1983) The carbonate-silicate geochemical cycle and its effect on atmospheric carbon dioxide over the past 100 million years. *American Journal of Science* **283**, 641–83.

Berner, R.A. & Maasch, K. (1996) Chemical weathering and controls on atmospheric O_2 and CO_2: Fundamental principles were enunciated by J.J. Ebelmen in 1845. *Geochemical and Geophysical Geosystems* **60**, 1633–7.

Berner, R.A., Petsch, S.T., Lake, J.A., *et al.* (2000) Isotope fractionation and atmospheric oxygen: implications for phanerozoic O_2 evolution. *Science* **287** (5458), 1630–33.

Bhattacharya, D. & Medlin, L. (1998) Algal phylogeny and the origin of land plants. *Plant Physiology* **116**, 9–15.

Bienfang, P.K. (1992) The role of coastal high latitude ecosystems in global export production. In: *Primary Productivity and Biogeochemical Cycles in the Sea* (eds P.G. Falkowski & A. Woodhead), pp. 285–97. Plenem Press, New York.

Blankenship, R.E. (1992) Origin and early evolution of photosynthesis. *Photosynthesis Research* **33**, 91–111.

Broecker, W. (1998) *How to Build a Habitable Planet*. Columbia University Press, New York.

Broecker, W.S. & Denton, G.H. (1989) The role of ocean-atmosphere reorganizations in glacial cycles. *Geochimica et Cosmochimica Acta* **53**, 2465–501.

Budyko, M.I. (1982) *The Earth's Climate: Past and Future*. Academic Press, New York.

Canfield, D. & Teske, A. (1996) Later Proterozoic rise in atmospheric oxygen concentration inferred from phylogenetic and sulfur-isotope studies. *Nature* **382**, 127–32.

Cooper, T.G., Filmer, D., Wishnick, M. & Lane, M.D. (1969). The active species of 'CO^{2}' utilized by ribulose diphosphate carboxylase. *Journal of Biology Chemistry* **244**, 1081–3.

Cramer, W.A. & Knaff, D.A. (1990) *Energy Transduction in Biological Membranes: A Textbook of Bioenergetics.* Springer-Verlag, Berlin.

Crowley, T. & North, G. (1991). *Paleoclimatology.* Oxford University Press, New York.

Delwiche, C. (2000) Tracing the thread of plastid diversity through the tapestry of life. *American Naturalist* **154**, S164–77.

Dismukes, G., Klimov, V., Baranov, S., Kozolov, Y.N., DasGupta, J. & Tyrushkin, A. (2001) The origin of atmospheric oxygen on Earth: the innovation of oxygenic photosynthesis. *Proceedings of the Nautical Academy of Science USA* **98**, 2170–75.

Dugdale, R. & Wilkerson, F. (1998) Silicate regulation of new production in the equatorial Pacific upwelling. *Nature* **391**, 270–73.

Ekart, D., Cerling, T., Montanez, I. & Tabor, N. (1999) A 400 million year carbon isotope record of pedogenic carbonate: implications for paleoatmospheric carbon dioxide. *American Journal of Science* **299**, 805–27.

Falkowski, P. (1997) Evolution of the nitrogen cycle and its influence on the biological sequestration of CO_2 in the ocean. *Nature* **387**, 272–5.

Falkowski, P., Barber, R. & Smetacek, V. (1998) Biogeochemical controls and feedbacks on ocean primary production. *Science* **281**, 200–206.

Falkowski, P.G. & Raven, J.A. (1997) *Aquatic Photosynthesis.* Blackwell Scientific, Oxford.

Falkowski, P. & Rosenthal, Y. (2001) Biological diversity and resource plunder in the geological record: casual correlations or causal relationships? *Proceedings of the Nautical Academy of Science USA* **98**, 4290–92.

Falkowski, P., Scholes, R.J., Boyle, E., *et al.* (2000) The global carbon cycle: a test of our knowledge of the Earth as a system. *Science* **290**, 291–6.

Fensome, R., MacRae, R.A., Moldowan, J.M., Taylor, F.J.R. & Williams, G.L. (1996) The early Mesozoic radiation of dinoflagellates. *Paleobiology* **22** (3), 329–38.

Goldman, J.C. (1992) Potential role of large oceanic diatoms in new primary production. *Deep-Sea Research* **40** (1), 159–68.

Harper, H.E., Jr & Knoll, A. (1975) Silica, diatoms, and Cenozoic radiolarian evolution. *Geology* April, 175–7.

Hoffman, P.F., Kaufman, A.J., Halverson, G.P. and Schrag, D.P. (1998) A Neoproterozoic snowball earth. *Science* **281** (5381), 1342–6.

Holland, H.D. (1984) *The Chemical Evolution of the Atmosphere and Oceans.* Princeton University Press, Princeton, NJ.

Holland, H. & Rye, R. (1998) Paleosols and the evolution of atmospheric oxygen: a critical review. *American Journal of Science* **298**, 621–72.

Imbrie, J., Boyle, E.A., Clements, S.C. *et al.* (1992) On the structure and origin of major glaciation cycles. I. Liniar responses to Milankovitch forcing. *Paleoceanography* **7**, 701–738.

Isozaki, Y. (1997) Permo-Triassic boundary superanoxia and stratified superocean: records from lost deep sea. *Science* **276** (5310), 235–8.

Kasting, J.F. (1990) Bolide impacts and the oxidation state of carbon in the Earth's early atmosphere. *Origins of Life in the Evolutionary Biosphere* **20**, 199–231.

Kasting, J.F. (1993) Earth's early atmosphere. *Science* **259**, 920–26.

Kasting, J.F. (1997) Planetary atmospheres: warming early Earth and Mars. *Science* **276** (5316), 1213–15.

Knoll, A.H. (1989) Evolution and extinction in the marine realm: some constraints imposed by phytoplankton. *Philosophical Transactions of the Royal Society, London* **325**, 279–90.

Kump, L., Kasting, J. & Barley, M. (2001) Rise of atmospheric oxygen and the 'upside-down' Archean mantle. *Geochemical and Geophysical Geosystems* **2**, 2000GC000114.

Lee, J.W., Tevault, C.V., Owens, T.G. & Greenbaum, E. (1996) Oxygenic photoautotrophic growth without photosystem-I. *Science* **273** (5273), 364–7.

Lenton, T. & Watson, A. (2000) Redfield revisited. 2: What regulates the oxygen content of the atmosphere. *Global Biogeochemical Cycles* **14**, 249–68.

Medlin, L.K., Saez, A.G., Engel, H. & Huss, V.A.R. (1994) Molecular biology and systematics. In: *The Haptophyte Algae* (eds J.C. Green & B.S.C. Leadbeater), Vol. 51, pp. 393–411. Clarendon Press, Oxford.

Mojzsis, S., Harrison, T. & Pidgeon, R. (2001) Oxygen-isotope evidence from ancient zircons for liquid water at the Earth's surface 4,300 Myr ago. *Nature* **409**, 178–81.

Moldowan, J.M. & Talyzina, N.M. (1998) Biogeochemical Evidence for Dinoflagellate Ancestors in the Early Cambrian. *Science* **281** (5380), 1168–70.

Mulkidanian, A.Y & Junge, W. (1997) On the origin of photosynthesis as inferred from sequence analysis – a primordial UV-protector as common ancestor of reaction centers and antenna proteins. *Photosynthesis Research* **51**, 27–42.

Olaizola, M., Ziemann, D.A., Bienfang, P.K., Walsh, W.A. & Conquest, L.D. (1993) Eddy-induced oscillations of the pycnocline affect the floristic composition and depth distribution of phytoplankton in the subtropical Pacific. *Marine Biology* **116**, 533–2.

Pace, N.R. (1997) A molecular view of microbial diversity and the biosphere. *Science* **276** (5313), 734–40.

Park, R. & Epstein, S. (1961) Metabolic fractionation of C^{13} and C^{12} in plants. *Plant Physiology* **36**, 133–8.

Peryt, T., Hoppe, A., Bechstadt, T., Koster, J., Pierre, C. & Richter, D.K. (1990) Late Proterozoic aragonitic cement crusts, Bambui group, Minas-Gerais, Brazil. *Sedimentology* **37**, 279–86.

Petit, J.R., Jouzel, J., Raynaud, D., *et al.* (1999) Climate and atmospheric history of the past 420,000 years from the Vostok ice core, Antarctica. *Nature* **399**, 429–36.

Petsch, S.T., Eglinton, T.I. & Edwards, K.J. (2001) ^{14}C-dead living biomass: evidence for microbial assimilation of ancient organic carbon during shale weathering. *Science* **292**, 1127–31.

Raymo, M. & Ruddiman, W. (1992) Tectonic forcing of the late Cenozoic climate. *Nature* **359**, 117–22.

Redding, K.K., Cournac, L., Vassiliev, I.R., Golbeck, J.H., Peltier, G. & Rochaix, J.D. (1999) Photosystem I is indispensable for photoautotrophic growth, CO_2 fixation, and H_2 photoproduction in *Chlamydomonas reinhardtii*. *Journal of Biology and Chemistry* **274**, 10466–73.

Rosing, M.T. (1999) 13C-depleted carbon microparticles in >3700-MA sea-floor sedimentary rocks from West Greenland. *Science* **283**, 674–6.

Rothman, D. (2001) Global biodiversity and the ancient carbon cycle. *Proceedings of the Nautical Academy of Science USA* **98**, 4305–10.

Schlessinger, W.H. (1997) *Biogeochemistry. An Analysis of Global Change*. Academic Press, New York.

Schopf, J. (1993) Microfossils of the early Archean Apex Chert: new evidence of the antiquity of life. *Science* **260**, 640–46.

Sigman, D. & Boyle, E. (2000) Glacial/interglacial variations in atmospheric carbon dioxide. *Nature* **407**, 859–69.

Summons, R., Jahnke, L., Hope, J. & Logan, G. (1999) 2-Methylhopanoids as biomarkers for cyanobacterial oxygenic photosynthesis. *Nature* **400**, 554–7.

Tabita, F. (1999) Microbial ribulose 1,5-bisphosphate carboxylase/oxygenase: a different perspective. *Photosynthesis Research* **60**, 1–28.

Urey, H. (1952) *The Planets: Their Origin and Development*. Yale University Press, New Haven.

Williams, R. & Frausto da Silva, J. (1996) *The Natural Selection of the Chemical Elements*. Clarendon Press, Oxford.

Woese, C.R. (1987) Bacterial evolution. *Microbiology Review* **51**, 221–71.

Zouni, A. *et al.* (2001) Crystal structure of photosystem II from *Synechococcus elongatus* at 3.8 A resolution. *Nature* **409**, 739–43.

Chapter 13
Marine Productivity: Footprints of the Past and Steps into the Future

Victor Smetacek, Marina Montresor and Peter Verity

13.1   Introduction

Textbooks on marine phytoplankton fall into one of two categories: those dealing with properties of organisms (taxonomy, phylogeny, morphology, physiology) and those devoted to system-oriented biological oceanography. There is a gap between these categories where there should instead be comprehensive, organised information on the biology of individual species and their adaptation to specific environments based on regional and seasonal distribution patterns of their populations. A comparison with the situation in neighbouring disciplines highlights this gap: ecologists, including freshwater planktologists, have specifically studied the autecology of dominant species (life cycles and behaviour) in order to understand their occurrence in the context of their environment. This information provides the basis for a quantitative understanding of the population biology of individual species and the causes of their fluctuations in time and space. One must know the trees to understand the forests. Marine chemists follow the same approach. Biogeochemical cycles are constructed on knowledge of the chemistry of individual elements in sea water in relation to sources, sinks and hydrography. Chemists even use the terms 'species' and 'behaviour' to describe the various chemical forms of elements and their interaction with the environment.

The scanty knowledge on the autecology of marine phytoplankton species cannot be blamed on methodological problems as such work has indeed been carried out successfully on marine protists: for instance, on the calcifying species *Emiliania huxleyi* as well as on various Foraminifera because of their relevance to palaeo-oceanography; similarly, various toxic species are well-studied because of their nuisance value. So why has the systematic study of the ecology of phytoplankton species relevant to marine productivity stagnated in the past decades? Steemann Nielsen's scientific biography is a case in point: he published several interesting monographs on the biogeography of tropical species of the genus *Ceratium* (Steemann Nielsen, 1934; Steemann Nielsen, 1939a, b) and speculated on their ecology based on the environmental data at his disposal, but he discontinued this auteco-

350

logical work after introducing the ^{14}C method in the early 1950s (Steemann Nielsen, 1952).

An impressive body of data has since accumulated on rates of primary production in the world ocean but our knowledge of the genus *Ceratium* remains meagre. It is a cosmopolitan genus which contributes significantly to global productivity; although not toxic, exceptionally large blooms have caused much economic damage in some areas. Yet we do not understand the reasons underlying seasonal population build-up and decline of the various species and what factors, intrinsic or extrinsic, trigger their growth or stop it in the course of the year. Indeed, we are still ignorant of the interplay of forces that drive phytoplankton species succession – from annual to geological time-scales. The attempts at explanation of this most fundamental aspect of pelagic productivity tend to be based on hindcast interpretation and fail to explain, let alone predict, its dynamics at the species level.

In this chapter, we show that the dichotomous development of marine phyto-plankton science which resulted in the autecological gap can be traced back to the founding fathers and the dynamics of their interaction. Mills (1989) has described the intellectual struggle that marked the birth and development of biological oceanography during its early years. Here we will focus on the philosophical backgrounds of the protagonists and the schools they established that led to the dichotomy. We leave it to our readers to decide how much intellectual progress has been made in the intervening century.

13.2   Footprints of the past

From 'philosophical dirt' to food of the sea

Johannes Müller pioneered the systematic study of minute organisms collected with hand-held nets in the 1840s. He demonstrated the net collections to his students and used the word 'Auftrieb' to characterise this community. The term held sway for 40 years during the period when the organisms were being busily described by many scientists, including Müller's students, as part of the Linnean drive to name and classify all organisms. 'Auftrieb' (literally up-drift or up-drive) is a widely used German word for which 'buoyancy' is the closest equivalent in English. For Müller, who dragged his net through surface water, it denoted 'that which collects in the surface'. Today the word is used in marine science exclusively for 'upwelling'. Müller jokingly referred to the 'Auftrieb' as 'philosophical dirt' as he saw no useful purpose in their investigation other than the scientific study of form and anatomy (Hensen, 1926).

In 1854 the 20-year-old student Ernst Haeckel was introduced to marine life on the island of Heligoland by Müller. Haeckel was deeply impressed by the symme-trical beauty of marine animals (his first love was jellyfish) and became an ardent zoologist. He acquired his postdoctoral degree (Habilitation) with a monograph on the Radiolaria (Haeckel, 1862), completed in Messina. At about this time he embraced Darwin's theory of evolution and eventually became its influential champion in Germany; the timing is probably significant in shaping Haeckel's world view, as we shall see below. In 1866, shortly after becoming Professor of Zoology at Jena University, Haeckel published two volumes on the general morphology of organisms in which he proposed the existence of a 'biogenetic basic law': an underlying order expressed in unifying patterns in the visible form of organisms. He called this principle an 'organic crystallography' (Haeckel, 1866). In this view, evolution was the unfolding of inherent properties that led from simple to complex forms, whereby function was of secondary importance. His radical thesis that ontogeny recapitulates phylogeny, first formulated in 1872 (Breidbach, 1997), follows from this view. To prove his point, Haeckel 'bent' his drawings of vertebrate embryos (Richardson *et al.*, 1997). His drawings of Radiolaria show how accurate a draughtsman he was, so his belief rather than his eye must have guided the hand that drew the embryos.

Haeckel was a gifted draughtsman but also vigorous biologist who made many field trips, also to the tropics, and studied animals in their natural surroundings. He coined the term 'ecology' and described functional relationships between benthic organisms. So his philosophical obsession with form to the negligence of function (Breidbach, 1997; Breidbach, 1998), which had a strong influence on the thinking of his age, appears strange in a man of such intellectual calibre and wide-ranging knowledge. It is tempting to suggest that plankton – the 'philosophical dirt' – was to blame. Until the late 1880s marine biologists wondered about the sources of nutrition of the deep-sea benthos and considered input of organic matter from the rivers and coast to be the major food supply. Although it was known that many organisms of the 'Auftrieb' were indeed algae, their role was not suspected at that time, presumably because the coarse-meshed nets in use caught only a small fraction. Since the 'Auftrieb' lived in a homogeneous medium with uniform nutrition, the wide diversity in intricate forms must have appeared as pure luxury with little if any functional significance. For Haeckel it was proof that evolution proceeded, even in the lowest forms, without environmental forcing, but simply as an expression of a natural law guiding morphogenesis and speciation. In this world view, natural selection is not a driving but rather a constraining force, setting limits to expression of the Biogenetic Basic Law which relentlessly creates forms of increasing complexity and sophistication without the necessity of function. Haeckel neglected to provide a mechanistic explanation for evolution (Breidbach, 1998).

Haeckel took his message to the public with books, lectures and art work and was favourably received in learned circles. As a successful fund-raiser and populariser of science throughout Europe, he had considerable influence on philosophy and the arts. His assertive nature is evident from his style of writing and his strong beliefs and anti-religious sentiments guided his actions. The series of meticulous drawings and paintings of organisms, particularly planktonic protozoa, published at the turn of the century under the title *Kunstformen der Natur* (Haeckel, 1904), were an instant best-seller whose magic has endured until today (Haeckel, 1998). The plates are an unabashed celebration of the symmetry of natural form as a principle unto itself: form is beauty and function profane. Not surprisingly, the arrangements of radiolarian skeletons, which particularly inspired art nouveau, are the central motif in this endeavour (Lötsch, 1998). It was a time when natural science was being integrated into evolving Western civilisation. Thanks to Haeckel's efforts, appreciating the beauty of plankton became a cultural experience.

Although Haeckel coined the terms 'protist' and 'ecology', it was Victor Hensen, a professor of medical physiology at Kiel University, who launched the science of protistan ecology with his much-cited but rarely read monograph published in 1887 (Hensen, 1887) in which he coined the term 'plankton' to replace 'Auftrieb'. He was the first to devise quantitative methods to measure phytoplankton productivity. He set the scales – the square-metre water column – and coined the currency by converting cell counts into ash-free dry weight (particulate organic matter). As a physiologist he acutely appreciated the need for accurate measurements and his 1887 monograph was merely intended to introduce the quantitative method he had painstakingly developed to study plankton. Although this monograph launched quantitative marine science, it has not been translated into English; indeed much of it consists of boring tables and details of his counts, presented as proof of the accuracy of his method, which are not worth the effort. But there are some gems in the introduction and discussion that reveal the thought behind the vision and are translated below. They show that Hensen, single-handedly and solely on the basis of net catches studied under the microscope, developed a quantitative, conceptual framework of the functioning of pelagic ecosystems that has since become self-understood but was a revolutionary feat at that time. Unfortunately, he was, like Haeckel, a man of firm beliefs which he stubbornly adhered to throughout, despite later evidence to the contrary. These detracted from his standing and the memory of his role in establishing quantitative planktology has faded, whereas that of Haeckel, who conveyed the beauty of plankton, still thrives (Smetacek, 1999). As we shall show next, Hensen's big picture was essentially correct but many of his details were wrong, whereas in Haeckel's case it was the other way round.

Hensen was a successful and respected physiologist, an active practitioner and believer in the chemistry-based branch of biology which, at that time, considered itself on the brink of unlocking the secrets of life's mechanisms. Hensen had been politically active before accepting the university chair and called for development of a scientifically based management of fisheries, then a major branch of the north German economy and plagued by fluctuating fish stocks. After joining the university he decided to develop this field of research himself. He approached the sea as a concerned scientist but examined it with the eyes of a physiologist trained to work quantitatively. He reasoned that fish stock size, and ultimately yield, had to be estimated, and hypothesised that the distribution of drifting fish eggs, which were amenable to sampling, might well prove to be a means to this end. Before commencing his surveys he satisfied himself that the fish eggs would, after spawning, be homogeneously distributed by the currents and waves as to make the effort worthwhile. He next went to great pains to design and test methods at sea to quantitatively collect the fish eggs from an entire water column, using a net of standard mesh developed to sift flour. This net also collected phytoplankton more efficiently than earlier ones and Hensen implies in his writings that their role dawned on him while searching under the microscope for fish eggs in masses of phytoplankton – primarily diatoms and *Ceratium* spp.

In the opening sentences of his monograph Hensen introduces his revolutionary viewpoint by stating that the organisms of the Auftrieb: 'have, apart from their interest for systematics and anatomy, no doubt great importance for the entire metabolism of the sea. This paper attempts to approach this metabolism.' The term 'plankton' is introduced and delineated in a few sentences, after which Hensen reasons that this plankton should be able to make full use of the light impinging on the sea since it can maintain itself in the surface layer. He next considers the sources of food to the sea and discounts the role of coastal input by pointing out the distances and areas involved but also the poor quality of this food which should be difficult to digest for primitive animals. For Hensen, it follows that: 'The plankton can generally grow well everywhere in the sea, it provides living food, and appears to be an excellent source of nutrition.' Not surprisingly, this grand view of a functionally self-sufficient ocean, radically different from that of the morphologists, was developed by a physiologist who understood metabolism and could comprehend how a comparatively thin soup of unicellular algae can nourish the rich marine animal life. In a later paper he likened the plankton to 'blood of the sea' (Hensen, 1911).

Hensen was acutely aware that his hypothesis had to be tested with accurate methods, and stated in the conclusion of his introduction that this monograph was:

'...merely the description of a methodology. It is more difficult for me than perhaps believed, to speak out this word, since not only would I have liked to provide more than just a methodology, rather, I must shut my eyes almost with force in order to say that it is only a methodology. Sometimes it appears that more has been achieved. This appears so because the results I have acquired are mostly the sole, in any case the most probable [thing] that can be said about the object in question; however, in almost each case I must acknowledge that renewed investigations are most desirable.'

The translation is as close to the original clumsy German wording as English grammar allows. Curiously, the only mention of nutrients is in the penultimate sentence:

'With regard to earlier references than my studies, I can only offer that Murray has stated that 16 tons of carbonate would be found under a square mile surface and 100 fathoms depth.'

Apparently, Hensen considered all nutrients to be as abundant as inorganic carbon.

The first third of his monograph is devoted to methodology and comprises a detailed description of his nets and the statistical significance of individual catches. This is followed by an account of sampling procedure (including how to deal with the ship), sample treatment on board and analysis in the laboratory. The section ends with 'Journal of a cruise'. Hensen was aware that the plankton would have to be uniformly distributed over long stretches for individual hauls to be representative. He compared catches from vertical and horizontal net hauls and regarded deviations from expectation to be due to difficulties inherent to the sampling techniques, e.g. due to sample spillage during rough weather.

The body of the monograph, termed 'Application of the methodology', presents and discusses the details of the data collected from various cruises carried out during various seasons in the Baltic and North Seas. The primary data are individual counts of all the species of plankton recorded, grouped according to dates or region and presented in numerous tables containing huge numbers scattered through the text. Hensen differentiated 23 systematic categories, starting with fish eggs, copepods, decapod larvae, and ending with unidentified cysts, diatoms and algae (comprising filamentous cyanobacteria and presumably *Phaeocystis*). The dinoflagellates are dealt with between tintinnids and radiolaria and are sometimes referred to as animals or plants, occasionally plant-animals. Twenty pages are devoted to the diatoms and dinoflagellates whereas the various zooplankters and protozoa are together accorded 30 pages. Since most of the larger planktonic species from European waters had already been described, Hensen's detailed examinations of many catches under the microscope yielded few new species but much additional information on their biology, particularly of the phytoplankton. He also outlined succession patterns from the different regions and added sometimes extensive notes (e.g. on life cycle stages) to his species-by-species description of the annual cycles.

In the five-page discussion at the end of the monograph, Hensen finally takes up the question posed in the introduction:

'It was my original intention to gain an estimate of the plankton production in the sea. A prerequisite for this is to ascertain the quantity in which plankton actually occurs. If we accept that this has been achieved in the above, one must ask what is to be done next? In order to estimate the production, one must either know the rate of destruction of the material present or the rate of its production; it would be best however, to know both. The most direct and only correct way to estimate production is to follow plant growth, or more correctly, the production of those organisms which are able to build their bodies from inorganic material. This was not possible for me to do, further, as will be shown, not as much is to be gained from it as might appear at first glance.'

With this last remark Hensen was referring to the difficulty of separating growth from grazing rate and went on to consider the food requirements of the copepods.

He performed incubation experiments using concentrated plankton in 4-litre glass bottles suspended at the side of the ship (to maintain turbulence) and calculated from counts before and after incubation that a single copepod consumed 12 *Ceratia* per day. He had derived the content of organic matter per *Ceratium* from his net catches and using this factor 'we come to an annual production of 133.35 grams organic substance per square metre surface'. This amount was consumed by copepods alone and if the rations of all the other zooplankton were considered, the production channelled directly to grazers would be much higher. He next examined annual cycles of organic matter derived from volume measurements of net catches and added up the differences between peaks and troughs. He again stressed that this was a minimum figure because his monthly sampling will have certainly missed peaks. The figure he obtained was 4162.1 cm^3 over 228 days which converted to 17.7 or 14.8 g organic matter depending on which factor, both derived by Hensen, was used for the conversion of catch volume to dry weight of organic matter. Hensen added this figure to the copepod consumption and concluded that the absolute minimum total annual production would be 150 g per square metre surface (equivalent to about 75 g carbon m^{-2}). Ironically, the first estimate of marine production was derived from feeding experiments and, although based on wrong premises (see below), was reasonably accurate: the first complete annual cycle of primary production measured with the ^{14}C method in 1973 in Kiel Bight recorded 150 g carbon m^{-2} (Smetacek *et al.*, 1984).

Hensen apparently did not attempt to measure growth rate of the algae in grazer exclusion experiments. His rough impression of population growth rates derived from following the rate of species replacement in the course of succession, gave him sufficient 'feel' for relevant time-scales in the plankton to run his mental quantitative model.

In order to convey the significance of his estimate, Hensen compared it with the productivity of cultivated meadows. With help from the agricultural department of the university, Hensen estimated the yield of hay to be 179 g organic matter, only 20% more than his minimum estimate of plankton production, and went on to conjecture that the yield from the sea was potentially much higher than that from the land. This analogy launched the agricultural paradigm of marine productivity studies which, as we shall see later, not only secured funding for the undertaking, but also profoundly influenced the direction of research. For Hensen it was more than a mere fund-raising gambit. He pointed out that natural landscapes were too heterogeneous (structured at small spatial scales) to be adequately measured but the sea was homogeneous, and like agricultural fields, amenable to quantitative analysis and estimation of yields. Clearly he anticipated mathematical models of functioning food webs.

Hensen summarised his vision of how plankton productivity studies were to be pursued in the penultimate paragraph:

'All that I have given is primarily an attempt at methodology, but the insight gained does go beyond it inasmuch as it renders observable the regulated wax and wane of all components of the plankton, and renders recognisable a mutual interdependence in occurrence of these components, so that an understanding of the life-processes in the marine desert has been brought closer than was thought possible so far. It has been well established that certain larvae appear at certain times, that some dinoflagellates are numerous in autumn, and more such details, but from where these primitive forms appear and where they go and what binds the larvae to this or that time of the year, what role the seasons play in the production of the sea and particularly, how and what the sea actually produces, are questions whose solutions could not be thought of up to now. Now these questions appear not only tractable to a certain degree but acquiring the answers becomes the decisive demand of various branches of marine science.'

Clearly, elucidation of the processes driving the annual cycle was perceived as the central goal of quantitative plankton research.

The monograph ends with the clumsily worded message:

'I hope that this will attract the diligent and self-denying assemblage of our youthful co-workers and, should this be the case, it will well be possible to acquire the funds that research to subjugate the sea unfortunately requires.'

Hensen was a brilliant visionary, unfortunately without the gift for words. He was nevertheless able to convince his colleagues and within a decade the new science launched by him was flourishing at Kiel University (Mills, 1989). He was also a successful fund-raiser and organiser of expeditions. Financial support came from the State and the business community whom he was able to convince that his line of research would yield scientific prestige and economic profit in the near future. Hensen not only launched the science of marine productivity, he also anchored it firmly in the realm of applied science, devoted to the study of function with clear deliverables.

Hensen's footprints

We have cited Hensen's monograph at length, not only for reasons of historical interest and to pay credit to an almost forgotten pioneer, but to gain perspective on the intellectual underpinnings of the science launched by him and its subsequent development. Hensen's monograph is square one. It projects the conceptual framework of the pelagic system developed by a mind trained in the quantitative logic of physiology but guided by yield-oriented agriculture and imprinted by extensive visual exposure to concentrated plankton samples. Closer examination of his monograph reveals a mind vacillating between defence of his approach and scientific objectivity. On the one hand he called for more and better observations, on the other he defended his beliefs with surprising stubbornness. Apparently he was confident that his holistic view of the metabolism of the sea was consistent. This situation is symptomatic of the course of development of plankton ecology, because the object of study, namely the pelagic ecosystem, is revealed piecemeal by the introduction of new methods and cannot be observed as a whole with the sense organs in the way terrestrial systems, whether fields or forests, can. Since the behaviour of the organisms and their interactions cannot be seen, but have to be inferred, there is an omnipresent danger that assumptions based on wishful thinking, dictated by the method or the approach, are regarded as fact by the investigator. Hensen's legacy provides an excellent example of the interaction between the reasoning mind and the mind enamoured of the method.

Hensen's estimate of annual production is remarkably accurate considering the inadequacy of his methods. It seems that his estimate was designed to conform with expectation and was not just fortuitous. We might be doing him great injustice but we suspect that he consulted his colleagues from the agricultural department before quantifying the results of his feedings experiment. Had he obtained figures an order of magnitude more or less, he would have rejected them as implausible. This would not have been a problem because he was aware of the limitations of his crude feeding experiments. In contrast to the plankton counts, the experiment (using four bottles, two of which were controls) is described perfunctorily in the discussion; it was not repeated and no data are shown. One is led to suspect that the question of how much the sea produces was solved before the measurements were obtained. So his view of plankton dynamics would not have changed had he been confronted with the ^{14}C data set measured in 1973. Which begs the question as to what we have learned from the modern productivity data.

Hensen had also developed some opinions on the other question he posed: what the sea produces. Again, his opinions are worth recounting because they throw light on current thinking on the subject. Hensen stressed in his monograph that diatoms were of little food value for zooplankton because they consisted mainly of silica walls and contained little nutritious plasma. The diatom cysts (spores) in contrast contained concentrated plasma and were a valuable food source. He reasoned that these would tend to sink out of the water column and hence provide the food supply of the benthos. Since most of his phytoplankton catches consisted of diatoms, and dinoflagellates were only important in autumn, his denial of their importance as food for copepods seems strange, since he did not mention what sustained the zooplankton in the absence of *Ceratium* spp. He claimed to have examined the guts of many copepods and found a diatom frustule on only one occasion. The guts contained amorphous matter but its origin is not referred to, certainly it was not the remains of diatoms. Why he did not perceive the food source of the copepods as a problem is difficult to explain. Apparently, it was simply overlooked. The easiest (and correct) explanation would have been that his nets missed a part of the plankton and this was what the copepods were eating. But defence of the accuracy of his method was more important and when Lohmann, who followed in Hensen's footsteps, discovered nanoplankton some years later, Hensen was not pleased because his method was exposed as not quantitative (Mills, 1989). Had the unknown food source of the copepods been perceived as a problem, he should have welcomed Lohmann's finding. Defending the method was more important than assimilating new findings in a quantitative coherent framework.

Hensen's other follower, Brandt, measured the chemical composition of diatoms and showed that they must be nutritious (Mills, 1989); but Hensen was not persuaded. A lively debate raged on the role of diatoms as animal food at the turn of the century. In a paper entitled 'The diatoms and their fate', Frenzel (1897) claimed that 'diatoms represented nothing but fodder for bacteria and building material for their own progeny'. Karsten (1899), in a monumental monograph on the diatoms of Kiel Bight, contradicted Hensen and Frenzel in no uncertain terms and documented many instances of diatoms being eaten by pelagic and benthic animals. Hensen remained stubborn and added a postscript at the end of Karsten's paper:

'I may well be allowed to add the following statement here. I would be pleased if I could let myself be convinced by the above reports that diatoms represent a significant food source of higher organisms. I am prevented from doing so by the fact that I have so far found the amount, the larvae and the egg production, e.g. of the copepods, to be more often in inverse rather than in direct relation to the frequently observed mass development of diatoms.'

Rauschenplat (1900) published a paper 'On the food of animals from Kiel Bight' again stressing the role of diatoms but Hensen was not to be swayed. In his final manuscript he reiterates his views on the unpalatability of diatoms (Hensen, 1926).

Ironically, we now know that it is the *Ceratium* spp. that are reluctantly, if at all, eaten by copepods. Their disappearance (in contrast to diatoms) in Hensen's feeding experiments was most likely due to the strong turbulence introduced by the ship's movements. Live *Ceratium* spp., in contrast to diatoms, are highly sensitive to turbulence and rough handling and are likely to have succumbed to these factors rather than the depredations of copepods. His grazer-dominated conceptual framework led him to estimate production from assumed feeding rates. In his view, measuring the growth rates of phytoplankton, i.e. diatoms, would not have been worth the effort.

In the following decades Hensen's opinion faded and diatoms came to be considered the pastures of the sea. Yet the issue is more complex and Hensen's argument in his postscript cannot be brushed aside. Hardy (1956) independently came to the same conclusion as Hensen and suggested that zooplankton avoided not only diatom but also *Phaeocystis* blooms, a phenomenon he termed 'grazer exclusion'. He seemed to imply that it was the density of bristly diatoms and mucoid *Phaeocystis* that deterred the zooplankton. More recently Miralto *et al.* (1999) and Wolfe *et al.* (1997) suggest that chemicals are to blame. It is now accepted that diatoms and haptophytes (the group which includes *Phaeocystis* and coccolithophorids) dominate blooms in the sea because they are grazed less than the other groups comprising the regenerating or microbial food web. We will return to the issue of differential grazing later.

Hensen also raised another issue pertaining to the plankton: 'I believe I should briefly draw attention to how many questions central to Darwin's theory are connected to research on these lower forms.' Presumably, he was referring to the proponents of biogenetic law who considered themselves the champions of Darwinism in Germany and were also claiming that evidence for ongoing speciation was provided by the many transition stages between species of plankton. Hensen stressed that since very little was known about the plankton, their 'proofs' were pure speculation. 'That all organisms are related I believe certainly enough as I am as convinced as possible that a creation or construction by anything other than the workings of natural laws is completely excluded.' But he pointed out that transition forms were exceptional and that many related species coexisted without any signs of merging into each other. Since he had examined multitudes of samples and counted astronomical numbers of plankton cells he felt in a position to make such a statement.

'Nothing seems more hopeless, considering the uniform conditions of the sea, where only the law of chance mediated by thorough mixing reigns, than attempting to prove the evolution and survival of that infinity of species on the basis of Darwin's theories.'

Hensen's arguments are convoluted but he makes it clear that understanding the forces that shaped the many species and forms of plankton and regulated their succession patterns required dedicated study. His perspective was that of a function-oriented physiologist, acutely aware of the dearth of knowledge, but confident that this could be overcome. The paradox of species diversity in a sea of uniformity outlined by him, now known as Hutchinson's paradox, is still being debated (Huisman and Weissing, 1999).

It is remarkable that Hensen developed a reasonably accurate, quantitative framework of the plankton despite the crudity of his methods and ignorance of the role of nutrients. As a physiologist he was well aware of their importance but assumed they were replete, at least partly influenced by his belief that plankton was uniformly distributed. Wishful thinking was involved because he was painfully aware that spatial heterogeneity would have jeopardised the utility of his method and his hopes of proving his grand vision of the metabolism of the sea. His obsession with accuracy led him to count every identifiable object, empty shells and all, so each sample took hours to days to count. No wonder he wished that each value be representative of as large an area as possible – for him it eventually became the entire ocean. Despite overwhelming evidence to the contrary, gained from an expedition he organised that covered the entire North Atlantic, he clung to his views and tried to reason away the glaring differences in plankton biomass between cold and warm seas (Mills, 1989).

Hensen's interest was focused on the seasonal cycle presumably because he felt that elucidating the processes underlying temporal change was intellectually more rewarding than accounting for regional patterns. It is tempting to suggest that he viewed the plankton as a heavily grazed meadow with diatoms being the 'thistles'. Although ignoring the role of nutrients, he was able to perceive a regenerating system in balance: If copepods ate 12 *Ceratium* spp. a day, then 12 *Ceratium* spp. must be being produced every day for each copepod to survive. Since the plankton caught by his nets was not enough to feed the zooplankton, he reasoned that the sum total of this 'regenerated production' (the ration consumed by copepods) would be many times larger than the 'new production' (the sum of all the biomass peaks in an annual cycle). This 'new production', because it was mostly diatoms, was not eaten by zooplankton but became 'export production' and fuelled the benthos. He fondly believed that this situation prevailed over the entire ocean. The regenerating food of the copepods was, as mentioned above, not considered an issue.

Hensen acknowledged deficits in knowledge at the level of organism interactions. His obsession with uniformity prevented him from reversing the paradox of species diversity he posed: it did not occur to him to turn the paradox around and examine the possibility that diversity was an expression of an as yet unperceived heterogeneity. However, these were mere details, an impediment to his efforts to achieve accuracy in a sea of biological complexity. His overall aim was to accurately quantify processes above the level of the organisms to prove that phytoplankton production nourished the sea – a hypothesis which grew and hardened into a grand vision long before the data were acquired.

Haeckel's kicks

Not surprisingly Hensen's monograph was sharply attacked by Haeckel (Mills, 1989) in language no editor would allow today. He accepted the term plankton, which he later extended by coining holoplankton and meroplankton, but ridiculed everything else. The intensity and insulting tone of the attack suggest that Haeckel's dislike of Hensen's monograph was provoked by the deprecatory remarks on Darwinian theory and the futility of applying it to the plankton. It is also very likely that the many boring tables repulsed Haeckel who would have preferred to see aesthetic illustrations of organism morphology in biological literature. Haeckel criticised the substance of Hensen's revolutionary insights on metabolism of the sea by doubting whether the plankton could feed the marine animals. He based his attack on the claim, in direct opposition to Hensen's views, that plankton was patchily distributed and that painstaking counts of individual samples were a waste of time, and expeditions directed at the quantitative study of plankton a waste of money. Possibly, an element of envy was also involved, induced by Hensen's success in raising funds. Competition for funds, the accepted reason for unfair attacks in science, could not have played a major role, because Hensen had opened up new resources inaccessible to Haeckel's brand of science.

Haeckel's attack was indeed unfair and had far-reaching consequences. As a remarkably wide-ranging biologist who defined ecology as 'the economy of nature' (Haeckel, 1873), he could not have intellectually rejected the ecological concept of metabolism of the sea. Indeed, he called for the development of chemical methods to study the plankton (Mills, 1989), implying that it was too early to commence the systematic study of plankton ecology. So he derided Hensen's cell counts as futile and intellectually unrewarding and in the process denigrated the entire approach within the zoology community which held Haeckel in high esteem. Thus was the budding science of biological oceanography literally scolded out of mainstream biology, at least in Germany. It is important to point out that this was not necessarily a clash between irreconcilable scientific viewpoints, because within his university, Hensen was able to convince his colleagues, including the renowned Professor of Zoology Karl Möbius, who had been an active proponent of the food-from-the coast view. Möbius accompanied Hensen on expeditions, worked on the zooplankton and demonstrated how mainstream zoology could well have naturally integrated with production biology. Instead, most marine biologists outside Kiel continued working on morphology and systematics of organisms without much regard to ecology.

This was probably only a minor skirmish for Haeckel who fought battles on other fronts with equal ferocity (Lötsch, 1998). It is not even mentioned by Breidbach (1997) who critically examined Haeckel's central role in disconnecting the study of morphology from physiology, an unfortunate segregation within zoology which has lasted until today. Since his cherished method had been ridiculed as senseless quantification, Hensen was driven into stubborn defence of a homogeneous distribution of plankton for the next three decades (Hensen, 1926), in opposition to the data and interpretation of his colleagues in Kiel who followed in his footsteps but did not share his views. This attitude led to disagreement over interpretation and relevance of results obtained from joint expeditions which would have slowed research progress. Strife was commonplace within the ranks of his colleagues in Kiel, and although it was part of the scientific culture of the day, one does get the impression that individual beliefs were unusually rigid even in the face of rigorous measurements to the contrary (Mills, 1989). Polarisation of position rather than assimilation of new findings seemed the norm in the birthplace of biological oceanography.

Development and demise of the Kiel school

Karl Brandt, who succeeded Möbius to the chair of zoology in 1887, took the agricultural paradigm further by examining nutrient availability and applying Liebig's law of the minimum to the study of plankton productivity. He developed firm views about the role of nitrate in determining production according to which warm seas should be poor in nitrate and phytoplankton (Mills, 1989). He, the father of marine biogeochemistry, disagreed strongly with the results of the first exhaustive